INTEGRATED CONSTRUCTION INFORMATION

OTHER TITLES FROM E & FN SPON

Architectural Management
M.P. Nicholson

Introduction to Building Procurement Systems
J.W.E. Masterman

Construction Conflicts Management and Resolution
P. Fenn and R. Gameson

Construction Contracts
Law and management
J.R. Murdoch and W. Hughes

A Concise Introduction to Engineering Economics
P. Cassimatis

The Idea of Building
S. Groák

Investment, Procurement and Performance in Construction
P. Venmore-Rowland, P. Brandon and T. Mole

The Management of Quality in Construction
J.L. Ashford

Management, Quality and Economics in Building
P. Brandon and A. Bezelga

Profitable Practice Management
For the construction professional
P. Barrett

Project Management Demystified
Today's tools and techniques
G. Reiss

Risk Analysis in Project Management
J. Raftery

Spon's Budget Estimating Handbook
Spain and Partners

Value Management in Design and Construction
S. Male and J. Kelly

Effective Speaking
Communicating in speech
C. Turk

Effective Writing
Improving scientific, technical and business communication
2nd Edition
C. Turk and J. Kirkman

Good Style
Writing for science and technology
J. Kirkman

Journals

Construction Management and Economics
Editors: R. Bon and W. Hughes

Building Research and Information
Editor: A. Kirk

For more information on these and other titles please contact: The Promotion Department
E & FN Spon, 2–6 Boundary Row, London, SE1 8HN. Telephone 0171 865 0066.

INTEGRATED CONSTRUCTION INFORMATION

Edited by
PETER BRANDON
AND
MARTIN BETTS

Department of Surveying
University of Salford, UK

Taylor & Francis
Taylor & Francis Group

LONDON AND NEW YORK

Published 1995 by Taylor & Francis
2 Park Square, Milton Park, Abingdon, Oxon OX14 4RN
52 Vanderbilt Avenue, New York, NY 10017, USA

First issued in paperback 2020

*Taylor & Francis is an imprint of the Taylor & Francis Group,
an informa business*

First edition 1995

© 1995 Taylor & Francis

Typeset in Great Britain by Saxon Graphics Ltd, Derby

A catalogue record for this book is available from the British
Library

Library of Congress Catalog Card Number: 94-69928

ISBN 13: 978-0-367-57972-2 (pbk)
ISBN 13: 978-0-419-20370-4 (hbk)

Contents

Contributors

Professor Brian Atkin *Department of Construction Management and Engineering, University of Reading, Whiteknights, Reading, RG6 2BU, UK.*
Brian Atkin is Professor of Construction Management at the Department of Construction Management and Engineering of the University of Reading. His interests include the application of IT within the construction industry, about which he has written many papers and several books. Brian Atkin is a Director of Atkin Research and Development Limited, a company specializing in construction industry research, development and training.

Godfried Augenbroe *Faculty of Civil Engineering, Technical University of Delft, PO Box 5048, 2600 GA Delft, The Netherlands.*
Godfried Augenbroe is Associate Professor in the Building Engineering group of the Faculty of Civil Engineering, Delft University of Technology. He heads the Computer Integrated Building unit. Recent work in this area is focused on the European COMBINE project, which he coordinates.

Martin Betts *Department of Surveying, University of Salford, Salford, M5 4WT, UK.*
Martin Betts is a Senior Lecturer in the Department of Surveying of the University of Salford. He is the Director of Postgraduate Studies within the Department and the Coordinator of the CIB W78 working commission entitled IT in Construction. His research interests are in the management of IT for construction.

Professor Björn Bindslev *Architect Kongelige, Danske Kunstakademi, Denmark.*
Björn Bindslev is Associate Professor at the Architect's School, the Royal Danish Academy of Fine Arts. He developed the 'CBC system', the first computer-aided project management system in Denmark. He became a Doctor of Technology at the University of Lund, Sweden, in 1973. He has practised as a cost consultant and project manager since 1960.

Professor Bo-Christer Björk *Department of Building Economics and Organisation, Royal Institute of Technology, Stockholm, Sweden.*
Bo-Christer Björk is Professor of Information Technology in Construction at the Royal Institute of Technology (KTH) in Stockholm, Sweden. Before joining KTH in 1993 he worked for thirteen years as a researcher in the Technical Research Centre of Finland (VTT). His research focuses on techniques which facilitate the integration of computer applications in construction.

Professor David I. Blockley *Department of Civil Engineering, University of Bristol, Bristol, UK.*
David Blockley is Professor of Civil Engineering and currently Dean of Engineering at the University of Bristol, England. His research interests include civil engineering systems; risk, safety and hazard engineering including human and organizational factors; uncertainty engineering, artificial intelligence and process modeling

Professor Peter Brandon *Department of Surveying, University of Salford, Salford, M5 4WT, UK.*
Peter Brandon is Pro-Vice Chancellor for Research at the University of Salford. He was formerly Chairman of the Department of Surveying. He was the lead researcher behind the development of the ELSIE system, the world's first commercially available expert system for construction. His current research interests cover knowledge-based systems, integrated databases and virtual environments.

James Breuer *Fluor Daniel, Mining and Metals, 1444 Alberni St., Vancouver, BC, Canada, V6G 224.*
James Breuer received his BS in Civil Engineering from the University of Illinois at Urbana and his MS and Eng degree in Construction Engineering and Management from Stanford University. He currently works for Fluor Daniel as a Construction Engineer on the Kennecott Smelter Modernization Project in Utah.

K.C. Choi *Bechtel Corporation, PO Box 193965, San Francisco, CA 94119-3965, USA.*
K.C. Choi is an Information Systems Manager within the Bechtel Corporation based in San Francisco.

Dr Grahame Cooper, *Information Technology Institute, University of Salford, Salford, M5 4WT, UK.*
Grahame Cooper is Senior Lecturer in the Information Technology Institute of the University of Salford.

Anne-Marie Dubois *CSTB, BP 209, 06904 Sophia Antipolis Cedex, France.*
Anne-Marie Dubois is a Research Engineer at the Computer Integrated Construction Division of CSTB in Sophia Antipolis, France. She specializes in Building Information Modelling in national and European projects, e.g. COMBINE, and in the application of STEP in the building industry.

Professor Charles M. Eastman *Graduate School of Architecture and Urban Planning, University of California, Los Angeles, USA.*
Charles Eastman is Director of the Center for Design and Computation at UCLA. He is a faculty member in the Departments of Architecture and Design and has taught courses in Computer Science. An expert in geometric modeling and engineering databases, he has consulted for major US and foreign companies in these areas for over twenty years. He is on the

editorial boards of several journals, including *Computer Aided Design, Research in Engineering Design,* and *Automation in Construction.*

Professor Martin A. Fischer *Center for Integrated Facility Engineering, Stanford University, Stanford, California, CA 94305-4020, USA.*
Martin Fischer is Assistant Professor in the Department of Civil Engineering at Stanford University. He focuses his teaching and research efforts on the role and use of IT in project management.

Professor Lars M. Giertz *Wadköpingsväg 15, 16151, Bromma, Sweden.*
Lars Giertz is an architect, a researcher and a documentalist. He was Vice-President of the International Council for Building Documentation in 1950 and Chairman of the International Building Classification Committee between 1950–60. He was a Professor at Haile Selassi University in Addis Ababa in 1961 since which time he has been engaged internationally.

Sven Hult *Technical Research Centre of Finland, Espoo, Finland.*
Sven Hult is a former research assistant at the Laboratory of Urban Planning and Building Design of VTT in Espoo, Finland. His background and research interests are in the area of telematics.

Pekka Huovila *Technical Research Centre of Finland, Espoo, Finland.*
Pekka Huovila is a civil engineer engaged as a Senior Research Officer in the Laboratory of Urban Planning and Building Design at VTT in Espoo, Finland. He has experience of civil engineering projects in Africa and as scientific attache in France.

Professor C.W. Ibbs *University of California, Berkeley, CA 94720, USA.*
Bill Ibbs is Professor of Management of Technology in Construction at the University of California at Berkeley.

Professor Ingvar Karlén *Unit of Informatics and Systems Science, Royal Institute of Technology, Fiskartorpsvägen 15A, Stockholm, Sweden.*
Ingvar Karlén is an Associate Professor in the Department of Design Methodology of the Royal Institute of Technology in Stockholm.

Lauri Koskela *Laboratory for Urban Planning and Building Design, VTT, PO Box 209, SF-02151 Espoo, Finland.*
Lauri Koskela is Senior Researcher at VTT Building Technology in Espoo, Finland. His main research area is lean construction.

Stephen Lockley *School of Architecture, University of Newcastle, Newcastle, UK.*
Stephen Lockley is an architect and also a lecturer at Newcastle University. He is engaged mainly in research into computer based information systems for the construction industry, information and process modelling.

Kevin J. Lomas   *De Montfort University, The Gateway, Leicester, LE1 9BH, UK.*
Kevin Lomas is Reader in Building Simulation in the School of the Built
Environment at De Montfort University, Leicester. He is also the Director
of the Environmental Computer Aided Design and Performance group
which specializes in thermal, lighting and air flow simulation.

Professor Noel H. McDonagh   *Visiting Professor, University of Salford,
Salford, M5 4WT, UK.*
Noel McDonagh is a retired Senior Partner of Mulcahy, McDonagh and
Partners quantity surveyors. He is the former president of the quantity
surveyors division of the Royal Institution of Chartered Surveyors and the
former co-ordinator of the CIB W74 working commission concerned with
information coordination for the construction industry. He is also a former member of the main board of the CIB and is currently a visiting professor of the University of Salford.

Paul Newland   *University of Portsmouth, Portsmouth, UK.*
Paul Newland is a member of the Centre for New Media Research at the
School of Art, Design and Media of the University of Portsmouth. He has
a doctorate in understanding information transfer to design professionals.
His present research interests include: learning strategies for interactive
new media, hermetic and alchemic knowledge in art and design, potential
and pitfalls of new media, the use of virtual reality and the world wide
web, interactive statistcal representation of medical data, and developing
new ways of disseminating any field of research in new media formats.

Rivka Oxman   *Faculty of Architecture and Town Planning, Technion, Israel
Institute of Technology, Haifa, Israel 32000.*
Rivka Oxman is Senior Lecturer in the Faculty of Architecture and Town
Planning at the Technion, Israel Institute of Technology. Her main areas
of research are design research, computer modelling of information,
knowledge and process in design, artificial intelligence and design cognition and the construction of design support systems. She is currently
developing the 'Design research computer laboratory' at the Technion.

David G. Platt   *Department of Civil Engineering, University of Bristol,
Bristol, UK.*
David Platt is a member of the Civil Engineering Systems Research Group
at the University of Bristol, England. His current research interests are
business process modelling, in particular, its relationship to organizational
culture and human factors.

Professor James Powell   *The Graduate School, University of Salford, Salford,
M5 4WT, UK.*
James Powell is Lucas Professor of Informing Design Systems and
Director of the Graduate School at the University of Salford. His interests
include team building, design information transfer and multi-media
design.

Patrice Poyet *CSTB, BP 209, 06904 Sophia Antipolis Cedex, France.*
Patrice Poyet is Head of the Computer Integrated Construction Division at CSTB in Sophia Antipolis, France. He has participated in many engineering research projects including ATLAS, COMBINE, and is the author of numerous papers and journal articles related to his areas of professional interest.

Professor Victor Sanvido *College of Engineering, Department of Architectural Engineering, The Pennsylvania State University, 104 Engineering Unit 'A', University Park, PA 16802, USA.*
Victor Sanvido is Associate Professor of Architectural Engineering at Penn State. He heads up the Computer Integrated Construction Research Programme and is the director of the Partnership for Achieving Construction Excellence (PACE), a university–industry collaboration aimed at developing leaders for the industry. He has several years of international construction experience.

Rowan Shulver *De Montfort University, The Gateway, Leicester, LE1 9BH, UK.*
Rowan Shulver was a research assistant in the School of the Built Environment at De Montfort University researching the integration of building design systems.

Stewart Thornton *De Montfort University, The Gateway, Leicester, LE1 9BH, UK.*
Stewart Thornton is a Senior Lecturer in the Information Systems Department at De Montfort University, with a special interest in object-oriented techniques.

Antony Thorpe *Department of Civil Engineering, Loughborough University, Loughborough, LE11 3TU, UK.*
Antony Thorpe is Senior Lecturer in Construction Management at Loughborough University of Technology. His research interests are the application of IT to communication processes in construction particularly EDI, automatic identification techniques and broadband communications.

Stephen Vincent *Scott Wilson Kirkpatrick, Basingstoke, UK.*
Stephen Vincent is Head of Information Systems at Scott Wilson Kirkpatrick. He is a civil engineer who has had to come to terms with information technology, and he takes a particular interest in the sustainability of new working methods and systems.

Alastair Watson *Computer-Aided Engineering Group, Civil Engineering Department, The University of Leeds, Leeds, UK.*
Alastair Watson is Senior Lecturer in Computer-Aided Engineering, Department of Civil Engineering, University of Leeds. His research interests span many areas but currently centre on the problems of information exchange and integration.

Foreword

*Professor Noel H. McDonagh, Visiting Professor, University of
Salford, Salford, M5 4WT*

INFORMATION SYSTEMS RESEARCH IN CONSTRUCTION

Much scientific research is currently being undertaken in information systems relevant to construction. There may be a tendency in the intensity of
the research itself, to become detached from the context within which the
research was originally promoted or conceived or is currently being
undertaken. The purpose of this Foreword is not so much scientific as
rather to set or consider the context for the various excellent scientific
chapters that follow. This Foreword is in the nature of a contextual or
reflective overview of the attitudes, trends and developments that may
tend to influence the direction of current and future building research
with particular reference to information research relevant to the building
process.

Many descriptions and qualities may be ascribed to our world today. It
has many incomprehensible characteristics. One such omnipresent characteristic is the factor of change. To say that we live in a time of considerable and dramatic change is a very obvious truism. But change is not new.
It has been said that one of the few eternal objective realities is change. It
could be said that change is inevitable and continuous. Change can also
be seen as being a necessary condition of existence. What distinguishes
change from time to time is its nature and rate. In our time and into the
future, change is comprehensive, immediate and its incidence is at an
ever-accelerating rate. It is pervasive, persistent and insistent.

In all that we do in life, we must face the challenge and influence of
change. How should we react to change? We can resist it – a sterile and
non-productive exercise leading to inevitable disillusionment and frustration. We can embrace it with such vigour and mindless enthusiasm that
we lose our sense of direction and our sense of purpose. What we should
do in an intelligent and balanced manner is to accept the reality of change
while at the same time preserving or conserving our continuity of objective. Change and continuity are not mutually exclusive concepts. In a
scientific context, they must be regarded as essential complementary

concepts which guide us towards deepening and broadening our knowledge base, developing new and improved techniques in applying our greater knowledge, and in finding better solutions to the problems we may be addressing in our scientific and research work.

Patterns of change relevant to our consideration are occurring within society as a whole, within technology, within construction practice and within construction research. (By construction practice in this context I mean the entire activity from design to production.) The rate of change is influenced by both technological development in other fields of human endeavour and the self-evident explosion in the availability of and access to information combined with even more efficient computerized information and communication technologies and systems.

This confluence of change and its rate – based as it is on information and communications – must make us pause and reflect on its potential effect upon the nature and direction of ongoing research activities concern with information related to the building process – its generation, exchange, conversion, use, application, processing, storage and, above all, its integration and its management. Those involved in the development of integrated approaches to the information problems of the construction industry will not only reflect on these matters, but we must also ask and inform ourselves as to what is happening in other areas such as building documentation and building economics, and consider those studying the organization and management of construction, and those considering modern technologies as well as maintenance and modernization of buildings. What are other users of information doing and thinking about?

CHANGING PATTERNS OF SOCIETY

The last decade has witnessed a significant change in economic thinking and attitudes world-wide. A growing public intolerance with levels of personal and indirect taxation combined with other major macroeconomic factors have caused many governments – not only those in Western Europe – to be more energetic and decisive in controlling levels of public expenditure and, in particular, in limiting the extent of public borrowing. The contraction in public spending and borrowing has affected both revenue or current spending and capital spending. This has resulted in there being less money available for public capital building programmes. While demand for certain programmes of public capital building investment may have diminished perceptibly in areas of Europe because of the combined effects of population trends and the post-war building boom, elsewhere the needs have not diminished. Indeed, even in the most developed countries there is continuing demand in the areas of housing, health and welfare as well as for maintenance and modernization of existing public building stock. In these circumstances society is demanding that more units of construction be provided per equivalent unit of expenditure, or that the equivalent number of units be produced for fewer units of expenditure – all without any reduction in standards. This is referred to as obtaining better value for money in all areas of public building capital expenditure. Private sector investors in industrial and commercial construction projects are equally insistent on the need for better value. This can only be achieved through greater efficiency of the building process. Society as a whole is seeking that degree of improved efficiency.

The pressure for control of public and private spending applies as much to current or revenue spending as it does to capital spending. Many of today's capital projects result in future revenue liabilities in operating, running, maintenance, modernization and replacement costs, as well as interest charges and debt repayments. There is now far greater pressure than ever before for the application of techniques to ensure that these future or continuing costs are also controlled and minimized. Hence the demand is not only for control of initial capital costs nor for the control of future costs, it is for both. Society is demanding the development of skills and techniques to achieve the optimum balance between initial and future costs.

Society then, requires both increased efficiency in performance from the building process as well as the development of the necessary skills and techniques to sustain continuing growth in efficiency. Both society and the direct public and private sector clients within it are now well enough informed to be able to justify their demands. They are equally aware that building practice has the tools available to increase its efficiency and to develop the required new techniques. They can appreciate that the building process incorporates not only building practice, which, as I have said before, includes design and construction, but also includes the promotion, development and use of buildings – just as much as it includes building practice.

Therefore, 'consumers' see themselves as part of the building process and are confident of their own competence to play a full role within the process. In doing so, they wish to see feedback from the use of new designs; they wish to see new designs that will respond to the criteria for economy and efficiency in initial and continuing costs; they wish to see standards of construction that will respond fully to design criteria; and they wish to limit the risk factors traditionally associated with construction projects. They are aware of the role that information and management information systems play in the realization of these objectives. Hence the desire to see an integrated approach to the various information systems required to meet these objectives and which will form the basis of comprehensive and integrated overall management of projects and programmes, and therefore control over the resources consumed by the same.

CHANGING PATTERNS OF TECHNOLOGIES

There is no need to dwell on the nature and extent of general technological development within today's world. It is self-evident and all around us. It is so obvious and pervasive that we take much of it for granted or fail to be impressed by it. This is probably equally true of computer technology. It was only a few short years ago that the first desktop calculating machines were available, and shortly before then the only computers generally available were mainframes requiring specially equipped and environmentally controlled computer rooms to house them. How quickly we have passed through the various computer generations – probably so quickly that it has been difficult for the human population to comprehend or assimilate. The development in software has been no less impressive and the currently available combination of microcomputers with standard office products, database and spreadsheet techniques and packages, provides an extremely broad range of applications appropriate to many of the requirements of the building process. The more sophisticated CAD and CAM systems and the developing expert systems open new horizons for the future. All of these combined with the advances in other communication technologies indicate considerable changes in the traditional skills, interaction of skills, and the methodologies applied by the respective skills in the future.

These changes are particularly relevant to the construction process and may herald a new era of efficiency provided there is consistency between the information systems adopted by the different sectors of the process and the separate skills within each sector. The potential of the new technologies will be fully realized and optimized only if frameworks or principles are developed and adopted for comprehensive, integrated information systems which permit consistency and ease of exchange to and between different users of the same information in different combinations for different purposes and across the whole range of projects and programmes. Full efficiency will not be achieved through the adoption of well designed but incompatible systems handling respectively the differing needs and purposes of each user.

Any studies of the information patterns of the construction process indicate the common information base for many of those involved and of the different purposes for which the information is required and applied. New technologies have the potential to accommodate a comprehensive integrated approach to the complexity of information handling within the construction process, with consequent development of improved and more efficient management of the overall process and the sectors and skills within it: in the words of a former President of CIB, Dr R.N. Wright, 'expectations are presented for computer integrated building practice'. These expectations must be considered in the context of both the general historical ultra conservative attitudes of the construction industry towards computer and computer-related technologies and techniques, and the fact that new generations of computer-literate graduates and trainees are being recruited into the industry.

CHANGING PATTERNS OF CONSTRUCTION PRACTICE

Insofar as construction practice is basically a service industry, it must be responsive to the demands of the market-place. As indicated earlier, the market-place is demanding higher levels of performance and competence, greater guarantees of the quality of the completed project, greater control of initial capital cost, more accurate forecasting of continuing future costs and elimination or limitation of the risk factors apparently inherent in construction projects. These demands for increased efficiency will only be met through improved management, improved methodologies, increased productivity and improved techniques. A key factor in many of these areas is the manner in which new technologies are adopted and applied.

It is true to say that the building process is information intensive, both in the information it generates and exchanges, and in the information it absorbs from sources outside the direct process. The quality of the information generated, the efficiency and format of its transfer, and its subsequent application and use are key elements in promoting effective management – it is the source of much of the management information upon which effective management is based. There is – and will be more so in the future – a growing awareness within construction practice of the significance of information as a touchstone for increasing efficiency; that the key to success for improving the effectiveness of information lies in the adoption and application of consistent information systems. There is also the beginning of awareness that a critical path of information effectiveness lies not only in the quality of information, but also in its form, content and presentation at the point of exchange. There is a realization that there should be some degree of consistency in these areas. There is a growing understanding of the role of databases as an effective management tool. Therefore, relevant information systems should not only be consistent with the skill or firm, they should be consistent and compatible across ranges of skills, firms, users and projects. How to achieve this is the big question.

CHANGING PATTERNS OF CONSTRUCTION RESEARCH

A consequence of government efforts to control public expenditure has been a general cut-back in the public funding of research programmes. This trend applies as much to construction research as to any other sector. The demand from those providing the funding is for more relevant, result-oriented, high quality research capable of immediate application with measurable benefit or effect. While researchers and research managers may regard this as an unattainable dream and a doubtful approach in principle, nevertheless, the pressure is increasing for greater dialogue and mutual support between construction research on the one hand, and practice and use on the other. Construction research is required to contribute to the overall management and efficiency of the construction process and the quality of the end product.

Much construction research in the past has dealt with technological and physical aspects of building construction, with performance criteria and with regulations and standards. There is now increasing demand for process-oriented research: on procurement systems, the management and organization of the overall process, the role of information within the process management, the performance of buildings as systems themselves, the organization of feedback information from use to design, the co-ordination of information generated within the process information injected into the process from outside, the relevance of integrated database development and so on, and the relevance of new technologies to the resolution of many of these issues. Overall combination of these issues and the relevance of computerized information and communications technologies can be regarded as the field of information management. This is now an important area of construction research worthy of commanding funding support.

The source of funding in itself is a major question. Traditionally, most construction research has been funded from government and other public sector sources. However, as these funds have become scarcer and more restricted, alternative or supplementary sources must be sought. These may need to be generated from the private sector and the industry itself.

COMBINATION OF CHANGING PATTERNS

The main common thread or theme present when considering the changing patterns of society, technology, practice and research is the emerging perception of the importance and interdependence of building use, construction practice and research. Construction research is a major support to practice and building use; practice is a supplier to building use; and building use is a major component of the overall process. All are essential parts of any national economy. Collaboration between use, practice and research is an important prerequisite in addressing the problems of:

1. increasing efficiency;

2. developing and applying new techniques;

3. optimizing the benefits of new computerized information and communication technologies;

4. optimizing the information flows within the process;

5. developing the frameworks for comprehensive integrated information systems basic to optimizing information flows;

6. establishing basic principles for integrated database development.

AREAS FOR REFLECTION AND THOUGHT

The need for effective integrated information systems is a recurring theme throughout this book; the relevance of the relationship between information systems and new technologies is another; the achievement of greater efficiency within the construction process is yet another, as is the role of management in achieving efficiency and the role of information in improving management. The book identifies the need for developing a conceptual framework for comprehensive integrated and co-ordinated information systems to optimize the benefits inherent in contemporary and future relevant technology. Much other research concentrates on other areas such as documentation, building economics, management and organization of the process, maintenance and modernization of buildings, and many other process-related areas. However, all this process-oriented work deals with areas that require development of appropriate supporting information systems. On the other hand, information-oriented research deals with information systems viewed separately from within and without the process. Much of the work is being undertaken in isolation; this probably reflects the manner in which the topics are being treated from a research point of view world-wide. Nevertheless, it seems logical that the genesis of an intelligent approach to developing a consistent structure or framework for comprehensive integrated information systems for the construction process, lies within the network of those who address directly information matters and wider process-oriented topics and subjects. A greater degree of cohesion and co-ordination is required between information-related and process-related research.

As well as the obvious benefits inherent in the sharing of scientific knowledge and experience, further benefits should be a deeper understanding of the context within which scientific work is being undertaken, a sharper focus and sense of purpose to the work, a greater cohesion in world-wide activity and a conservation of scarce scientific resources in terms of both finance and skilled personnel. Meetings provide the opportunity to share information between different generations of scientists and the potential for new generations to learn some of the lessons of the past first-hand in order to build on the accumulated wisdom, experience,

The field of integrated construction information: an editorial overview

Peter Brandon and Martin Betts

INTRODUCTION

This book contains 24 contributions covering integrated construction information from scientists in different parts of the world. The contributions involve 33 contributors from 17 different universities or research institutes in 10 countries. The result is a unique collection of state-of-the-art reports of leading research in a priority area for construction. Although there are several international conference series that provide fora for discussion on such matters and several journal publications which embrace integration; this volume is rare in containing the most recent contributions from many of the leaders in this field. There has been previous work documented on construction integration but this volume takes a more comprehensive view of the subject's past, present and future.

International collaboration

A particular feature of this book is that it draws from a very widespread international community. Business activity is increasingly adopting global perspectives and creating organizations that work in multiple countries and cultural environments. Academic and scientific research is similarly developing to the point where most of the leading contributors to this volume are often in daily contact with collaborators from elsewhere in the world about aspects of their work.

Integrated construction scientific endeavour is well advanced through media based around the Internet, such that there is research work in construction integration being pursued on a continuous 24 hour basis. Through the Internet, and other mechanisms, an increasing number of projects are conducted collaboratively between research centres as evidenced in the authorship of papers in this volume. Well established international research communities such as the *Conseil International du Batiment*

(CIB), have been active in this and other construction research arena for a long time and are growing in importance.

The people

This volume does not only reflect contributions from world academics. Work in this field from practitioners is also proving to be important to the development of the discipline as a whole. Even those chapters that have been written solely by academics will typically have drawn from research projects with extensive industrial involvement, and often sponsorship. The people behind this book are increasingly becoming concerned with forging a link between the fundamental theories of the issues of integration and the practical applications of the science to the problems of construction practice.

Most of the major research centres active in the field are represented here. Many of the people involved will have known each other and worked together before the production of this volume. However, the coming together on this initiative has formed a further common bond between them as well as documenting their respective work. This book represents a forming of a community of scientists in this field who in their contributions to this work have filled important parts of an overall jigsaw about the construction integration issue.

The long term view

Research for many is a life-time activity. We have deliberately sought contributions for this book that draw from people whose involvement in the field has been long-standing as well as those who have led the field more recently. The section on the history of construction integration, in particular, contains a series of contributions from the founders who have been active for forty years or more, yet are still making important and practically useful contributions today.

In a rapidly changing field such as this, where the influence of rapid advances in technology are so great, it is sometimes tempting to close our eyes to what has gone before and rush to apply the latest technological fix. We feel that the opportunity that we have been given to publish such a book requires us to try to take stock of how the field of integration has evolved in a broader time frame.

Often in science and technology, the worth of a contribution is only fully acknowledged some time later when its relevance can be applied with improved technology or in an emerging context. It is vitally important, in our view, that those looking to the latest technologies, to solve today's perception of the construction integration problem, should look back to the well-established progress that has been made in thinking

about the general principles of classifying construction information produced some time ago.

IMPORTANCE OF INTEGRATION

This then brings us to the question of why a book about integration should
be published at all. For so many contributing scientists and practitioners
from around the world to be making this their life's work demonstrates that,
at least to them, this must be perceived as an important problem. But what
relevance does it have to the practical activities of the construction industry?

The integration problem has many roots and origins. They key to its
current importance is that construction as a form of investment has to
compete for funds with other economic activities on the basis of its ability
to make a value-adding contribution. At present the ability to add value is
being curtailed by poor performance within construction with regard to
productivity and quality. The explanation for the poor productivity and
quality attainments is widely based but a key issue is the high level of
fragmentation.

Fragmentation refers to the fact that the different participants in the
design and construction of buildings have been trained, and gained experience, through different educational and professional paths, are
employed in distinct organizations, have distinct value systems and methods of working, and have derived separate information management systems. Information management, because of this, is highly fragmented.
Advances in management and technology in other sectors have enabled
dramatic advances in productivity, quality and business performance
which have been denied to construction partly for this reason.

Considerable opportunity therefore exists for construction to achieve
dramatic improvements in performance in productivity and quality once
the crucial stumbling block of integrated information can be achieved.
Given that the construction industry typically represents 10% of an economy, by measures such as value added and total employment, the importance of solving the integration issue for economic advance generally, is
considerable.

This problem has long been recognized. Recent advances in information technology are seen by many as a rare opportunity for a substantial
solution to be found. Until now our efforts to solve the problem have been
piecemeal. The bringing together of a book like this represents an attempt
to form an overall view of where we are now.

From construction manufacture to integrated construction information

In the past we have been tempted to always think of ourselves in construction as being different. We might say that the only thing that makes

different disciplines the same is their view that they are different! Because we in construction have considered ourselves to be different in the past has caused us to look for unique solutions to our problems from within our own concepts, terminologies and theories. The world, and scientific and managerial thought is changing considerably, and there is a much greater encouragement for us to look at the parallels between sectors and areas of thought. The growth of interdisciplinary and multidisciplinary activities within scientific communities and the more widely diversified corporate approaches to business both illustrate this.

There is a current tendency throughout the world to view construction as a particular type of manufacturing process. This causes us to look to manufacturing industry's developments as a model for construction. One of the major developments in manufacturing productivity and quality improvement from a technological viewpoint has been the development of Computer Integrated Manufacturing (CIM). As we have described, there is an opportunity in construction for integration to allow fundamental improvements in performance. Recent major investigations into construction industry performance in the UK, such as the Latham Committee, have suggested targets for industry cost reduction of as much as 30%. In arriving at the conclusion that significant improvements are possible, we are only restating in our context the way that CIM has allowed substantial improvements in performance elsewhere. The direct corollary of CIM to construction is Computer Integrated Construction (CIC) which is a term widely used in the chapters that follow. However, we believe the CIC as a practical methodology in itself will only begin to be realized once we have a clearer definition of how to integrate construction information. Thus Integrated Construction Information (ICI) is, in our view, a fundamental prerequisite of CIC and the achievement of manufacturing-like productivity and quality goals within construction.

THE PARTS OF THE BOOK

It is within the context of the above that the contributions to this book have been selected and assembled. The book is structured into five sections that deal with the background and rationale for integration, the history of our previous attempts to solve the problem, two sections which address the technologies of product and process modelling that are behind our current major drives to achieve a solution, and a final section describing how ICI can be applied to construction problems. As a whole they represent a complete pass through the subject of ICI.

All the authors of the chapters that follow were asked to place their own research onto each of the dimensions of the research map in Table 1.3. The individual results were aggregated to give some assessment of the balance of our overall activities. Analysis of this suggests we are concentrating our ICI research at present in the area of integration of the design function but our understanding of what integration means in terms of data, models, knowledge or goals is by no means consistent between us. We are also clearly utilizing a variety of technologies. We can illustrate this diversity in our approaches to ICI research in the following examples of chapters that follow.

Powell and Newland (Chapter 5) are approaching the issue of integration from the viewpoint of the designer. Their work is pitched at the design stage of projects and aims for integration of views of the building product. It is an activity in applied research that is taken through to a product development. The emphasis is on sharing information between people using multi-media. Lockley (Chapter 14), by contrast, focuses more on the owner or client function. The object of his work is more widely oriented to the whole construction industry with more basic research. The aspect of integration it is concerned with is sharing data between projects. Platt and Blockley (Chapter 16) are also basing their integration work on the design phase and function. The aspect of integration they are concerned with has more of a project emphasis. Their work is more conceptually based and utilizes process modelling technology.

Integrated Construction Information map

Scientific progress often takes the form of individuals or research groups advancing specific technologies and theories in isolation and periodically taking stock of the overall scene and interrelationships. Part of the latter is about meta level analysis of where our discipline is going and of what its constituent components are. For our efforts at analysing the meta level of our discipline in a practical context we can describe this as the drawing of an Integrated Construction Information map. The parts of this book are themselves likely elements within this map although the different technologies that are being developed are other dimensions of it. We might start to draw such an Integrated Construction Information map as in Figure 1, which is depicted as an incomplete jigsaw puzzle.

A picture like this starts to provide for us a structure for the field of ICI. We can see that two of the dominant aspects of the research methods we use are the information technologies by which we seek to achieve integration and the modelling approaches we use in developing and applying these technologies. The latter has been a key means of structuring this volume into parts, whereas the former is addressed slightly differently by each of the chapters that follow. The third dimension to the research field is the types of data structures which, from the examples in the map, embrace both IT-related work and technology-neutral classification systems.

The three pieces of the jigsaw, completed so far in the lower part of the puzzle, are concerned with the application of ICI. They relate to different

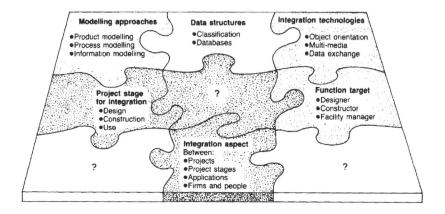

Figure 1   A map of Integrated Construction Information

aspects of the target for the work. Two of these, namely the project stage at which integration is targeted and the function target, are parameters that we use commonly to differentiate aspects of the construction discipline. The integration aspect dimension is one that is more particular to construction integration and refers to the different aspects of construction information that can be integrated. It suggests we can look to integration information used within different projects, within different stages of the same project, within different IT applications, and between different firms and people.

We feel that in compiling this book, we have put together some key pieces of the jigsaw. The map that is formed by completing the puzzle allows us to take a good view of how the ground of construction integration appears. The work is, of course, incomplete and there are some important pieces of the jigsaw that have yet to be filled in. At this stage we may even be unaware of the nature of the pieces of the puzzle for which we are looking. However, knowing what the remaining parts of the puzzle are, is made easier when those pieces we do have are put together in this way. It is in this light that the contribution of the book should be viewed.

These relationships within our work are fundamental and it is only when our work is collected together in this way that we can clearly see the overall picture. The gaps, in terms of parts of the map of our field, that are only partly addressed appear to be integration research concerned with the stages of construction and use of buildings. Work with a constructor, supplier or facility manager function base, that is corporately targeted, and that utilizes a broad range of integration technologies is also missing.

It will be interesting to see whether the next major collection of scientific thought on integrated construction information has brought us nearer to the attainment of a fully integrated industry by drawing from and inter-relating some other aspects of this research map and filling in further important pieces of the jigsaw.

ACKNOWLEDGEMENTS

This work has been made possible by the combined efforts of the contributing authors to the chapters that follow, making considerable efforts not only in writing such informative chapters but also in ensuring the consistency and accuracy of their contributions. We are most grateful to them for their efforts. In addition, all contributed most ably to the workshop at Armathwaite Hall, in the English Lake District, where we met to present our work to each other and we developed the material from which this volume has grown.

Several others contributed to this workshop although their work does not appear as a chapter. We are most grateful to Geoffrey Hutton, Alan Day, Robyn Thorogood, Ghassan Aouad, Terry Child and Serena Ford for their contributions.

Finally, we would like to express our sincere thanks to the Engineering and Physical Sciences Research Council of the UK government who provided the funding not only for some of the projects reported in this book but also for the workshop that allowed it to be brought together.

What is Integration and Why is it Important?

Part One contains five chapters which collectively address the issues of what is meant by integration and why it is important. This is vital context for the more detailed technical issues that are raised by chapters in the later parts.

Chapter 1 is a combined effort by Betts at Salford, Fischer at Stanford and Koskela at VTT in Finland. In itself, a contribution by three authors from leading centres in different parts of the world illustrates the way research in this field is progressing with international collaboration assuming ever greater importance. This chapter raises the issues of the purpose and definition of integrated construction information.

The purpose of integration is discussed in the context of emerging production management and business philosophies such as lean production and strategic IT exploitation. Both of these potentially provide new business change environments in which information integration will be encouraged. The definition of integrated construction information is then discussed through considering questions such as why, who, where, how and when integration will be realized. Alternative answers to those questions are put forward as potential points on a map of the integrated construction information field. This map can be used as a means for relating the contributions of all the other chapters in the book, as has been attempted in the Editorial Overview.

Chapter 2 by Fischer and Breuer both of Stanford, extends some of these issues further by attempting to define a whole range of managerial and organizational issues that impose upon information integration. These are then used by the authors as a framework for evaluating some of the technical efforts towards integration that have been made by the Center for Integrated Facilities Engineering (CIFE) at Stanford.

Chapter 3 extends our definition and understanding of information integration by focusing on the issue of data transfer. Here Thorpe relates data transfer technologies to a discussion of integration in the context of a four-stage model of how IT is exploited within organizations. The parallels to this model of UK construction experience are discussed, and the significance of data transfer technologies as construction companies enter the mature stage of IT exploitation is addressed.

Chapter 4 continues the debate of what integration is and why it is important by considering some different views. The fact that this chapter from Vincent is by a practitioner rather than an academic makes the viewpoint particularly valuable, given the focus of this part is on organizational and managerial issues.

The other important dimension that Vincent's chapter introduces is a human one and this is very much continued by Powell and Newland. Their chapter extends the definition of the integration problem by focusing on integration of interaction and sharing of meaning and understanding of information. Chapter 5 clearly shows that integrated information systems in construction will need to take note of the considerable variation in the way that information must be presented in order to be adequately received and thus for effective communication to have taken place. They are dealing with the integration of values.

These five chapters define what integration means from alternative viewpoints and establish why it is important from managerial, organizational, practical and human perspectives. It provides a broad conceptual framework by which to consider the past, present and future attempts at achieving integration that are covered in the next three parts.

CHAPTER 1

The purpose and definition of integration

Martin Betts, Department of Surveying, University of Salford,
Salford, M5 4WT, UK
Martin A. Fischer, Center for Integrated Facility Engineering,
Stanford University, Stanford, CA, 94305 4020, USA
Lauri Koskela, Laboratory for Urban Planning and Building
Design, VTT, PO Box 209, SF-02151 Espoo, Finland

1.1   INTRODUCTION

A problem with technology research generally and construction IT research in particular, is that researchers have rarely seen their projects in the light of the way they can be applied in practice, i.e. most integration research provides a technology push by developing a new technology and trying to push it onto an application. The purpose of integration research in a practical sense is often ill defined and how well the proposed solution helps to overcome a practical problem, or how easily the solution can be organizationally exploited in the evolving construction context, is seldom tested. Thus, integration research has produced many solutions looking for real world problems. An alternative approach to technology push as a model for innovation that has gained much acceptance in management disciplines, is to follow research and development in support of a strategy pull. That is, for seeing what the evolving business needs are and then developing a solution to the business problem.

A number of writers have commented on how IT can be exploited by companies to exploit strategic purposes including Porter and Millar (1985), Earl (1989) and Daniels (1991). Earl, in particular, stresses the need for the 'technology strategy connection' to be made and advocates the use of planning frameworks to assist in this. A combination of technology push

Integrated Construction Information. Edited by Peter Brandon and Martin Betts. Published in 1995 by E & FN Spon, 2–6 Boundary Row, London SE1 8HN.
ISBN: 0 419 20370 2

and strategy pull is likely to offer the best promise for successful implementation and application of computer-integrated construction (CIC).

In addition, while the achievement of CIC is recognized by some as a process of continuing improvements, most research projects are isolated short-term efforts. This temporal fragmentation has led to many ideas that have been developed to the prototype level but that have typically been abandoned before being tested in a practical context. These prototypes are seldom picked up and built upon by others.

Furthermore, researchers generally bemoan the lack of standardization in the Architect Engineer Constructor industry, but have themselves done little to agree on a standard terminology or objective for integration research. For example, there exists no generally accepted definition of the word integration in the context of CIC. Fischer (1989) defined integration as 'the continuous interdisciplinary sharing of data, knowledge, and goals among project participants'. Is this definition acceptable or appropriate for us all in our different contexts? There are likely to be many different definitions of integration which currently implicitly underpin our research efforts. Yet integration, with computers or otherwise, is not an intrinsic goal of benefit to construction organizations. It should be motivated by the specific improvement needs of the construction process or the business strategies of the participating firms.

What then, are the appropriate goals to be used in connection with improvement of construction processes? This chapter argues for a new, alternative view of CIC based on several recent approaches stressing process improvement and business strategy. After discussing these new views generally, the implications of them for CIC are analysed. The way that this is done is by posing the question 'why should we pursue integrated construction information?'

In many ways this question arises from taking a systems approach in construction as advocated by Armstrong (1985), who suggests that we should only move to the fourth stage of a systems approach, to operational programs, after having:

1. stated our ultimate objectives,

2. identified our indicators of success and

3. considered the alternative strategies for realizing them.

The argument here is that much of the CIC research is preoccupied with the implementation of operational programs. There is a need to shed light on the other three stages of a systems approach and this chapter will begin to do this.

1.2 WHY SHOULD WE PURSUE CIC?

This question may equate with the systems approach activity of identifying our ultimate objectives. The answer may appear to be obvious and related to the productivity and fragmentation issues raised. But in itself this is too simplistic a view and a deeper justification is required in the light of developments in management thinking. This is particularly so given that CIC, as a concept, originates as a parallel of CIM, which is applied extensively in manufacturing but usually in the context of different production management and corporate planning frameworks. For a justification for CIC then, some discussion of the recent developments in new production philosophies, competitive strategy and strategic IT planning are appropriate. These reflect part of the body of theory of organizational behaviour that we have seldom applied to our CIC research. They are increasingly being used as a justification for technological innovation in other sectors.

1.2.1 The new production philosophy

Since the end of the 1970s, many new approaches to production management have emerged within the manufacturing sectors. These include just in time (JIT) materials management, total quality management (TQM), time-based competition, value-based management, process redesign, lean production, world class manufacturing and concurrent engineering. Their significance for construction has been analysed by Koskela (1992) who found that many of the approaches view a common core from different angles. This common core is based on a conceptualization of production activities with the angle determined by the design and control principles emphasized. For instance, JIT materials management aims to eliminate wait times in the management of deliveries, production layouts and storage, whereas TQM aims to eliminate errors and related rework. Both apply to flows of work, material or information.

The sum of these different management advances is a new production philosophy. Regardless of what term is used to name it, it is the emerging mainstream approach practised, at least partially but increasingly, by major manufacturing companies in America, Europe and East Asia (Womack *et al.*, 1990).

The core of this new production philosophy is the observation that there are two types of activities in all production systems: conversions and flows. While all activities incur cost and time, only conversion activities add value to the material or piece of information being transformed into a product. Thus, the improvement of non-value-adding flow activities (inspection, waiting, moving) should aim to reduce or eliminate them, whereas conversion, or value-adding activities should be made more efficient. In design, control and improvement of production systems, both aspects have to be considered. Traditional managerial principles have considered only conversions, or all activities have been treated as though they were value-adding conversions. This distinction may seem slight but within manufacturing has been found to be of great consequence in effective production management in general and the successful implementation of CIM in particular. In some cases it has led to changes in business performance and strategy that has been of major significance to nations. Our argument here is that such a management approach may become of prime concern to the way construction is managed in the future in which case this issue would become a prime determinant of why we would aim for integrated construction information.

We can illustrate the basis of the new production philosophy and our current preoccupation with conversion activities in construction with a simple example drawn from Koskela and Betts (1993). This assumes a sequence of activities at the production stage of construction design to consist of those shown in Figure 1.1. First it has to be noted that the conventional description methods, such as generic flow-charting (used in Figure 1.1), do not show attributes of flow processes important to the new production philosophy such as waste (for example, amount of rework due to errors or omissions). Taking the preparation of the brief (Activity 1) as an example, making the brief is a value-adding activity. The needs of the owner do not exist explicitly, but have to be established and stated. On the other hand, any requirement which is missed in the beginning but occurs later in the process, is costly and produces waste. Thus, a systematized analysis of needs is extremely important; however, our conventional description methods do not suggest this in any way. The significance of avoiding having to restate needs is missing.

Second, value-adding and non-value-adding activities have not usually been distinguished, as they are in Figure 1.1. The exchange of files (Activities 5 and 6) provide an example. These are non-value-adding activities which should be reduced or eliminated. If we are aiming only to increase the efficiency of these then our research may be misguided. However, this is the goal which has been primarily adopted in much integrated construction information research. To what extent does it apply to other projects reported in this book?

This is obviously a very simplified example at a general level; as an example, it serves to illustrate the concepts of the new production philosophy. For a practical process improvement study involving CIC, the analysis should be taken to much more detail.

1.2.2   Strategic IT planning

In parallel with the emergence of the new production philosophy, a second major area of development in management thinking has been the concern with business strategy. Dominant in developments in

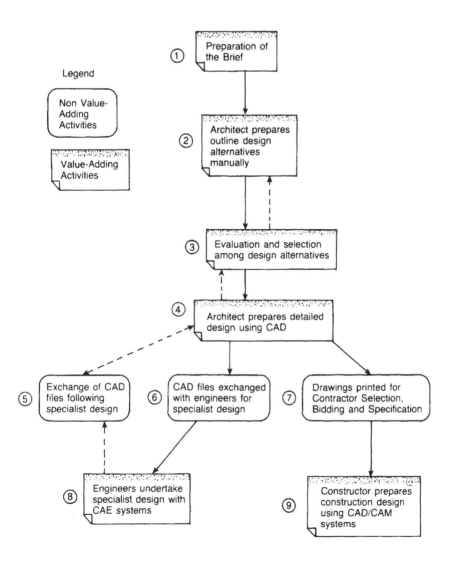

Figure 1.1 Construction design: an example of conventional analysis

competitive strategy and strategic planning in the 1980s was the work of Porter, whose advancement of a five forces model for analysing industry profitability potential, value chain for exploring competitive advantage and advocation of alternative competitive strategies have all been popularly adopted. More recently, Porter's work in the competitive advantage of national industry segments has resulted in a diamond of four interrelated components of a competitive environment which has been applied to the construction professions by Betts and Ofori (1994).

Other strategy writers have also recently advocated other business planning techniques such as benchmarking (*The Economist*, 1991) and the core competences approach (Prahalad and Hamel, 1990). The former is also seen as one of the principles of flow process design in the new production philosophy, illustrating the link between these two emerging management approaches. Core competences address the production and process issues more directly by looking to identify a basic production skill which can be applied broadly to out-compete rivals for whom that basic skill is not available. Their significance to our discussions here may lie in whether a technical ability in integrated construction information could result in a core competence that a construction organization would use as a strategic weapon. These strategic planning techniques and their relevance to construction enterprises have been discussed in detail by Betts and Ofori (1992).

In addition to the techniques of strategic planning, there have been other advances in techniques for identifying strategic information systems (Earl, 1989). Many of these techniques pose such questions as who, why, when and for what information systems are applied. Their significance for construction and their place within CIC has been considered by Betts (1992). They are also considered by Fischer and Breuer in Chapter 2. These techniques again take a conversion view of information systems with less concern about flows. But what relevance, if any, do these strategic IT planning techniques have to the new production philosophy and to an evaluation of the purpose of our research in CIC?

1.2.3   A production and strategy synthesis

The motivation for many process improvement activities and technological innovations is generally derived from strategic considerations, although there is some evidence for exploitation of technological opportunity. Most current strategic models do not acknowledge explicitly both the conversion and flow aspects of the new production philosophy. Many of Porter's early models most directly relate to conversions rather than flows and would therefore be out of line with the new production philosophy described. Porter argues that in particular, the generic competitive strategies of product differentiation, low cost leadership and focus are mutually exclusive alternatives, and points out the dangers of falling between any of them. The new production philosophy in contrast advocates that both lower costs and higher quality are attainable and indeed should both be aimed for in process improvement.

That is not to say that the strategic IT planning techniques are irrelevant to CIC but only that they do not show the complete picture. Applying CIC strategically using some of the above models for conversion activity improvements would be beneficial. They may not, however, bring the order of magnitude improvements that the new production philosophy has shown to be possible in other industries, which presumably is at the heart of our goals for integrated construction information.

The core competence approach is also restricted by virtue of its concern with only conversion activities. It is proposed by Stalk *et al.* (1992) that the approaches of core competence (referring primarily to conversion aspects) and process capability (referring primarily to flow aspects) could be used in parallel to chart the strategic improvement potential. Thus, in the case of building design, core competence primarily refers to the skills of designers and their design tools, such as calculating programmes, whereas capabilities refer to the properties of the actual information flows in design. Which of these do we primarily consider in CIC?

What we are left with from this review of developments in strategic planning is an understanding that strategy and production are interdependent. To implement process improvement without a mind to the strategic objectives is unlikely to result in innovations to suit the way the construction industry will be organized in the future. But by the same token, to develop competitive strategies without consideration of the way that production is carried out, and the scope for process improvement that exists, is unlikely to result in a strategy that is practically realizable. This applies most particularly to production-dominated activities such as construction. Our review of these two areas of management developments therefore can be combined to give a synthesized justification for technology applications. Any new developments in technology management must address the process improvement approach required by the new production philosophy and the requirements of the new thinking in business strategy. Taken together, we might describe this need for technology to contribute to the synthesis of these two approaches as 'evolving customer needs'. These are clearly important drivers to our efforts to attain integrated construction information.

Our consideration of these matters has been concerned with both supply-side and demand-side issues. This again brings us back to the idea of a technology strategy connection which we can illustrate here as an opportunities matrix. Organizations within construction may view a combination of technology and strategy issues, as shown in Figure 1.2. All in construction should identify the key demand-side issues, or 'evolving customer needs', that face them in the future and combine these with a consideration of the emerging technologies of a constructional, financial, organizational and behavioural type. These may be those technologies they can acquire or gain access to rather than develop internally. They should then see where the technology-push/strategy-pull forces act together.

In terms of our desire for integration of construction information through CIC, we can use some examples from our matrix as illustrations. We may assume that one of the key emerging technologies is object-

Demand Side / Strategy Pull

			Evolving Customer Needs			
			Buildability	Maintainability	Early Visualization	Automatic File Transfer
Supply Side/ Technology Push	Emerging Technologies	Object Oriented Systems	*Example 1*			
		Virtual Reality				
		Multi-media			*Example 2*	
		Broad-band comms				

Figure 1.2 Opportunities matrix for an AEC organization

oriented systems and that an evolving customer need for an integrated AEC constructor organization increasingly involved in 'build, operate and transfer' projects would be for early buildability and maintainability advice. This may give rise to a technology strategy connection opportunity that gives a specific purpose to one particular type of integration research.

A second example may see the emerging technology of broad bandwidth data communication links and of multimedia technology with the evolving customer need for value-added early visualization of internal office space alternatives for an owner–occupier of an office development. In addition, the traditional flow activities of the types shown in Figure 1.1 may need to be completely removed by automatic file transfer. Again, a specific technology strategy connection opportunity for integration research emerges but of a quite different nature. Both examples draw from a vision of a dynamic industrial viewpoint and research in response to change. Not all cells in the matrix will give rise to integration research scope and the detailed analysis with the matrix that would be required in an individual, practical context therefore gives rise to identification of where to research.

From this initial consideration of the current state of development in these dynamic fields it appears that the new production philosophy and the strategic concepts of core competence and capabilities offer scope for a combined strategy and process improvement approach which should be combined with our current technological preoccupation. It is to these ultimate objectives that our efforts in CIC may most purposefully be directed. Up till now most CIC research appears not to have been guided by any production philosophy or strategic model other than automating and integrating current practice based on an implicit assumption that automation is in itself desirable and that current industry practice is likely to remain.

The answer to our question of why we are aiming for CIC is thus not a generally applicable one. There will be different reasons for following different types of integration research for different beneficiaries at different points in time and in different construction contexts. This is the range of CIC research objectives that we must consider.

1.3 A GENERIC DEFINITION OF INTEGRATION

Having identified the principles that explain why we are aiming for integrated construction information in the context of emerging management philosophies we can now return to the second stage of a systems approach whereby we specify our indicators of success. This issue can be addressed by having a clearer and operationalized definition of what we mean by integration in the different contexts outlined above. The definition of integration mentioned above (Fischer, 1989) prompts us to ask several questions regarding CIC. Who integrates what; how and when one should integrate; and why one would choose to integrate. This leads to the development of the framework presented in Table 1.1. This framework was first proposed as an initial basis for discussion to define dimensions and levels of integration towards CIC (Fischer *et al.*, 1993). The relevance of the framework to us here in this book may lie in its explanation of the alternative definitions of integration that underpin the projects described in the other chapters.

With regard to who, we can imagine integration among individuals and departments leading to the integration of entire firms and projects, and ultimately to the integration of the entire construction industry. With regard to the question of what to integrate, as an initial step we might choose to focus on sharing data. This could then be expanded to include models, such as product and process models, knowledge about decisions, and project goals. Ultimately, data, models, knowledge and goals would all be shared. With regard to when we integrate, just a few applications within one phase and discipline might be a starting point which could then be expanded to include all applications from all disciplines and phases. Reasons why anyone might integrate or increase the level of integration are to stay in business, increase profit, market share, market size or to enter or even create new markets. That CIC offers these opportunities has already been demonstrated by such companies as OTIS (Cash and McFarlan, 1990) along with many others from outside of construction. Bradshaw (1990) also demonstrates a building materials supply company exploiting similar opportunities.

Table 1.1 Dimensions and levels of integration

	Low Integration	→	→	→	*High Integration*
Who?	Individuals	Departments	Entire organizations	Whole project life cycle	Entire industry
What?	Data	Models	Knowledge	Goals	All project information
When?	Islands of automation	Multiple applications in one discipline and phase	Multiple applications for multiple disciplines in one phase	Multiple applications for multiple discipline and phases	All applications in project delivery process
Why?	Survival, stay in business	Increase profit	Increase market share	Enter new market	Create new market

Fischer *et al.* (1993) attempted to list the various values of these four dimensions of integration to indicate increasing levels of integration. However, they found that it is not always necessary to reach the previous step to go to the next level of integration. For example, a firm could well opt to first integrate goals and then tackle knowledge or model integration. Yet a framework such as this is important to distinguish different forms and stages of integrated construction information and for all in practice and research to be able to cross-refer to different initiatives using a common understanding. Earl (1989) also argues that such a temporal framework is important in identifying sequential stages of progression in research and development.

This framework allows individuals, departments, companies, projects and industries to plot their current state of integration and to indicate efforts to increase the level of integration. Thus, the framework becomes a vehicle for comparison and for focusing development and implementation efforts. For example, two departments might differ in their capabilities of sharing project information, or a company might be interested in pushing its integration capabilities from level 3 to level 4 for the how and when dimension.

This framework also provides a generic and focused definition of integration. Generically integration can be defined as the sharing of something by somebody using some approach for some purpose. Obviously this is not a very useful definition. However, if one substitutes the vague expressions with values from the framework, one can create a definition that suits a particular purpose. For example, a firm might define integration as the sharing of data and models by departments using several applications pertaining to a number of disciplines and project phases for the purpose of increasing profit and market share. A government body may see integration as industry-wide data exchange between applications in multiple disciplines for industry survival. The advantage of the framework is that different definitions can be related to each other easily. The framework has five dimensions that are each presented in Table 1.1. independently. In reality, each dimension can be combined in a multidimensional way which we can illustrate by combining the dimensions why and who as in Table 1.2.

Through this framework and the generic integration definition that it allows, we have not specified a single indicator of success for our systems approach. We have given the range of different indicators of success for the range of CIC research objectives that we foresaw at the conclusion of Section 1.2.3.

1.4   RESEARCH MAP

The third stage of a systems approach is to consider alternative strategies. This is something we seldom do in our CIC research. It is much more

Table 1.2 An example of the interplay of two dimensions within the framework

WHO?

	Individuals	Departments	Enterprises	Projects	Industry
	Stay in a job	Department survival	Stay in business	Complete project	Industry survival
	Increase earnings	Increase profit contribution	Increase profit	Increase project success	Increase industry profitability
WHY?	Extend job authority	Increase department political role	Increase market share	Increase project scope	Increase industry share of economy
	Change jobs	Assume additional roles	Enter new market	Make project more widely useful	Extend into other sectors
	Create a new job	Create new roles	Create new market	Create project extension	Create new sectors and economic services

common for each CIC research group to plough its respective furrows in ignorance of what is happening in neighbouring fields and with no desire to learn from the better harvests of our close rivals. We do appear sometimes over-willing to look more distantly for instant technological solutions in the form of new ploughing machinery but once we have acquired it, inertia returns. The preoccupation of CIC researchers first with expert systems and subsequently with object-oriented systems illustrates this well.

Of great value within such a scenario may be an overview of different research approaches in the form of a research map of the field. Few of us appear to work towards the drawing of such a map or to be aware of how it shows the ground to be lying.

Table 1.3 is a first attempt to draw a CIC research map arising from some of the discussions of a group of researchers at a workshop in Finland (Fischer *et al.*, 1993). It shows many dimensions by which our CIC research can be defined and delimited. There are many others that are excluded here but may be relevant. The important questions each CIC research group should ask of itself are: Where are we on the map? Are we in the right place? Are we heading towards our intended target? Where are others? Who are our closest neighbours? What links are possible between us and our neighbours? The important questions to be addressed by the CIC research community collectively are: How do we cover all the ground on the map? How do we make it interrelate? What is on the other maps next to ours?

When some of these questions were put to the authors of the other chapters in this book, the following patterns emerged. Most of the work reported in this book is oriented towards the design stage and the design disciplines, suggesting that most others see that as the key to subsequent integration. The aspects of integration considered by authors was varied, suggesting that different authors have quite different interpretations of what integration means. Finally, this book describes projects embracing a range of integration technologies with object-oriented systems and data exchange standards well favoured.

We have described this concentration in more detail in the Editorial Overview. This gives examples of coverage of parts of the research map from the individual chapters that follow.

1.5 CONCLUSIONS

This chapter began by making a case for more thought to be given to the question of why we seek CIC and what constitutes integration. By following a systems approach we first identified objectives, then specified indicators of success and followed this with a consideration of alternative strategies. Only then would we adopt and implement an operational programme. Much current CIC research seeks to implement a particular

Table 1.3 CIC research map

	1	2	3	4	5
Project stage	Programming	Design	Planning	Construction	Use
Discipline target	Owner	Designer	Constructor	Supplier	Facility manager
Research directors	University/Research institute	Corporate R&D department	Software vendors	Professional institutions	Construction end users
Strategic target	Building product	Project	Enterprise	Profession	National construction industry
Research category	Basic research	Applied research	Product and process development	Dissemination	Implementation
Integration aspect	Sharing data between projects	Sharing data between life cycle stages	Sharing multiple data sources within an application	Sharing data between distinct applications	Sharing data between organizations and people
Deliverable	Concept, theory, methodology, framework, knowledge	Models	Tools for R&D	Prototype applications, guidelines, standards	Commercial products and services
Integration technology	Object-oriented systems	Hypertext	Data exchange standards	Virtual reality	Multimedia
Driving factor	Technology push				Strategy pull

operational programme without putting the work into a context of broad objectives and indicators of success. This first chapter does not contribute a solution to the integration problem nor does it describe a particular programme.

What the chapter has shown is that the prerequisites to a programme are:

1. deciding why we are doing it;

2. how far we want to go with it and

3. how we might get there.

These questions are addressed in the chapter by considering our current quest for CIC in terms of recent managerial philosophies, by considering the extent in terms of a multidimensional generic integration definition and by speculating on how we might get there by attempting to draw a research map. The contribution of the chapter therefore lies in the foundations it may give to the operational programmes research of others. It is these issues that much of the rest of this book is concerned with.

ACKNOWLEDGEMENTS

This chapter brings together recent collaborative work that has had a significant input from Matti Hannus of VTT and Jarmo Laitinen of HAKA in Finland, George Ofori at NUS in Singapore and of Yusuke Yamazaki of Shimizu Corporation. We acknowledge and thank their contribution to many of the ideas that have been assembled here.

REFERENCES

Armstrong, J.S. (1985) *Long Range Forecasting: From Crystal Ball to Computer*, 2nd edn, John Wiley and Sons, New York.

Betts, M. (1992) *Information Technology Planning Frameworks for Computer Integrated Construction*, Proceedings of International Workshop on Models Supporting Computer Integrated Construction, Technical Research Center of Finland (VTT), Espoo, Finland, October 5–9, 1992.

Betts, M. and Ofori, G. (1992) Strategic planning for competitive advantage in construction. *Construction Management and Economics*, **10** (6), 511–32.

Betts, M. and Ofori, G. (1993) Strategic planning for competitive advantage: The institutions. *Construction Management and Economics*, **12** (3), 203–17.

Bradshaw, D. (1990) Building blocks of efficiency. *Financial Times*, 30 August, p. 18.

Cash, J.I. and McFarlan, F.W. (1990) *Competing Through Information Technology*. Harvard Business School Video Series, Harvard, MA.

Daniels, C. (1991) *The Management Challenge of Information Technology*, The Economist Intelligence Unit Management Guides, London.

Earl, M.J. (1989) *Management Strategies for Information Technology*, Prentice Hall, London.

Fischer, M. (1989) *A Constructability Expert System for the Preliminary Design of Reinforced Concrete Structures*. Proceedings of the 6th Conference on Computers in Civil Engineering, ASCE, pp. 60–6.

Fischer, M., Betts, M., Hannus, M. *et al.* (1993) *Computer Integrated Construction Framework*, First International Conference on the Management of Information Technology for Construction, Singapore, August.

Koskela, L. (1992) Application of the New Production Philosophy to Construction. *CIFE Technical Report #72*. Department of Civil Engineering, Stanford University.

Koskela, L. and Betts, M. (1993) *Computer Integrated Construction in the Context of the New Production Philosophy*, First International Conference on the Management of Information Technology for Construction, Singapore, August.

Liker, J.K., Fleischer, M. and Arnsdorf, D. (1992) Fulfilling the promises of CAD. *Sloan Management Review*, Spring, 74–86.

Porter, M. and Millar, V.E. (1985) How information gives you competitive advantage. *Harvard Business Review*, July–August, 149–60.

Prahalad, C.K. and Hamel, G. (1990) The core competence of the corporation. *Harvard Business Review*, May–June, 79–91.

Stalk, G., Evans, P. and Shulman, L.E. (1992) Competing on capabilities: The new rules of corporate strategy. *Harvard Business Review*, March–April, 57–69.

The Economist (1991) Competing with tomorrow, 12 May, 67–8.

Womack, J.P., Jones, D.T. and Roos, D. (1990) *The Machine That Changed The World*, Rawson Associates, New York.

Wright, R.N. (1991) Competing for construction in the world arena. *Construction Business Review*, May/June, 36–9.

CHAPTER 2

Managerial issues affecting integration

Martin A. Fischer, Center for Integrated

Facility Engineering, Stanford University, Stanford,

CA 94305 4020, USA and James Breuer, Fluor Daniel,

Mining and Metals, Vancouver, Canada

2.1 NON-TECHNOLOGICAL ISSUES AND IT

Most of the integration research in civil engineering has focused on proposing technological solutions to overcome fragmentation in the AEC industry (Howard *et al.*, 1989). Yet there is an equally important aspect critical to the success of such IT solutions. Many non-technological or managerial issues need to be considered by companies as they adopt advanced information technologies. We performed a survey of the literature to summarize these non-technological issues, borrowing many of the ideas from the general field of information management, since not much has been written on the subject specifically for the AEC industry. We group the issues into three main categories: planning, organization and development and management.

2.2 PLANNING ISSUES

Planning issues are those that deal with the company's overall plan or strategy for the exploitation of information technology. Their purpose is to set the course of all IT efforts and to develop a blueprint for the development of individual applications. Some of the questions that need to be answered are: What can IT do for us? How do we find the right applications? Who should lead the effort? When should we do it? The answers to these questions should be broad enough to be applicable to changing situations, but useful enough to provide the company with a systematic method of improving its IT resources.

Integrated Construction Information. Edited by Peter Brandon and Martin Betts. Published in 1995 by E & FN Spon, 2–6 Boundary Row, London SE1 8HN.
ISBN: 0 419 20370 2

2.2.1   The role of IT

The role of IT in the commercial world has changed considerably over the last 30 years. The first applications were data-processing in nature, covering functions like accounting, payroll and complex numerical calculations. The second-generation applications were decision-support systems, whose function was to manipulate and filter information to facilitate decision-making. Finally, a third generation of applications evolved, strategic information systems which are critical to a company's business and provide it with an advantage over its competitors (Earl, 1989), or which improve the competitiveness of the industry as a whole (Porter and Millar, 1985). Although the AEC industry has been using the first two types of IT applications, it has not yet been able to use advanced information technologies in a strategic way.

To achieve this, companies will need to define the role that IT will play within their environments. Porter and Millar describe the potential applications of IT within a firm using the concept of the value chain which 'divides a company into the technologically and economically distinct activities it performs to do business' (Porter and Millar, 1985). The value chain is composed of activities and linkages between these activities. The value system, composed of many value chains, represents an entire industry. Companies and industries can use IT strategically, if they can find ways of applying it to either improve individual value activities, or alter the linkages between them to make the overall value chain or value system more efficient. The opportunities for IT increase as the information requirements of the value chain increase. This can be assessed by looking at several criteria, such as number of suppliers and customers, number of distinct product varieties, number of steps in a product life cycle, and duration of the product life cycle (Porter and Millar, 1985). As we can see, the AEC industry is a sector with very promising IT applications, since it has a high information intensity in all of these criteria.

2.2.2   Leadership

Until now, AEC companies have treated IT as a support function, where key decisions have been made by middle management, responsible for the entire IT project cycle from initial planning to implementation to management. As information systems become more strategic, however, companies must regard them as an integral part of their business just like any other technology or resource. Therefore, IT has to be included in the overall strategic planning process, which necessarily involves senior management.

Senior management's participation in the IT planning process is key for success. In fact, a survey of IT managers from different industries shows that the majority of them think that the single most important factor in determining the effective use of IT is the perspective on IT of Chief Executive Officers and other top managers (Index Group, 1989). Top management's involvement in IT will ensure that adequate resources are committed. It will help align IT strategies and goals with those of the company as a whole. Equally important, it will send a strong signal throughout the company that IT is taken very seriously (Bawden and Blakeman, 1990).

2.2.3   Identifying IT opportunities

As the scope of IT opportunities grows rapidly, the task of identifying those uses that will maximize benefits becomes more important. Researchers agree that the key is to compare and align the goals of IT constantly with those of the company (Porter and Millar, 1985; Earl *et al.*, 1988; Earl, 1989; Goldsmith, 1991; Betts, 1992). Porter and Millar recommend a methodical approach to finding IT opportunities by looking at ways to improve the value chain and to enhance the information content of the product. In the first case, the production process becomes more efficient; in the second, the final product is improved. This argument is similar to Tatum's more general discussion on the use of technology to achieve process or product innovation (Tatum, 1988).

Earl *et al.* (1988) prefer a more informal method of identifying IT opportunities. They propose an environment where all users take the lead in finding ways to use IT. They argue that user managers are 'most likely to identify and prosecute competitive advantage opportunities [because they are] close to competitive forces and can see IT threats and opportunities more clearly than information system specialists'. For this to occur, however, users have to learn about the technology's capabilities and limitations, and they have to embed IT in their everyday thinking. In the case of the AEC industry, potential IT users, namely, design engineers, project managers, estimators, site engineers, etc., should be aware of the capabilities of the technology so that they can identify and promote its application to information-intensive tasks.

2.2.4   Resource commitment

IT must be treated as an investment, rather than as an expenditure. This definition requires a change in perspectives in favour of a more long-term view of IT projects. Just as a company engages in an investment venture it projects the benefits over a certain time period, so it also has to realize that the benefits of IT will come gradually rather than immediately. Therefore, the AEC industry must understand that there is a lag between the up-front cost of IT and the benefits it offers (Björnsson, 1991; Gibson and Bell, 1992). Also, case studies have shown that the majority of successful applications were built on top of good existing IT infrastructures (Earl *et al.*,

1988). Therefore, companies must realize that benefits do not come overnight, but are the result of a long-term commitment.

It is also critical to include all resource requirements in the budgeting and funding process. This seems to be too obvious to even mention, but as companies start using more advanced types of IT, it can become a serious issue. Hodge (1992), for example, spends considerable time itemizing all the costs of a typical CAD station, a relatively widespread application, and concludes that software and hardware add up to only 34% of the total costs. The rest is comprised mainly by items such as technical support, training and maintenance. With newer and less-known technologies, we can expect these items to take an even larger share of the total cost. Since they are not easy to quantify, a good method of estimating them is essential.

2.2.5   Innovation vs. imitation

Researchers argue that advances in computer integration in the AEC industry are smaller than in other industries because of fragmentation in the industry and because the scale and complexity of facility delivery projects make integration a very difficult task (Howard *et al.*, 1989). We can add a third cause: intense competition and the project-driven nature of the industry make companies reluctant to invest in new types of IT, whose benefits are not immediate and many times not guaranteed. Companies who would rather invest in more traditional, commercially proven technologies, do not have the flexibility or vision to innovate with IT (Betts *et al.*, 1991). Although imitating other firms or industries is a highly valid strategy for technology transfer, it should not be the only one. Tatum (1988) tries to dispel the idea that 'in construction, we don't experiment' by pointing out the first-user advantages of technological innovation. As companies map their IT strategy, they should definitely consider the possibility of being IT innovators.

2.3   ORGANIZATIONAL ISSUES

Organizational issues are those related to the changes in organizational structure and attitudes that need to accompany the use of advanced IT. As a strategic tool, IT has the ability to alter radically the ways in which business is done. However, if implemented in a stagnant environment, IT will not provide the benefits it offers. Along with its introduction, companies must also embrace organizational changes that will be possible thanks to IT and, at the same time, are required to maximize the use of the technology.

2.3.1   A different perspective on IT

Two points regarding this issue stand out in the literature. The first is the need to integrate IT to all activities and levels in the company. Regarding this first point, Earl talks about the need to 'normalize' information management, in other words, treat information resources as any other resource (Earl, 1989). He describes how in highly successful companies, users and IT specialists work closely together on IT projects. Other researchers also stress the importance of active user involvement (Goldsmith, 1991; Betts *et al.*, 1991; Cash *et al.*, 1992; Wei, 1993). In fact, Betts *et al.* (1991) argue that the reason why the AEC industry is not able to use IT strategically yet is because companies view IT as a technological issue, rather than a business issue, and thus delegate it to IT specialists.

The second point is the idea that IT can alter the way business is conducted by offering opportunities that did not exist before. The main push in the AEC research community is to use IT to 'integrate' different parties and phases in the project life cycle (Howard *et al.*, 1989). In the future, applications will be geared towards user groups, rather than individual users (Applegate *et al.*, 1988). This implies a major shift in the use of information systems. Integrated environments will need to solve the issues of how to interface with users from different disciplines, and how to make different hardware and software communicate with each other (Fenves *et al.*, 1990).

2.3.2 Changes in organizational structure

Traditional organizational structures, both at the company and at the AEC project levels, have been chosen for good reasons. As projects became increasingly complex, the industry specialized into architects, design engineers, consultants, general contractors, specialty contractors, suppliers, etc. This division of labour allowed owners to have access to the expertise needed to build complex facilities. However, it also resulted in a fragmented industry, characterized by a linear facility delivery process with little concurrent exchange of information between interdependent parties. Researchers agree that advanced information technologies require changes in current organizational structures. Teicholz and Fischer (1994), for example, argue that the only way the industry will achieve computer integration is if a more co-operative atmosphere exists between the different parties involved. Fenves *et al.* (1990) say that current information flow structures must be modernized, since they do not lend themselves to integrated decision-making. Liker *et al.* (1992) studied six major manufacturing firms and concluded that the main reason for their ineffective use of CAD/CAM technology was a lack of 'social integration'. In essence, companies must redesign their processes to materialize the opportunities of IT.

2.3.3 Role of users vs. role of IT specialists

The importance of user involvement in all stages of the IT development process has been mentioned before in this article, specially with regards to providing leadership and identifying opportunities for the use of information technology. IT specialists, on the other hand, although not in complete control of all IT activities anymore, still must perform the very important tasks of technical support, advising and the actual design and construction of the information systems (Cash *et al.*, 1992). Although the degree of influence of each group will depend on the organizational nature of the company, it is important that the two enjoy fairly equal status, since they both play very critical roles.

2.3.4   The threat of IT

Because of the changing nature of IT, successful applications require continuous learning and adaptation from users. In many cases these conditions do not exist in companies, in effect creating barriers to the improvement of information resources. People feel threatened by technology because they think their jobs will be replaced or become automated and routine-like. Liker *et al.* (1992) describe a situation in a manufacturing firm where users felt that a 3D CAD system was lowering the quality of their jobs. Agai (1988) describes a similar situation with an office management system in an engineering/design firm.

These problems are solvable if companies establish proper methods of educating users about IT. At the same time, IT specialists must be aware of the limitations and concerns of users, thus facilitating co-operation between these two groups. Most importantly, senior management has to ensure that the entire company shares the same commitment and objectives regarding its information systems.

2.4   DEVELOPMENT AND MANAGEMENT ISSUES

While planning and organizational issues cover a broad scope, development and management issues deal with the development of individual applications and the everyday management of information systems. In this sense, some of the questions that a company should ask are: Should we develop IT resources internally or procure them from the outside? How centralized should the IT function be? What is the best system implementation method?

2.4.1   Make or buy

The decision of developing a system in-house or externally will depend on the type of application, size of the company, size and experience of the IT staff, and backlog, among other factors. In the AEC industry, except for large firms, companies typically buy IT applications from the market. This has become especially common since the 1980s, when the advent of the

PC allowed firms of all sizes to have access to cheap, mass-marketed software and hardware (Lackowitz, 1992). As information systems gain importance, however, companies will need to get more involved in the development process for two reasons. First, many applications based on knowledge-base systems will contain proprietary knowledge, which means that user participation will be essential. Second, as companies become more dependent on their information systems, they will need to become much more IT literate.

Some of the advantages of in-house development are the ability to create a customized product, the ability to maintain confidentiality and the ability to adapt quickly to changing system requirements. Some advantages of out-sourcing are lower cost, quick access to specialized skills and the larger variety of IT services available in the market (Cash *et al.*, 1992).

2.4.2 Centralization vs. decentralization of IT

The main question is whether to delegate control of information systems to users or whether to have a central information systems group manage several systems. IT organizations were centralized in the 1960s and 1970s because the highly technical systems needed a specialized staff to run them. In the 1980s, this changed with the widespread use of PCs and user-friendlier software. Some research suggests that this trend will continue. A recent survey of large firms in the United Kingdom, including AEC firms, shows that most companies prefer an IT structure of 'federal decentralization', where a central function is responsible for overall strategic planning and divisional units are in charge of development and management decisions (Hodgkinson, 1992). Other people feel that the control over IT resources will be re-centralized in the future. Reinschmidt (1993) thinks that AEC firms went too far in decentralizing their systems in the 1980s. Users often made the wrong decisions because of their lack of expertise in IT. They also spent large amounts of time maintaining and upgrading their systems, a task that could have been done more efficiently by specialists. As AEC markets become increasingly competitive, companies will start paying closer attention to IT investments and require that a central group oversee all major decisions (Reinschmidt, 1993).

2.4.3 Implementation method

The implementation process deals with the introduction of a new system into the firm's existing operations. Although it largely influences the success of an IT system, it is often not planned adequately. The selection of the best implementation method is a function of several variables, including the size and nature of the system, the differences between the old and new ways of performing the task the system is trying to improve, and the risks and consequences of failure. There are three main methods of system implementation: total conversion, parallel system operation and gradual transition (Bawden and Blakeman, 1990). The first involves switching from the old to the new system at one time. If everything goes well, it offers minimum disruption, but in general it is a fairly risky method. In the second method, the old and new systems are run concurrently, which allows for easier comparison and debugging, but can be very expensive. The third method involves implementing the new system gradually, either by pieces or by trying it on a small user group. It fits well with modular-type applications, but it can be very tedious.

The best implementation method depends on the characteristics of the application, rather than on the type of industry where it is implemented. Therefore, it is improper to select an 'ideal' method for the AEC industry. Instead, a company must look at the features of individual applications to select the appropriate method.

2.4.4   Results control system

A results control system describes the mechanisms that a company uses to control and measure the results of its IT systems (Cash *et al.*, 1992). It resolves the issue of accounting responsibility and relationships between the IT function and users. Earl identifies four major types: service centre, cost centre, profit centre and hybrid centre (Earl, 1989). A service centre means that the company does not charge users for IT resources. This method can promote the use of and experimentation with IT, but it can also be very expensive if resources are employed irresponsibly. A cost centre approach charges users for access to information systems. This method promotes a more efficient use of IT resources, but it can also discourage people from using them as they might consider them too expensive. The profit centre approach is similar to the cost centre method, with the difference that the IT function needs to make a profit on its services. The hybrid centre, a combination of any of the three methods previously described, can be particularly helpful in a company where different information systems have different requirements.

As with the implementation method, the selection of the results control system should be a function of the individual IT application, rather than a function of the type of industry. However, those companies in the AEC industry that are trying to find ways of using IT strategically should probably favour a service-centre approach, at least initially, since it is the one that provides the greatest incentives to experiment and innovate with the technology.

2.4.5   Training and education

Companies frequently invest heavily in information systems hoping to achieve major benefits, yet neglecting one key aspect of the IT process: the training and education of their workforce. Researchers agree that this trend has to change if companies want to exploit fully the advantages that IT offers (Earl, 1989; Bawden and Blakeman, 1990; Baker, 1991). In fact, one author views the lack of training and education as the major barrier to the use of IT in the AEC industry, suggesting that many people in the industry still need basic computer education (Baker, 1991). Not only should companies train their employees to use individual applications, but they should also develop an 'information management education' plan, where managers are taught about the capabilities of IT and the conditions for its successful exploitation (Earl, 1989).

2.4.6 Operational dependence

As IT systems become more critical and strategic to companies, then these companies become more dependent on the proper functioning of their systems. Therefore, particular attention has to be placed to make sure that information systems work properly both technically, i.e. avoiding crashes and breakdowns, and functionally, i.e. giving the right answer. This issue does not seem that difficult for traditional IT applications. The majority of current applications are stand alone and produce some kind of numerical or quantitative output. Therefore, it is relatively easy to keep systems running and to control the accuracy of their input and output. With newer and less-known technologies, this issue is not as obvious. First, the networking and interoperability requirements of integrated environments will add a degree of technical complexity, making quality control a more difficult task. Second, many new applications, such as those associated with artificial intelligence, will produce outputs that are qualitative and even judgemental in nature (Rosenman *et al.*, 1989). In these cases, the performance of companies will be greatly affected by the effectiveness of these systems, yet it will not be an easy task to ensure that the systems themselves perform adequately.

2.5 DISCUSSION

It is clear that the use of advanced information technologies in the AEC industry and thus the organizational achievement of integrated construction information, will demand more than the elimination of technological barriers. Along with acquiring technologies that work, companies must also deal with many planning, organizational and development and management issues. None of the solutions to these issues are revolutionary or counterintuitive. Many of the ideas discussed are based on common sense. Yet they can present serious obstacles to the successful implementation of IT, if they are not dealt with properly. Therefore, companies should not only consider them when they adopt IT, but academics should also keep them in mind when planning their research. Later chapters of this book give some consideration of these issues for the respective research initiative. Here, we examine IT research at CIFE in light of the issues presented so far in this chapter.

2.6   RECENT IT RESEARCH AT CIFE

CIFE was started in January 1988 as an industry affiliate programme in the School of Engineering at Stanford University. The main participating departments are the Departments of Civil Engineering and Computer Science. Currently, over 40 companies are members of CIFE. These companies are in Asia, Europe and North America, and participate in the AEC industry as owners, operators, architects, engineers, contractors, or software and hardware developers and vendors. CIFE's goal is to establish a research partnership between industry and university to help increase the quality of the products and services provided by the AEC industry. The main research focus so far has been on improved automation and integration over the facility life cycle. It is probably fair to say that one of the main challenges CIFE faces today is that of technology transfer to its member companies.

A significant number of research projects at CIFE have used artificial intelligence (AI) methodologies as the tool of choice for investigation. In the context of CIFE, AI has been a very useful tool for exploration or discovery, whereas in an industry context, AI should be a tool of delivery. This difference alone might explain our problems of technology transfer since we are very familiar with the use of AI for discovery but not for delivery. Let us, however, first summarize some research efforts at CIFE and then examine how well these projects have addressed the non-technological issues outlined in this chapter.

At CIFE, researchers have used AI tools in a number of ways to model aspects of the project delivery process. They have developed tools, methodologies and approaches to model AEC products (e.g. facilities), processes (e.g. construction schedules) and organizations (e.g. design teams). These tools support the generation and sharing of project information and the simulation of products, processes and organizations under specified loads and requirements.

In the style of traditional expert systems, some projects have captured heuristics applied by professionals in the solution of a particular problem. Examples of such projects are CAADIE and COKE. The CAADIE project automated the generation of bubble diagrams for the early phases of the architectural design of university buildings (Chinowsky, 1991). It captured the expertise of professional architects with respect to positive and negative adjacency requirements of rooms, noise, access and lighting requirements of required spaces. Given a set of rooms and requirements,

CAADIE automatically generates bubble diagrams and shows how well a diagram satisfies the stipulated requirements. An architect can then modify the diagram and receive instant feedback on how well the solution is doing with respect to the requirements. CAADIE thus automates the application of architectural layout knowledge. The COKE project automates constructability feedback to the preliminary design of reinforced concrete building structures (Fischer, 1991a). It captures construction knowledge that experienced designers, contractors and suppliers would use in critiquing the constructability of a structure.

AI tools have also been used to model products, processes and engineering organizations. For example, the COKE project used a symbolic model to represent the geometry and topology of a reinforced concrete structure (Fischer, 1991b). The primitive composite (PC) approach is a more substantial product modelling research effort (Phan, 1993). The PC approach provides a methodology to model individual concepts (the primitives) used in facility engineering as objects and supplies mechanisms to aggregate these primitives into composite objects that represent discipline, application and phase-specific views of the current state of a project. In the area of process modelling, the OARPLAN project has automated the generation of construction schedules from a product model (Darwiche *et al.*, 1989), and the IRTMM project automates the generation of maintenance schedules for power plants from a process model of the plant and from real-time sensor data (Jin *et al.*, 1992). All the projects described in this paragraph demonstrate a departure from classical heuristic AI systems to model-based systems based on symbolic, object-oriented models. In the OPIS project, Froese (1992) illustrated the usefulness of such object-oriented models to integrate project management activities. The VDT project applies object-oriented models to represent the structure and experience of an engineering organization and to simulate the performance of such an organization given certain tasks and tools (Cohen, 1992).

The main emphasis of the projects described so far has been on the modelling of project information for the purposes of data sharing and simulation. AI tools have also been used for knowledge sharing and for the rapid generation of product models. Bicharra developed the ADD system which automatically captures and documents the rationale used by a designer of HVAC systems in preliminary design (Bicharra-Garcia, 1992). When later queried about the reasons for a certain decision, ADD can explain the reasoning that was applied in making a particular decision. Tauber *et al.* (1992) developed the CONCEPTIMATOR system. Given client requirements, site conditions and user input, CONCEPTIMATOR rapidly generates foundation designs for buildings to determine the most cost-effective solution.

We would now like to examine whether the research projects summarized above have considered the managerial issues outlined earlier at all. It is probably fair to say that few of the projects were started and carried out with an explicit consideration of these issues. Instead, AI research projects at CIFE are usually put in a framework that focuses on the purpose of the research, the representation of 'things' in the system, the reasoning done by the system, the user interface(s) of the system and the testing of the system (Kunz, 1989).

2.7   CONSIDERATION OF NON-TECHNOLOGICAL ISSUES BY RESEARCHERS

While we do not think that it is necessarily bad or wrong not to consider the managerial issues raised in this chapter, we would still like to analyse how well the research projects at CIFE have addressed each of the issues described above.

2.7.1   The role of IT

Heuristic systems and (model-based) simulation help a company to provide services that might not have been possible before. In addition, product models allow easier sharing of project information and can therefore place a firm in a strategic position to tie into information networks with other participants in the project delivery process. The systems prototyped in the research projects described above clearly offer the strategic potential identified as important for the success of IT applications in AEC firms.

2.7.2   Leadership

In the United States, top management in construction companies is often removed from the day-to-day activities on projects, i.e. project managers are usually free to run projects as they see fit. This means that the strong leadership required for the success of IT solutions is inherently fragmented by the current organization of firms and projects. In addition, the projects described above offer solutions that are still in their infancy stages and do little to convince project managers to try them.

2.7.3   Identifying IT opportunities

Project managers are usually too busy and too distant from the IT community to learn about IT opportunities. By the same token, researchers are usually too far removed from the field to understand practical problems fully. Furthermore, not all research addresses an existing, identified practical problem. In other words, we face the classic R&D problem of poor or lack of communication between the affected parties. At CIFE, the company representatives that visit and provide input to research projects and learn about on-going and completed projects, often come from staff positions and face a tough internal selling job.

2.7.4 Resource commitment

The long-term view required for successful implementation and application of the systems summarized above is difficult to achieve owing to the project focus that is prevalent in construction in the United States. Japanese clients and contractors seem to be able to take a longer term view by building projects where learning about technology is more important than building them in the cheapest way that is currently known (*Engineering News Record*, 1993). Thus, they have the opportunity to invest in technology. To our knowledge, little research regarding this issue has been done, and we are not sure how one would come up with reasonable figures regarding the resource commitments required for successful application of the new technologies described above.

2.7.5 Innovation vs. imitation

CIFE researchers have usually failed to show (and quantify) the benefits of an innovative application of their ideas for construction companies. A manager in charge of implementing such tools is, however, very concerned about the bottom line. We wonder whether it is this fundamentally different way of looking at the world, i.e. the view of IT opportunities vs. the IT costs and pitfalls, that makes communication between the two parties difficult.

2.7.6 A different perspective on IT

Researchers have recognized that fragmentation of the IT tools used in the AEC industry is a problem. Many research efforts have proposed solutions to overcome the data fragmentation problem that exists because project data are used in various phases and by various disciplines.

2.7.7 Changes in organizational structure

There has been little research in the area of process redesign in construction. The introspection required for such a redesign does not come easy to most construction professionals, but – as shown by Koskela (1992) – is essential for the success of integration and automation efforts proposed by the research projects at CIFE.

2.7.8   Role of users vs. role of IT specialists

There has been little consideration of this issue at CIFE, except that user involvement is usually recognized as critical to the success of AI technology. However, how to sell such technology to IT specialists has not been addressed by CIFE researchers.

2.7.9   The threat of IT

In our experience, CIFE researchers have generally been enthusiastic about the opportunities generated by IT and not concerned with the threat of IT for potential users or non-users.

2.7.10   Make or buy

While knowledge publishing has been an idea for some time now (Feigenbaum *et al.*, 1988), the knowledge required for the systems proposed above is not yet available on the market. Companies thus face a significant investment in making their own AI tools. Given the structure of the AEC industry, these investments are not likely to occur in the near future. Object-oriented systems, however, offer more flexibility and modularity, and should reduce the effort required by individual companies to customize their AI tools (Levitt *et al.*, 1991).

2.7.11   Centralization vs. decentralization of IT

The VDT project mentioned above (Cohen, 1992) has started to model these issues in the context of a design project team. The other projects described above were typically not concerned with this issue.

2.7.12   Implementation method

This was not addressed by any of the research projects.

2.7.13   Results control system

Again, no research was done in this area. It is not entirely clear what the cost and value of the contributions of the various systems we described should be; how value and cost should be measured and charged. The one thing that is certain though is that existing ways of measuring progress, e.g. work-hours per drawing, will not be useful measures in the future and probably hinder the implementation of more advanced IT.

2.7.14   Training and education

While researchers do not specifically address the issue of training and education in firms, ideas, concepts and tools that were developed in CIFE projects have started to be incorporated in courses at Stanford University.

2.7.15   Operational dependence

As shown by the framework used for the development of applications at CIFE (Kunz, 1989), testing is an important aspect of system development. However, researchers probably do not (cannot) provide the level of testing required for industry acceptance since their systems are only prototypes or proof-of-concept systems.

2.8   CONCLUSIONS

As the discussion above shows, the more an issue relates to the actual implementation of IT, the less it has been considered by researchers. This is hardly surprising given the type of IT research done at CIFE so far. The question is, can or should IT researchers consider these issues in more depth? The least we can do is be aware of them. We could, however, think about shifting the focus of our research efforts to include more participation by industry and affected professionals. We could also argue that IT research at CIFE has not yet produced really useful tools and that we therefore need to do more research to improve the usability and applicability of these tools. Both approaches, however, require a vision of where we would like to be and what we would like to achieve. We have started to outline our vision at CIFE in a technical sense (Fischer and Kunz, 1993), but have not done so with respect to the issues presented in this chapter. It appears however that the market-place will increasingly demand a more holistic approach to research projects. This is clearly a fundamental issue for us to consider in our attempts to move towards greater practical integration of construction information.

ACKNOWLEDGEMENTS

We would like to thank the Construction Institute at Stanford University for the support of this research project.

REFERENCES

Agai, E. (1988) Automation is Management, in *Proceedings of Fifth Conference on Computing in Civil Engineering*, (ed. K.M. Will), ASCE, pp. 537–41.

Applegate, L., Cash, J., Mills, J. and Quinn, D. (1988) Information technology and tomorrow's manager. *Harvard Business Review*, 66(6), 128–36.

Baker, C. (1991) Building on IT. *Accountancy*, 6, 12–14.

Bawden, D. and Blakeman, K. (1990) *IT Strategies for Information Management* Butterworth Scientific, London.

Betts, M. (1992) *Information Technology Planning Frameworks for Computer Integrated Construction.* Proceedings of International Workshop on Models Supporting Computer Integrated Construction, Technical Research Center of Finland (VTT), Espoo, Finland, October 5–9, 1992.

Betts, M., Cher, L., Mathur, K. and Ofori, G. (1991) Strategies for the construction sector in the information technology era. *Construction Management and Economics*, 9, 509–28.

Bicharra-Garcia, A.C. (1992) Active design documents: A new approach for supporting documentation in preliminary routine design. *PhD Thesis*, Department of Civil Engineering, Stanford University.

Björnsson, H.C. (1991) *Information Technology Strategies in Construction.* Final Proceedings of the Technological Innovation in Construction Seminar, Sponsored by the Swiss National Research Fund and the National Science Foundation, USA, Zurich, April 3–5, pp. 104–114.

Cash, J.I., McFarlan, F.W., McKenney, J.L. and Applegate, L.M. (1992) *Corporate Information Systems Management: Text and Cases.* Irwin, Homewood, IL.

Chinowsky, P.S. (1991) The CAADIE project: Applying knowledge-based paradigms to architectural layout generation. *PhD Thesis*, Department of Civil Engineering, Stanford University.

Cohen, G. (1992) The virtual design team: An information processing model of design team management. *PhD Thesis*, Department of Civil Engineering, Stanford University.

Darwiche, A., Hayes-Roth, B. and Levitt, R.E. (1989) OARPLAN: Generating project plans in a blackboard system by reasoning about objects, actions, and resources. *CIFE Technical Report # 2*, Department of Civil Engineering, Stanford University.

Earl, M. (1989) *Management Strategies for Information Technology*, Prentice Hall, London.

Earl, M., Feeny, D., Lockett, M. and Runge, D. (1988). Competitive advantage through information technology: Eight maxims for senior managers. *Multinational Business*, Summer, 15–21.

Engineering News Record (1993) Building-by-numbers in Japan: A mechanical systems approach spurs productivity and slices erection time on a 20-story bank building, 230(9), 22–4.

Feigenbaum, E., McCorduck, P. and Nii, P.H. (1988) *The Rise of the Expert Company: How visionary companies are using artificial intelligence to achieve higher productivity and profits*, Times Books, New York.

Fenves, S.J., Flemming, U., Hendrickson, C. *et al.* (1990) Integrated software environment for building design and construction. *Computer Aided Design*, 22(1), 27–36.

Fischer, M. (1991a) Constructability input to preliminary design of reinforced concrete structures. *PhD Thesis*, Department of Civil Engineering, Stanford University.

Fischer, M. (1991b) Reasoning about constructability: Representing construction knowledge and project data, in *Artificial Intelligence and Structural Engineering*, Proceedings of Second International Conference on the Application of Artificial Intelligence to Civil and Structural Engineering, Oxford, September 3–5, 1991, (ed) B.H.V. Topping Civil-Comp Press, Edinburgh, pp. 105–12.

Fischer, M. and Kunz, J. (1993) Circle integration. *CIFE Working Paper # 20*, Department of Civil Engineering, Stanford University.

Froese, T. (1992) Integrated computer-aided project management through standard object-oriented models. *PhD Thesis*, Department of Civil Engineering, Stanford University.

Gibson, G.E. and Bell, L.C. (1992) Integrated data-base systems. *Journal of Construction Engineering and Management*, **118**(1), 50–9.

Goldsmith, N. (1991). Linking IT planning to business strategy. *Long Range Planning*, **24**(6), 67–77.

Hodge, C. (1992) True costs, in *Proceedings of Symposium on Computing in Civil Engineering and Geographic Information Systems*, (eds B. Goodno and J. Wright), ASCE, pp. 1095–100.

Hodgkinson, S.L. (1992) IT structures for the 1990s: Organization of IT functions in large companies. *Information and Management*, **22**, 161–75.

Howard, H.C., Levitt, R.E., Paulson, B.C. and Tatum, C.B. (1989) Computer integration: Reducing fragmentation in the AEC industry. *Journal of Computing in Civil Engineering*, **3**(1), 18–31.

Index Group (1989). *Critical Issues of Information Systems Management for 1989*. Index Group, Boston, MA.

Jin Y., Kunz, J., Levitt, R. and Winstanley, G. (1992) *Design of Project Plans From Fundamental Knowledge of Engineered Systems*. Proceedings of AAAI 1992 Fall Symposium: Design from Physical Principles, Cambridge, MA, pp. 149–54.

Koskela, L. (1992) Application of the New Production Philosophy to Construction. *CIFE Technical Report # 72*, Department of Civil Engineering, Stanford University.

Kunz, J. (1989) Concurrent Knowledge Systems Engineering. *CIFE Working Paper # 5*, Department of Civil Engineering, Stanford University.

Lackowitz, G.W. (1992) Acquisition issues, in *Proceedings of Symposium on Computing in Civil Engineering and Geographic Information Systems*, (eds), B. Goodno and J. Wright ASCE, pp. 1031–5.

Levitt, R.E., Axworthy, A. and Kattajamäki, M. (1991) Automating engineering design with design++. *Nikkei AI Journal*, Winter, 118–27.

Liker, J.K., Fleischer, M. and Arnsdorf, D. (1992) Fulfilling the promises of CAD. *Sloan Management Review*, Spring, 74–86.

Phan, D.H.D. (1993) The primitive-composite approach: A methodology for developing sharable object-oriented data representations for facility engineering integration. *PhD Thesis*, Department of Civil Engineering, Stanford University.

Porter, M. and Millar, V.E. (1985) How information gives you competitive advantage. *Harvard Business Review*, July–August, 149–60.

Reinschmidt, K. (1993) *Experience with Computer-Integrated Construction at Stone & Webster Engineering Corporation*. Presentation at Construction Congress III, ASCE, San Francisco, February 28–March 2, 1993.

Rosenman, M.A., Balachandran, B.M. and Gero, J.S. (1989) The place of expert systems in civil engineering. *Civil Engineering Systems*, **6**(1–2), 11–20.

Tatum, C.B. (1988) Technology and competitive advantage in civil engineering. *Journal of Professional Issues in Engineering*, **114**(3), 256–64.

Tauber, E.R., Levitt, R.E., Oralkan, G.A., Reinberg, F.C. and Walsh, T.J. (1992) The Conceptimator: An Expert System for Conceptual Cost Estimating of Building Foundations. *CIFE Working Paper, # 14*, Department of Civil Engineering, Stanford University.

Teicholz, P. and Fischer, M. (1994) A strategy for computer integrated construction technology. *Journal of Construction Engineering and Management* **120** (1), 117–31.

Wei, T. (1993) Personal conversation at Stanford University, February 1, 1993.

CHAPTER 3

The role of data transfer

Antony Thorpe, Department of Civil and Building Engineering,
Loughborough University of Technology,
Loughborough, LE11 3TU, UK

3.1 INTRODUCTION

The ability to achieve wide scale integration of construction information and data, often referred to as computer integrated construction (CIC), is finally approaching reality due to continued advances, and cost reductions, in computing hardware and software technologies. In addition enabling methodologies relating to information modelling, requirement analyses, interface design, CASE tools, etc., are also now developed to a point where they can be applied successfully and with demonstrable benefits.

Integration is a confusing and ill-defined term which Augenbroe (1991) noted can mean 'almost anything'. The concepts and meaning of integration and CIC are discussed elsewhere in Part One by Vincent (Chapter 4) and by Betts, Fischer and Koskela (Chapter 1). If the full benefits of integration are to be achieved it is important to encompass within the framework of CIC all methodologies and technologies which enhance the capture, transmission, analysis and processing of construction data and information in all its guises.

This chapter reviews the stages of development of IT within the contracting sector of the UK construction industry and discusses its current integration problems. The use of currently emerging data transfer standards such as STEP and, in particular, EDIFACT, in the integration process are discussed, and it is proposed that these protocols and other communication technologies, such as multimedia, could form the bonds between the integrated system 'molecules' of the future construction industry.

3.2 THE GROWTH OF IT IN CONSTRUCTION

The following four stages in the development and growth of IT in construction are presented as a typical pattern. It does not represent any par-

Integrated Construction Information. Edited by Peter Brandon and Martin Betts. Published in 1995 by E & FN Spon, 2–6 Boundary Row, London SE1 8HN.
ISBN: 0 419 20370 2

ticular company and it does not imply that all companies have followed this pattern. However, it does represent a not uncommon route by which contracting firms have approached the integration of construction information.

3.2.1 Stage 1: early 1960s to late 1970s

The use of computer technology in the UK construction industry began in the early 1960s. At this stage the systems were designed primarily to take advantage of the complex numerical computations that could be achieved with the new technology (e.g. finite element analysis). Computers were seen mainly as sophisticated calculators. The cost of permanent data storage prevented storage of large volumes of data and the development of information systems. By the 1970s the larger construction organizations had purchased mainframe computers. These were primarily for accounting and financial management purposes. Smaller organizations bought time via computer bureaux.

3.2.2 Stage 2: early 1980s to mid 1980s

This phase was initiated by the introduction of low cost personal computers which resulted in a rapid expansion of computer facilities. The cost of such facilities dropped to a level which made the purchase of hardware affordable to management at all levels within construction organizations. During this phase of computer development the adoption of computer systems was seldom co-ordinated within organizations. A wide variety of hardware platforms and operating systems were often adopted, with little or no consideration given to standardization.

The growth in affordable computer hardware resulted in a demand for software. Many companies developed software in-house to provide for their own functional requirements. There was also a growth in the availability of packaged software for applications such as accounting, estimating and other construction management functions. The integration of computer systems and software was largely unconsidered at this stage.

3.2.3 Stage 3: mid 1980s to late 1980s

Construction organizations, seeking to contain the rapid expansion of computer usage, increased centralized control of IT departments. Larger organizations who had established data processing departments in the past re-enforced their control. Those organizations that were new to computer usage achieved centralized control through central IT departments, usually based at the head office. The larger construction organizations recognized the need for corporate IT strategies and for direction and vision from top management.

The benefits of system integration were recognized during this stage partly as a result of the success achieved by other industries, such as the retail and banking industries. The change from localized control of IT development to centralized control allowed companies to standardize their hardware platforms and operating systems. This standardization of equipment was considered essential to achieve systems integration.

Construction organizations began to initiate integration of their computing systems. Although new software development tools emerged and were marketed as the solutions to system integration, there was seldom the opportunity to develop new fully integrated systems across the range of computing applications. Most computer systems within construction organizations remained at varying stages of their life cycles, thus preventing the development of wide scale integrated systems. The majority of integration achieved by companies at this stage was within the head office of the company, between major applications, using one-off linking programs by utilizing ASCII file export/import facilities on standard software.

3.2.4 Stage 4: early 1990s

The fourth, and current, stage of development has seen a growing trend for the reversion of control of computer systems from central IT departments to localized groups of end users. This localization of IT responsibility is commonly referred to as down-sizing. The reasons for returning to localized IT control extend beyond the availability of cheaper more powerful personal computers to include the dissatisfactions in the service provided by central IT departments. The central IT departments have not been removed but have been greatly reduced in size with modified objectives concerned with the guidance of IT strategy rather than the control of all IT operations.

The integration of systems within companies remains a primary objective. Open systems technology is now available to facilitate hardware integration and neutral file formats are available for software integration. However, there is a realization that achieving fully integrated computer systems within the fragmented and ever-changing structure of a contracting company is far more demanding than originally perceived.

3.3 MATCHING CONSTRUCTION TO THE STANDARD GROWTH MODEL

A standard growth model for the development of data processing within organizations was first suggested by Gibson and Nolan (1974). Their four-stage model depicts the growth of data processing from initial investment to mature operation (Figure 3.1). There are notable similarities between the experience of the UK construction industry and the four-stage model

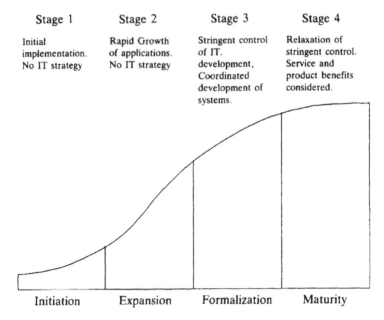

Stage 1	Stage 2	Stage 3	Stage 4
Initial implementation. No IT strategy	Rapid Growth of applications. No IT strategy	Stringent control of IT. development, Coordinated development of systems.	Relaxation of stringent control. Service and product benefits considered.

Initiation   Expansion   Formalization   Maturity

Figure 3.1 Four-stage model of data processing growth (Source: Nolan, 1973)

as proposed by Gibson and Nolan. The first stage of computing in the construction industry follows closely the first stage within the model, characterized by the *ad hoc* nature of computer system development, with no common IT strategy between departments.

The second stage of the model maintains the approach of little coordination in IT development, but a rapid growth in IT investment. This was partially experienced in the construction industry with the adoption of a wider range of applications. However, the rapid expansion, as presented by the model, was not experienced, although there was an overall increase in the use of computing by contractors during the 1980s (Vickery, 1988).

The third stage of the model is characterized by the widespread formation of central IT departments, which was witnessed in the late 1980s within the construction industry.

It would appear that the UK construction industry has experienced shortened second and third stages of development, while not experiencing the associated high levels of expenditure, and is now moving into the fourth stage. Two possible hypotheses for this are:

1. *Delay hypothesis*: The traditional reticence of the construction industry to accept change and the lack of suitable packaged software may have delayed the onset of the third stage. This delay then allowed management to identify the need for centralized IT control before rapid growth of IT spend occurred.

2. *Observation hypothesis*: It has been noted that the construction industry lags several years behind, and lacks the commitment of, other industries in the use of computing technology (Atkin, 1990; CICA, 1990). It is therefore likely that top management within contracting organizations observed the trend of companies within other industries of adopting a co-ordinated IT strategy. This allowed them to identify the benefits of a co-ordinated IT strategy before the rapid growth in IT investment occurred.

Construction industry organizations are currently approaching/entering Stage 4 of the four-stage model. This is indicated by the trend of decentralization or down-sizing of IT, thus relaxing of the stringent centralized control of the third stage.

3.4   CURRENT INDUSTRY TRENDS: NEW BARRIERS TO INTEGRATION

Large construction organizations generally comprise several trading groups or divisions, which are commonly further subdivided into functional or regional centres. Each of these units will have their own computing requirements. The majority of the technical and commercial computing undertaken by these units has become an integral part of their everyday operations. It is therefore necessary to have effective computer systems at this functional level. However, the use of central IT departments for the development of systems and applications has resulted in end-user dissatisfaction.

The sources of end-user dissatisfaction are argued by Thorpe (1992) to include:

1. long development times for systems;

2. too little user input in new systems design;

3. lack of empathy between computing and construction staff;

4. lack of the right solutions to the real problems; and

5. too costly solutions.

The majority of these problems are caused by poor or ineffective communication between the end-user departments and the IT department responsible for developing the software.

In many organizations the central IT departments have failed to provide individual departments with the effective computer systems they require. This, combined with their increased costs, has prompted senior management to seek new solutions for IT control.

The term 'down-sizing' is used to describe the practice of installing applications on smaller hardware systems usually in locations close to the end users. This term has arisen as many organizations have transferred their portfolio of applications software from large centralized mainframe computers to smaller linked minicomputer and microcomputer systems. Down-sizing implies not only changes in the location of computer hardware but also changes in IT control and management.

3.5   DOWN-SIZING – REASONS FOR ADOPTION

Down-sizing consists of the establishment of localized IT departments, which develop and maintain computer systems for a small section of the company. There are three driving forces behind the down-sizing of computer systems:

1. *Business drivers*
   (a) Pressure to reduce cost;
   (b) organization and decision-making decentralization;
   (c) a 'users-know-best' attitude; and
   (d) increasingly competitive and changing business environment which requires more flexible and responsive systems.

2. *End-user drivers*
   (a) Increased computer literacy/increased expectations;
   (b) user friendliness of PC software; and
   (c) frustration with existing approaches to systems development.

3. *Technical drivers*
   (a) Increased power of down-sized hardware;
   (b) graphical user interfaces, e.g. Windows;
   (c) more powerful and flexible application packages;
   (d) faster development tools;
   (e) down-sized systems require less support staff; and
   (f) cheaper hardware and software.

The decentralization of IT changes the role of existing IT departments. The modified role of the new central IT department is to:

1. contribute to IT strategy;

2. determine policies and standards;

3: procure hardware and software cost effectively;

4. future watch and ensure that organizations take advantage of new technological developments;

5. provide IT training;

6. provide technical advice; and

7. vet outside suppliers.

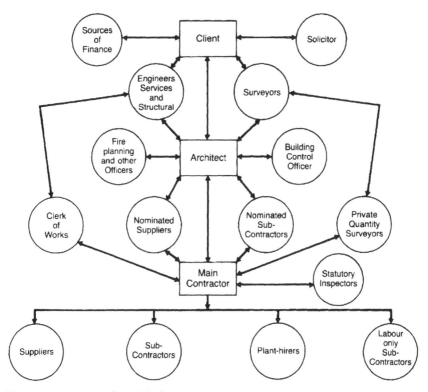

Figure 3.2 A general model of construction industry communications (Source: Wix and McLelland).

However, the local control of IT resources and systems can pose considerable integration problems. The purchase and development of incompatible and isolated systems, providing localized and single function benefits has, unfortunately, once again become a possibility.

3.6 DATA TRANSFER AND INTEGRATION

A Construction Industry Institute study of its membership in 1988 found that all respondents emphasized the importance of integration within their computer systems (Choi and Ibbs, 1989). Many of the companies interviewed in the study considered that the next major advance in the construction industry will take place when project, design and facility computing systems are integrated. If this advance is to occur it is important to consider all the parties involved in the construction process, the flow of information between them, and the nature of these flows.

Studies of communications and information flows within the construction industry have been undertaken by many researchers (for example,

Higgin and Jessop, 1965; Guevara, 1979; Ndekugri and McCaffer, 1988). Figure 3.2 shows a general model of construction industry communications suggested by Wix and McLelland (1986).

While new systems may be developed using the latest CASE/4GL environments to achieve internal integration, there will still be the need to communicate with other, possibly older, systems which do not possess the same attributes and facilities. Thus, communication protocols are required which will enable these transfers to occur using internationally recognized standards.

3.7   OPEN SYSTEMS

There are two possible methods for achieving the transmission of data and information between computer systems within a company:

1. develop individual translation software for each pair of applications within the company; and

2. adopt a neutral file transfer communication strategy.

The selection of an integration strategy is dependent on the number of systems to be integrated, as this affects the number of translators required and hence the cost of achieving integration.

Figure 3.3 illustrates the number of translators required for direct application connection in comparison with the number of translators required using a neutral file strategy. The two approaches require equal numbers of translators to interconnect three computer systems; however, as the number of systems increases there is an increasing advantage in adopting the neutral file strategy.

The use of a neutral file format alone does not allow integration. The standardization of the hardware and software that undertakes the communication process also has to be standardized. The Open Systems standard was developed to achieve this. To eliminate the need for multiple translators on every computer system, the International Standards Organization (ISO) has developed a framework for the interconnection of systems. The generic title for this work is Open Systems Interconnection (OSI). The development work commenced in 1977 and resulted in the production of a reference model, detailed in ISO IS 7498. This model defines a standard architecture for computer communications using a seven-layer structure.

OSI only deals with the way in which systems may communicate, it does not provide details of implementation. It is at this level that neutral file standards provide the means of describing data items and groups which form messages.

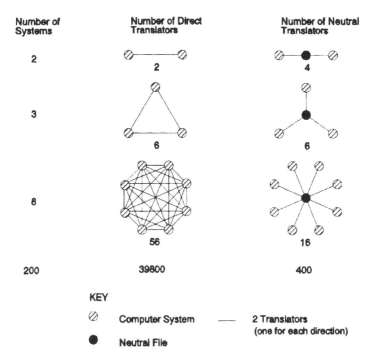

Figure 3.3   Comparison of direct and neutral file transfer

3.8   THE USE OF ELECTRONIC DATA INTERCHANGE (EDI)

EDI can be defined as:

> The electronic exchange of structured and normalised data between computer applications of parties involved in a (trade) transaction
> *Hoffman, 1987*

Currently, there is growing interest in the applications of EDI in the construction industry. Several contractors in the United Kingdom are using EDIFACT message standards to transfer trading cycle data, including invoices and purchase orders.

The EDIFACT (Electronic Data Interchange for Administration, Commerce and Transport) standard was ratified by the United Nations in 1987 and has already been adopted by many EDI users for international interchange of data. EDIFACT is based on earlier European and American standards for data interchange called UN/TDI and ANSI.X12 respectively (HMSO, 1988). The syntax for structuring and encoding electronic documents for interchange between computer systems is defined in international standard ISO 9735. EDIFACT is the basis for agreement on a common structure for documents, such as those for trading, so that they can be interchanged electronically. The need for an EDI standard, such as EDIFACT, was recognized in the 'Building IT 2000' report which states that standards must be agreed and observed if there is to be growth of communications in a global market (George, 1991).

In 1992 56 major construction-related companies were surveyed with regard to their plans for the implementation of EDI (Lewis, 1992). Of the respondents, 75% indicated that their companies proposed adopting EDI within the next five years. If this take-up of EDI occurs there will be increasing pressure on all construction organizations to adopt this technology to trade efficiently.

Taking the example of a regional office in a down-sized construction company, Figure 3.4 details how EDI can form part of a contractor's existing computer system. All EDIFACT trading cycle messages shown in Figure 3.4 are currently available to the construction industry through EDICON, the UK construction industry group. The bill of quantities message and valuations messages are currently being developed and tested by EDICON and EDICONSTRUCT (the French construction EDI group), respectively.

To achieve the transfer of electronic messages by EDI, each application requires an EDI interface. The use of these interfaces need not be limited to the transfer of the standard EDIFACT messages, but may be used to transfer any data coded in the EDIFACT syntax. Sections of existing EDIFACT standard messages could be used to form internal EDI messages or alternatively new in-house messages could be designed. The EDIFACT standard is therefore an ideal neutral file format for the transfer of text-based data between offices/site within a large contracting company.

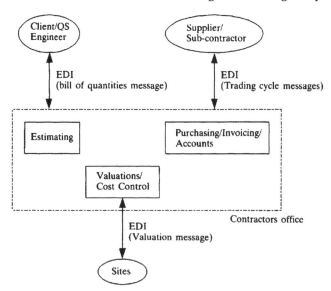

Figure 3.4 External communications of a contractor's office

As all companies within an industry will need to exchange very similar information internally, it is possible to create a standard directory of internal EDIFACT messages for each industrial sector. These messages would not form an industry standard for internal data interchange as this would be too constrictive to individual companies. However, the directory would greatly reduce the development time of integration translation software as suitable messages, which require very little or no modification, would be readily available.

The use of the EDIFACT standard as a neutral file format has the following benefits:

1. No additional hardware costs

2. Short development time

3. Compatibility with existing software

4. A well-defined set of directories is available for internal file standard development

5. It allows freedom of computer systems/applications selection within regional offices

The adoption and use of EDI should enable the transfer of information, and hence the integration of systems, across commercial boundaries. However, EDIFACT is not suitable in its current form for technical information exchange and this has been addressed by the development of PDI (Product Data Interchange).

3.9   THE USE OF PRODUCT DATA INTERCHANGE (PDI)

PDI can be defined as:

The electronic exchange of structured and normalised data between CAD applications, covering the whole life cycle of a product, making use of the explicit distinction between logical and physical data
Neuteboom, 1993

PDI may be regarded as technical data transfer, as opposed to EDI which is concerned with commercial data transfer. The thrust for generic PDI standards resulted from the inability to transfer CAD data files between different systems. It was recognized that it was insufficient to merely transmit the physical data from application to application. What users required was a formal and standardized description of the data at a logical level.

This was provided by the development of the EXPRESS data description language while the physical file can be exchanged using the STEP physical file format. This is a major difference to the EDI/EDIFACT developments.

PDI and EDI have developed independently to address separate identifiable requirements. However, the linking of the two approaches may be possible via a neutral project model as proposed by Neuteboom (1993) and shown in Figure 3.5. This approach depends on the development of the methodologies required for the neutral project model, currently being investigated in projects such as the EC funded ATLAS and COMBINE projects. Undoubtedly these developments will further enhance the communications and integration facilities available within the construction industry.

3.10   OTHER MECHANISMS FOR INTEGRATION

As well as the continuing data transfer and project-centred integration developments, other emerging mechanisms and technologies will play important roles in the total CIC framework.

3.10.1   Automatic identification

Automatic identification is required if we are to complete the loop of EDI/PDI between contractors, suppliers and sites. The missing (computerized) link is the recording of the flow of goods from suppliers/subcontractors to site. Automatic Identification, and in particular Bar Coding, has revolutionized the retail and marketing industries. Its use in construction

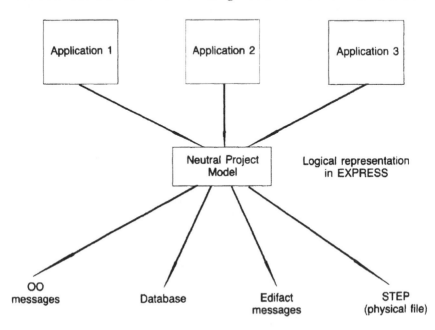

Figure 3.5   Linking EDI and PDI (after Neuteboom, 1993)

is currently minimal; however, work is under way in the United States (Bell and McCullouch, 1988), Finland (Lehtonen, 1993) as part of the RATAS project, and the United Kingdom (Alkaabi, Baldwin and Thorpe, 1992) to produce a model for the generic labelling of goods in the construction industry. Obviously great benefits could accrue if the labelling system can incorporate segments from the EDI and PDI data, describing both the product and the transaction, in a machine-readable form.

3.10.2 Broadband communications

Video techniques together with voice and data transmission are opening up new areas for improved communications and integration within construction. The European Community funded BRICC project (Broadband Integrated Communications for Construction) is currently investigating linking the construction 'players' via an ISDN network to provide instantaneous and interactive voice, video and data communications. A typical system is illustrated in Figure 3.6.

The systems being developed will allow computer-supported co-operative work (CSCW) enabling several people to work remotely, at the same time, on identical screen material. In addition, object-oriented or model-based 3D CAD packages will enable different professionals to view and discuss the same material by using the metaphors they understand, or require, the most (Leevers, 1993).

The use of multimedia communication between sites and remote experts is expected to emerge as an efficient and dynamic form of co-ordination and decision-making once an accessible broadband network is in place. This is predicted for the EC by 1998.

3.11 CONCLUSIONS

The construction industry is currently approaching/entering Stage 4 of the Gibson and Nolan model which is characterized by the relaxation of stringent central control over IT. In the case of construction this has been recognized by the decentralization, and hence down-sizing, of IT within contracting organizations. However, the computing applications and the level of integration achieved are not as advanced as expected by Stage 4.

It is widely recognized that the next advance in construction integration will involve the integration of the design and project systems. The integration of systems on a local level has been achieved in several contracting companies, using specifically written translation software. However, in a down-sized contracting company there may be many computer systems which are incompatible.

The adoption of EDI in the industry is on the threshold of expansion, and shortly contractors will come under increasing commercial pressure

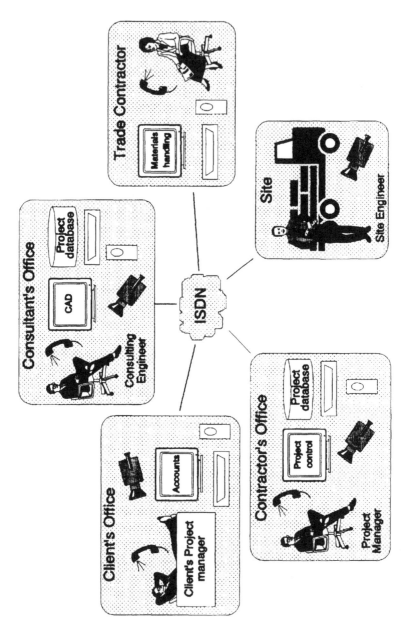

Figure 3.6 Typical construction-related ISDN network (after Leevers, 1993)

to adopt EDI. The use of the EDIFACT EDI standard as a neutral file format could therefore be used as a mechanism for system integration, as contractors will have to develop EDI interfaces for their existing applications regardless of their integration strategy. Similarly, PDI offers a neutral file format with the additional benefit of the formal data definition language, EXPRESS. The merging of PDI and EDI via a neutral project model appears to offer an achievable way forward.

Finally the ability to label goods in an intelligent and computer-readable form which links back to their product model and the transaction details, together with the ability to communicate not only to the site, but also to specific individuals within the site at the point of production, will undoubtedly herald a new wave of integration opportunities, benefits and problems.

ACKNOWLEDGEMENTS

I gratefully acknowledge the help provided by my researchers, Alan Couzens, Tony Lewis and Chris Carter, and for the financial support, for the work described in this chapter, provided by the Science and Engineering Research Council (SERC) of the United Kingdom, and by BICC plc.

REFERENCES

Alkaabi, J., Baldwin, A.N. and Thorpe, A. (1992) Bar Coding for Construction: A Feasibility Study. Internal Paper, Department of Civil Engineering, Loughborough University of Technology, UK.

Atkin, B. (1990) *Information Management of Construction Projects*. T.W. Crow Associates and Crow Maunsell Pty. Ltd, Management and Project Consultants, 50 Margaret Street, Sydney, Australia.

Augenbroe, G. (1991) *R&D View on Integration in the Building Industry*. Proceedings of the CIB seminar – The Computer Integrated Future, Eindhoven.

Bell, L.C. and McCullouch, B.G. (1988) Bar Code Applications for Construction. *Journal of Construction Engineering and Management*, **114**(2).

Choi, K.C. and Ibbs, C.W. (1989) Cost Effectiveness of Computerisation in Design and Construction. The Construction Industry Institute, Austin, TX, *Source Document 50*, August.

CICA (1990) Building on IT – for the 90s. The Construction Industry Computing Association and Peat Marwick McLintock (PMM), Guildhall Place, Cambridge, CB2 3QQ.

George, J. (1991) Building IT 2000. The Building Centre Trust.

Gibson, F. and Nolan, R.L. (1974) Managing the four stages of EDP growth. *Harvard Business Review*, January/February, 76–88.

Guevara, J.M. (1979) Communication in Construction Companies. PhD Thesis, University of Illinois, Urbana, IL.

Higgin, G. and Jessop, N. (1963) *Communications in the Building Industry*, Tavistock Publications, London.

HMSO (1988) EDI and X.400 Study. *A Vanguard Report*, Her Majesty's Stationery Office, London.

Hoffman, W.J. (1987) Electronische ge gevensuitwisseling tussen organisaties Tutein Notteius, *EDI Handboek*, ISBN 90-72194-09-8.

Leevers, D. (1993) Multimedia in Construction. *CICE '93*, Institution of Civil Engineers, London.

Lehtonen, H. (1993) *Goods Labelling Project*. Proceedings of the joint EDICON/EDIBUILD Conference, National Motorcycle Museum, Birmingham, UK, April.

Lewis, T. (1992) EDI in the Construction Industry. Internal paper, Department of Civil Engineering, Loughborough University, UK.

Ndekugri, I.E. and McCaffer, R. (1988) Management information flow in construction companies. *Journal of Construction Management and Economics*, (6).

Neuteboom, J. (1993) *Bridging the Gap between EDI and PDI*. Proceedings of the joint EDICON/EDIBUILD Conference, National Motorcycle Museum, Birmingham, UK, April.

Nolan, R.L. (1973) Managing the computer resource: A stage hypothesis. *Communications of the ACM*, July.

Thorpe, A. (1992) The Laing IS Experience. Internal paper, Department of Civil Engineering, Loughborough University of Technology, UK.

Vickery, N. (1988) Flying into the Future. *Construction Computing*, Winter, 18–19.

Wix, J. and McLelland, C. (1986) Data Exchange between Computer Systems in the Construction Industry. Building Services Research and Information Association.

CHAPTER 4

Integrating different views of integration

Stephen Vincent, Scott Wilson Kirkpatrick, Basingstoke, UK

4.1 SETTING THE SCENE

The word 'integration' has become very widely used to describe the desirable concept of freely exchanging information between different participants in the construction process, yet actual examples of integration are relatively limited and localized. Grand schemes for the flow of information between the stages of the construction process were a well-established topic of discussion over 20 years ago, when computers first caused the various participants, or 'actors', to study how the design process worked in order to utilize new technology. During the early days of computerization, many practical people took a deep interest in the impact of information technology, and everyone expected rapid results.

Over the years, information technology has become a specialist discipline in its own right; computer software has become the province of the specialist vendors selling to the immediate market; and research has dispersed into a wide range of complex localized topics, many of which are looking some years ahead to viable solutions (Dupagne, 1991). At the same time, most of the best practical and professional brains in the construction industry have been focused upon new working methods and organizational structures, as greater separation between managerial and professional skills has become essential for commercial success.

Integration is both a simple concept for a managerial mind to understand, and a monumentally complex wish to fulfil for the professional and research minds who are grappling with the subject. This chapter is the result of studying this problem from many different angles, in an attempt to chart a logical route forward. An important theme throughout is to get back to looking at the real information system first, then look at technology and its effects. Information technology has been an important stimulus for understanding the system, but it is dynamic and transient, and will continue to be so for many years to come. Lives depend on getting con-

Integrated Construction Information. Edited by Peter Brandon and Martin Betts. Published in 1995 by E & FN Spon, 2–6 Boundary Row, London SE1 8HN.
ISBN: 0 419 20370 2

struction skills right. Sustainable changes are likely to outlive several generations of computer hardware. This point can be well illustrated by Part
Two of this book, where historical integration efforts that pre-dated the IT
explosion can be seen to have followed a somewhat similar path to our
current efforts.

4.1.1 Who wants integration?

Changes will only occur if both someone has a strong interest in achieving
them, and they can be achieved. Figure 4.1 is an analysis of what appears
to be happening at the moment. At the highest level, national governments and international bodies have observed the effects of the integrated
use of information systems upon other industries, and have therefore
identified the potential for improvement in the construction industry. In
the European Community the essential need for common standards is a
driving force in integration initiatives, to create an open market where different countries and actors all work to consistent and compatible methodologies. Strategic concepts come from above.

The many different firms actually involved in the process are all driven
by commercial interests; they will invest where proven benefits will
improve their financial returns. This applies equally to planners, designers, constructors and operators. In localized areas, for instance, the
exchange of computerized drawings, standards have been developed but
as yet they are either rather localized in their application, or very limited
in the depth of the information transferred. Initiatives in steelwork design
and fabrication and other areas may extend a little further, but they still
only cover parts of the construction industry.

Meanwhile the users of construction projects will continue to complain
whenever the results are not perfect. Why was the latest knowledge not

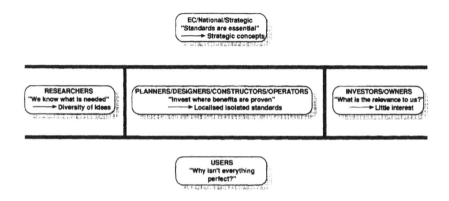

Figure 4.1   The forces driving integration

applied? Why do things not work? Why must extra expense be incurred? Who is responsible?

Researchers have a key role to play; they are the people with ideas. But they must fight for funding in a world which does not always value original thinking, and new concepts and ideas are not easy to communicate to those outside specialist research communities. The best people in industry are too busy with the commercial world to discuss and explore academic ideas. As a result there is an increasing gulf between industry and academia, especially in the region of grand ideas rather than localized applications. Hence researchers do their best to understand what is needed and get on with their work. An apparent lack of focus and common purpose leads to increasingly diverse ideas, without an effective framework of regulation.

Of course, there is another group with a vested interest in improving the situation, the investors and owners who take the financial risks and reap the results. They pay for a product, and employ others to work out the best way of doing it. Hence talk of integration seems of little relevance, and generates little interest. They do, however, wield immense power over the whole construction industry; if they saw benefits in commercial returns through reduced costs or improved quality, things would happen.

4.1.2 Concepts, objects and data

Integration is sometimes quoted as having happened when any substantial amount of information has been exchanged in digital form. Exchanging electronic drawings may actually offer little improvement over exchanging paper drawings; it still requires a skilled eye to interpret them. For instance, almost every type of analysis involves a human process of simplification to convert the problem into a suitable mathematical form based on geometric elements, volumes or visual appearance.

Although many advantages are quoted for computerized integration, large complex projects have been designed, constructed and operated throughout history without the need for information technology. Perhaps we sometimes fail to realize the full extent of the information which we are handling and how it is being communicated. It is all too easy to concentrate on the obvious when it is placed in front of us without thinking any further.

Human communication is very sophisticated and complex, and Figure 4.2 seeks to separate communications into three different levels for comparison with computerized communications. At the highest level, humans converse in concepts. There are no constraints on what a concept can involve; it does not need to be consistent, logical or complete, as long as it conveys a meaning of some sort. The understanding of a concept often varies between different people depending on their knowledge and background, requiring a two-way interaction to correct any misinterpretations. Concepts are the crucial components of political discussions and humour,

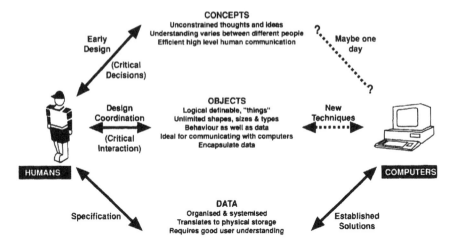

Figure 4.2   What level are we working at?

and are often well illustrated when translating between languages. Much of the early design of any construction project is done in terms of concepts which allow efficient complex communications. Most of the critical decisions which will determine the success of a project are also made while communication is still in the form of concepts. Chapter 5, by Powell and Newland, develops this argument further.

Many human communications occur in the form of objects. A distinction is drawn here between concepts and objects in that objects must be logical and definable 'things'; it must be possible to write down a complete definition. Objects can take any form; there is no restriction upon their shape, size or type. Language includes many individual words which each convey a whole wealth of experience and information, including behaviour as well as just size and shape. The understanding of what a particular object is, is also fairly consistent between different individuals, although the emphasis on which aspects are important may vary.

At a more basic level, communications often include data; independent facts and figures which require some form of knowledge to apply them. Data are usually organized and systemized, and translates well to physical methods of storing information.

Looking at how we use computers, most established solutions are in terms of data, requiring good human user understanding to utilize them. This frequently generates a barrier between the technical specialist who can fully interpret the computerized method, and management and professional specialists who only understand the subject.

Object-oriented techniques have been the subject of research for some time and are mentioned extensively later in this book. They are ideal for improving communication with computers, allowing information to be encapsulated into understandable units for communication. Real applications are still in their infancy though.

As for concepts, computers emanate from a stable of logical solutions, and concepts break all of the rules. Computerized techniques have quite some way to go to catch up with the human brain.

4.1.3 Models and integration

In the past, a lot of effort has been devoted to defining exchange standards, so that information can move between different organizations. However, there has been little widespread acceptance of any standards except at rather basic levels of data exchange as Chapter 3 by Thorpe has demonstrated. All parts of the construction process rely on models of what is happening which are formed in the human minds of those involved, to provide hooks upon which thoughts about all of the information involved can be hung. A model is any representation of something which is real; it is used by different people to describe both representations and processes.

A model may just be a mental picture of something which is real, but it can also be a virtual model of something which does not exist yet, or a process which has not yet happened. Models are formed from past experience, and allow simulation and prediction of the future.

An understanding of how information can be represented is one of the keys to communication, and hence to integration of information. Figure 4.3 illustrates a method of assessing the effectiveness of any model for

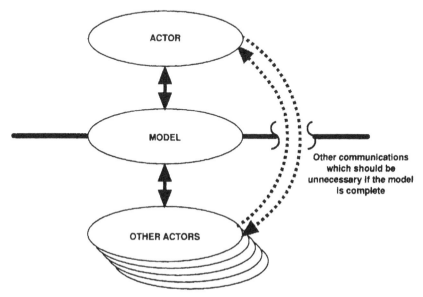

Figure 4.3 Models as an integration tool

communication. If a model is complete, it forms a complete method of communication. Hence a measure of how effective a model is can be determined from studying how much other communication is needed in addition to the model itself. Everything which is not properly defined is a potential source of errors through misunderstandings.

4.1.4　Where does the model end and the world begin?

Most current modelling techniques only model a very small proportion of the parameters and effects which are evident in the real world. For instance, the structural engineer reduces most structures to an assembly of points, lines and planes, and then actually analyses this gross over-simplification of the real complexity of structural behaviour; the mechanical services engineer divides a building into numerical volumes for analysis; the financier converts the complexities of the cash flows on a project into a simple graph. Everybody has just enough information to do their job.

Many of the actors also have localized information which is of no relevance to other actors: the reinforcing bars inside concrete only interest the structural engineer and the contractor; predictions of water flows are only useful to the drainage engineer. So integrating all of the information which does exist together is also an unnecessary exercise.

Figure 4.4 indicates the current mechanism of integration, the human mind. Information is gathered from many diverse sources, meetings, drawings, regulations, maps, analyses and experience, and is linked together by a unique process of thought to create an individual model of the project relevant to the needs of a specific person. The quality of the result will always depend to a large extent upon human insight as well as logical systems and processes.

Where integration can assist is by enabling each individual to assimilate more information more effectively, to see the effects of changes which others may make more rapidly, and to allow more complex simulations to predict the effectiveness of solutions.

4.2　PRACTICAL VIEWPOINTS

4.2.1　The life cycle and knowledge feedback

Construction projects go through a life cycle from an initial idea to eventual demolition, and it is possible to analyse the integration of information from the viewpoint of looking at the stages of this process. This is one of the parameters of an integration definition offered earlier by Betts, Fischer and Koskeler.

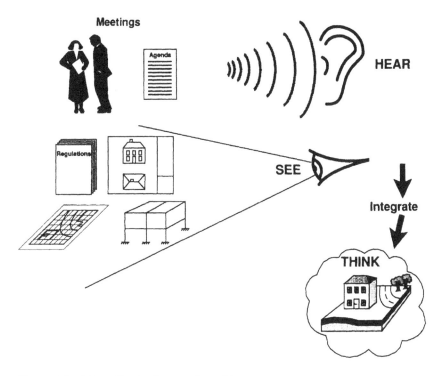

Figure 4.4 Integration by human thought

Figure 4.5 illustrates a representation of the life cycle and the flows of information which occur during it. Central to this particular representation is the development of a product model as an integration focus throughout the life cycle. The potential existence of such a model is currently a concept rather than something which can actually be achieved in practice in the near future. Such a model would, however, provide continuity of the existence of information about the project throughout the life cycle. At present the nearest thing to such a model on most projects is a collection of drawings, documents and human knowledge which is usually distributed between a number of locations.

Looking at the process, it starts with the appraisal of an idea when the project is nothing more than a concept, progresses through design where specifications are generated and operation parameters are determined, is constructed whereby the specification is transformed into a physical form, and then it moves into a period of use during which it must be operated and maintained. At present, most projects mainly involve the information flows shown on the left-hand side of Figure 4.5. Experience from previous projects is assimilated through human thought and the definition of codes of practice to influence and improve the design of new projects. If an

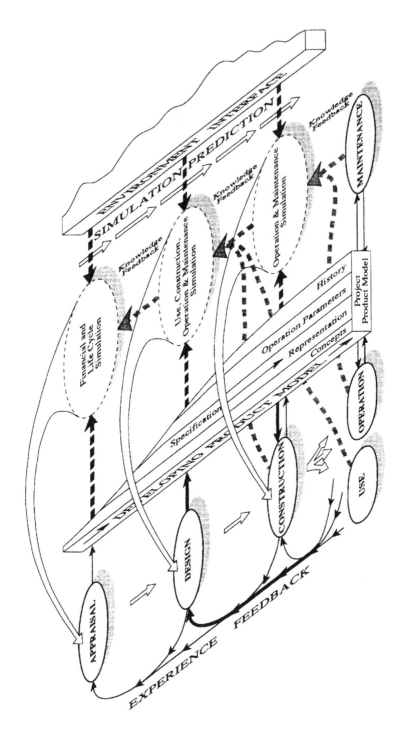

Figure 4.5 Simulation and feedback

effective model were to exist, more comprehensive simulations of later stages of the life cycle could be carried out during the earlier stages, to predict and avoid future problems. Knowledge could then be fed back in a more formalized and logical form to become part of these simulations.

4.2.2   Cost, time and value

The previous part of this chapter concentrated upon stages and processes familiar to the engineer, architect or contractor. However, the key players without whom projects would not exist are the owner and the user. They have a very different view of the process.

To the owner, the project can usually be summarized as a financial quantity; at any point in time it is either costing money or making money. This is illustrated in Figure 4.6. Some types of owner employ professional people to create the project, and then charge users to use it.

However complex the process may appear to those deeply involved in doing it, the owner can often simplify it by a process of calculation into basic financial terms. What is of great interest to the owner, though, is a prediction of future costs and future income. The ability to make good predictions will often be critical to the success of the project, and does rely on a deep and complete understanding of all of the processes involved.

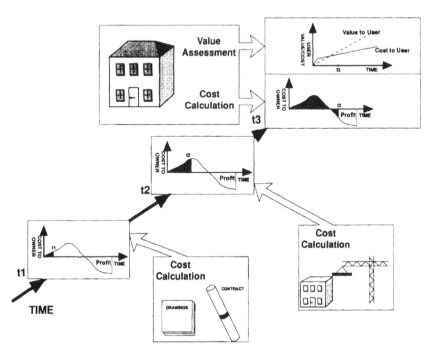

Figure 4.6   Value and profit

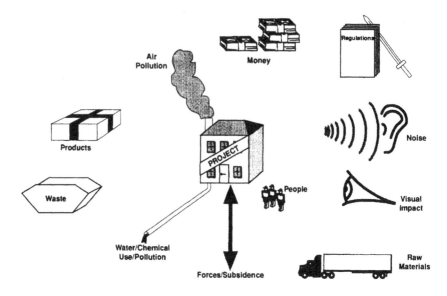

Figure 4.7   Harmony and the real world

The user has a slightly different view. In most cases there must be a balance between the intrinsic value of the project to the user, and the financial cost of payments to the owner. The 'user' may actually be a large number of individuals paying small sums, or a single organization paying for long-term use. The value of the project to the user is rarely purely financial; there is usually some element of choice which is influenced by other factors.

4.2.3   Environmental harmony

Construction projects can never be completely isolated from the environment in which they are constructed. There will always be interaction with the ground below, the air above and the people in and around it. Figure 4.7 indicates some of the factors with which any project must achieve some form of harmony. Regulations constrain what can be done; taxes may affect its viability. People living nearby will be concerned about its visual appearance and any noise produced. Raw materials need to be brought in from outside; products need to be transported away. Waste materials and any liquid pollution must be correctly disposed of.

Achieving harmony with all of these environmental influences is not always easy, but a project cannot be designed and built successfully without taking them into account. The theoretical simulation processes in the life cycle stages in Figure 4.5 include interaction with an environment of external information outside the project itself. In effect, one cannot just

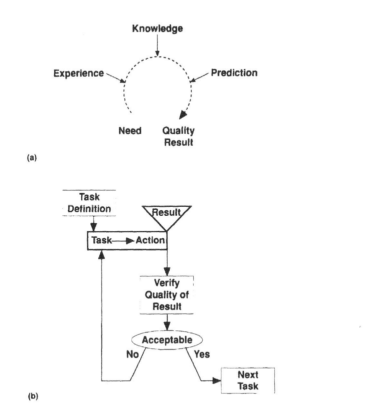

Figure 4.8 Routes to quality: (a) traditional quality control: (b) modern quality control

draw a neat clear line around the edges of a project, and work inside it. Full integration involves including these rather less tangible factors.

4.2.4 Striving for quality

Another viewpoint with a strong following at the moment is the pursuit of quality. Procedures to ensure that quality standards are maintained are being introduced throughout industry, and every aspect of the construction process is progressively becoming the subject of a detailed analysis of working methods.

It is interesting to reflect whether this is such a new phenomenon. Many older professionals can remember a time when an understanding of quality was regarded as an integral part of normal working practices, without the need to monitor and record its existence. Figure 4.8(a) illustrates traditional quality control which depended on the right mixture of experience, knowledge and perceptive prediction.

Rapid changes in working practices, systemization of working techniques to reduce skill levels, and the high cost of staff time have all had their effects upon the construction industry. A modern approach to quality control is illustrated in Figure 4.8(b). Work is broken down into tasks, each task has defined working procedures and an end result, and a specific check to verify the quality of the result is a documented part of the process.

Integration of information in the construction industry must take account of the need for quality control. To be effective, quality control procedures must be part of normal working practices, rather than an add-on extra. Quality checks and documentation are part of the process and the information to be integrated.

4.3   EVOLVING A SOLUTION

4.3.1   Pawns in the process

The first lesson to be drawn from the points raised above is that no single person or organization involved in the construction process is ever in a position to know every piece of information about a major project. Every actor has a role to play, referring to certain information and drawing conclusions to convey to other actors. This again illustrates a major dimension of the map of integrated construction information parameters drawn earlier by Betts, Fischer and Koskela.

Human processes have developed as the result of a long evolution of ideas, and are often more efficient than technologists imagine. Major construction projects were being built around the world long before computers, or telephones, were even dreamt of. As the next part of the book will show, some measure of integration was also being achieved without computers.

Grand schemes to computerize and integrate all aspects of other industries have frequently run into complex technical problems. Construction poses even more problems than most other industries because of the number and diversity of different firms and organizations who must communicate during the different stages of a project.

Figure 4.9 indicates a fairly simple subdivision of the different actors involved in the construction process. It is not exhaustive, and many of the actors shown actually represent a number of firms or organizations. They are all within a communication environment where specific links are established and maintained, and they need to communicate in terms of concepts, knowledge, facts and supposition. Each deals with their own sample of information about the project and about the environment into which it is constructed.

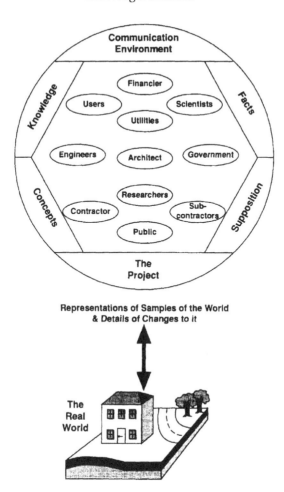

Figure 4.9 Viewpoints and communication

4.3.2 Sustainable islands

Within this complex communication domain, specific groupings of actors and activities who regularly need to communicate, become established for any project. They will exchange information, and the need for localized integration of information within such groups will evolve as a natural result.

Sustainable islands of information integration are beginning to emerge in different parts of the construction industry. Driven by commercial pressure, they utilize whatever technology is readily available. Academic interests will progressively observe and improve the facilities available to work within these islands of integration, if industry can define appropriate objectives.

4.3.3   Windows of understanding

The question then arises, why do we need more? The answer is that many parts of the process will be enhanced by an improved knowledge of other parts. For instance, most environmental work, such as the assessment of visual impact, is based upon human interpretation of drawings and documents which were really prepared for other purposes. More direct and immediate access between different actors will assist and improve the process.

Traditionally, people own their own information, they take responsibility for organizing, managing and updating it. Moving towards a more general integration of information by gathering it together in a single location could disrupt this established principle. Observing the rapid advances in computer technology at the moment suggests an alternative. Allow each actor, or group of actors with a common interest, to maintain the information for their own activity in the form in which it can most easily be updated and used. Then provide a means of dynamic interchange of information with other activities. This approach tends to support and strengthen the human processes which already exist.

This activity-centred approach is illustrated in Figure 4.10. An activity may extend to a whole group of firms or a whole stage in the construction process, or it may be rather localized in one small firm. The viewpoints and activities described in the above parts of this chapter give some idea of the diversity of potential activities. The 'who' dimension of the earlier framework presented by Betts, Fischer and Koskela is also relevant here.

Integration is then achieved by establishing and communicating through links which are dynamically established with other activities when required. Translation during communication is almost inevitable if freedom of implementation is allowed for each activity; on the other hand, complete standardization is unlikely to be feasible. However, if suitable interface standards are established as part of product model research, translation should not be too difficult a task.

To the human involved in the activity, it should appear through the interface with the computer that there is a model of the data needed for the activity and that there are facilities which can act on this model. A process structure and model structure behind the interface provide access to data and real facilities. Data is either local to the activity, and is then owned and maintained as part of it, or is remote and accessed through translators. The activity contains both knowledge of other activities to assist in finding other data, and local knowledge in the form of algorithms or rules which form analysis facilities.

In effect, over a period of time, the local real existence of data and facilities will be extended by the apparent existence of additional virtual data and facilities which are actually located elsewhere. As the translation techniques which support integration become more sophisticated, the

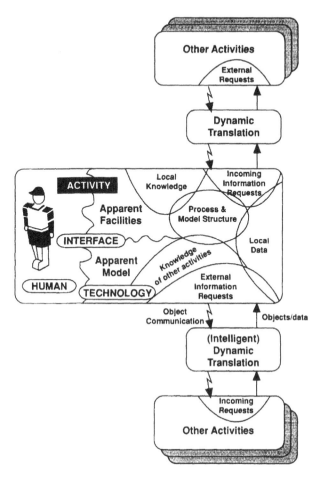

Figure 4.10 Supporting the human process

difference between local and remote data should become less evident to the system user. One consequence of this approach to integration is that there is no attempt to extend access to specific information unless there is a demand to do so. New actors can also become part of the integration whenever required, without the need to define new data structures or working methods; the integration is dynamic and adaptive.

The different roles of commercial and academic activity in developing integration also become easier to understand. Commercial activity generates information and ways of handling it to suit immediate needs, while academic research progressively formalizes and develops exchange mechanisms which will allow sophisticated, dynamic and direct access to remote information between different areas of commercial activity.

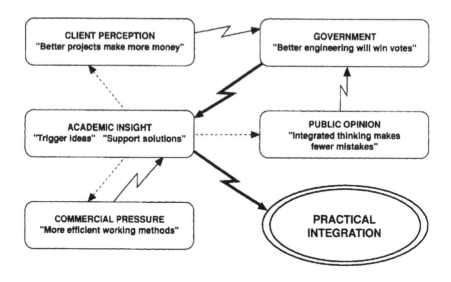

Figure 4.11   Who influences who?

This solution makes productive use of many new technologies, and is intended to be equally applicable on future new computer systems. It also supports the human processes rather than trying to introduce dramatic changes. With a little imagination, it can be seen that Figure 4.10 could represent much of the human process for an activity, even without information technology. Local islands of integration can be set up now; future translations between activities will provide additional capabilities if and when they occur.

4.4   MAKING IT HAPPEN

4.4.1   Mechanisms of change

Going back to the comments raised by Fischer and Brewer in Chapter 2 and repeated at the start of this chapter, describing the constraints on progress illustrated in Figure 4.1, an alternative view of how changes might occur is shown in Figure 4.11. Commercial links between industry and academia are not currently producing very much progress. Industry concentrates on relatively short-term horizons and benefits, and academics tend to become involved in more immediate practical problems as a result. Long-term academic thinking does not always gain much support. Nevertheless, the academic community does have time to think, and read, and understand problems which industry is too busy to concentrate on.

This academic insight must be used to trigger ideas and generate enthusiasm to follow them.

There are two other groups who might lobby for more rapid progress. Project owners, or investors, will understand that better projects are likely to make more money. The public do not like seeing mistakes made and money wasted, and might understand that integrated thinking makes fewer mistakes. Government might listen to these two groups, and this could result in sufficient support for the academic community to achieve solutions which will result in more widespread practical integration.

4.4.2 A common understanding

The first step towards achieving a more general acceptance of the need for integration will be to understand how all of the different and diverse research avenues currently being followed will eventually fit together, to present a common understanding which can be widely understood. The solution presented in this chapter is to allow a considerable degree of freedom for different user groups to progress integration concepts independently, while steering them towards the need to plan for future communications with other activities.

ACKNOWLEDGEMENTS

The views presented in this paper have been derived from a combination of extensive discussions with many researchers and a practical involvement with the development of many construction projects over the past few years. It is not straightforward to attribute specific elements of this experience to individual published documents, and an extensive list of references would not provide a proper opportunity to acknowledge the vital contribution to the content of this paper of discussions with practical people who have not published their experiences. I would like to extend my thanks to the many colleagues, researchers and friends who have contributed to the development of these views over the past few years.

REFERENCE

Dupagne, A. (1991) Computer Integrated Building. *Strategic Final Report*, Esprit II, Exploratory Action No. 5604, CEC, D.G. XIII, December.

CHAPTER 5

An integrating interface to data

James Powell, The Graduate School, University of Salford, Salford,
M5 4WT, UK
Paul Newland, University of Portsmouth, Portsmouth, UK

5.1 INTRODUCTION

Construction is becoming extremely competitive as the industry strives for survival. In time of recession, where there is little prospect of growth, the demands for profitability produce the drives to cost cutting, improved efficiency, total quality and value for money. However, as de Bono (1993) points out, regenerating any business towards a successful future will require more than better competition, a change will be needed in attitude of mind, strategy and concept design towards systemic thinking based on human value integration. We believe integrating values will become the new trend in successful business thinking. Therefore future information systems must design and offer integrated values, so that

> the producer will integrate not just with the customer, but with all the complex values of the customers' lifestyles. For, we all now live in a complex world with many values, where a *building* is no longer just a piece of engineering. But how does the value that the *building constructor* offers integrate into the complex life values of the buyer or consumer? These are going to be the important integrating values in the future.

Values created by the human perceptions of all involved in the construction process.

In the final analysis, in any construction context, it is people's value that will drive change and it is people who will transform data into workable actions. People will work with the new decision support tools in the hope that they will increase their connectivity to the appropriate databases, infor-

Integrated Construction Information. Edited by Peter Brandon and Martin Betts. Published in 1995 by E & FN Spon, 2–6 Boundary Row, London SE1 8HN.
ISBN: 0 419 20370 2

mation and people. Therefore, integrated construction information systems or databases will only be truly integrating when the interfaces to them are properly designed to enable more coherent decision-making and, in particular, support the necessary actions of all construction professionals. For no matter how good the formal integration of disparate and complex databases, if they cannot be fully accessed and used by decision-makers then they are of little real value to them. In this respect, information systems have to recognize the nature of the individuals, groups and teams likely to use them, and be able to respond positively to their very different sorts of interrogation. In particular, they have to be intelligent in recognizing and presenting information in a style and mode that enables the deep understanding so necessary for good decision-making.

5.1.1 Existing information transfer to construction designers

Once information is accepted by a designer, patterns of understanding embedded in the information connect. This brings the information under the control of the designer. Therefore, the major task of an information disseminator is to ensure that when information is sought, such information is in a form that is readily acceptable to any designer. We would contend that at present those who perceive themselves as disseminators of information to designers feel that this acceptability mainly relates to the attractive presentation of information. However, in the following review of research we suggest that attractive presentation alone is not sufficient to gain designers' interest.

Almost 20 years ago the apparent lack of impact of technical literature upon architectural designers' decision-making motivated both the National Bureau of Standards (NBS), Washington, USA and the Building Research Establishment (BRE), UK, to commission major studies to assess the reasons. Goodey and Matthew (1971) reported to the BRE that their findings suggested presentational style was a key factor for encouraging the transfer process of available literature. Their concluding recommendations emphasized the requirement for brevity, clarity and the importance of visual illustration and vocabulary of an architectural nature. Burnett (1979a,b) for the NBS, endorsed these findings with respect to American designers and suggested that when the format had been decided, the information should appear consistently within that format.

Once familiarity with a format is attached, the architect will avoid using a formal index relying instead on memory, browsing and place recognition to locate information. Knowledge of the formats themselves also serve as a memorized guide to content and thus to the relevance of the resources for the problem at hand; while facilitating quick access in this way, consistent formats also promote a consistency in the quality and scope of information on the items described.

Despite the enormous effort to reformat information to be consistent with the detailed recommendations of the above researchers, many, not least those at the BRE responsible for making relevant changes, were intrigued to find that improved presentation did not seem to markedly improve technology transfer. Attractive and more relevant presentation alone does not seem to gain the confidence, acceptance or interest of the design profession to an extent where they actually use information any more readily.

In 1982 MacKinder and Marvin made a follow-up study for the BRE in order to give them a rigorous understanding of the reasons for this communication failure. Their findings confirmed the importance of the previous presentational recommendations, but went on to highlight the unwillingness of architects even to consult technical information, whether in exemplary formats or otherwise. There is no doubt that better presentation did help designers in memorizing information efficiently and exploiting it effectively. However, we believe that in the first instance designers must be prepared to expose themselves to, and have accepted such information, if they are to make it part of their designing.

In one sense this appears trivial, but, unfortunately, creating exemplary formats does not appear to serve as a means to attract initial attention to information. Formats may appeal, they may even be beautiful, but if designers have a predisposition to ignore information there is clearly no way the format of information presentation by itself will overcome these prejudices. This view has led us to contend that instead of research attention being placed solely on information presentation, greater consideration should be given to understanding the self-informing strategies designers use to facilitate their own design process. With such knowledge, information proffered to designers could then be appropriately portrayed to match designers' actual self-informing strategies in the hope that they would then choose to perceive the information as being relevant and useful.

5.1.2 The relationships between designers and information

The relationship between architectural designers and information has been brought into focus by two questions posed by Cooper (1988):

1. Is building design an information-dependent activity?

2. What role does the precision and manipulation of information play in legitimizing architects as a separate occupational group?

The simple answer to the first question is that, quite clearly any design activity ought to be, and in at least one sense is, information dependent. For the key aptitude of designers is their ability to select useful distinctions (information) in order to formulate a boundary appropriate for a required solution. In other words, designers are able to unfold a solution by constructing a boundary, or boundaries, from distinctions which need to be accounted for in the designed solution. Even to conceive of solutions requires dependence on a selection of appropriate information from that held within the architect's cultural heritage. Each designer requires some kind of self-informing strategy to viably operate in the world.

Interestingly architects, and designers in general, are perceived to have the ability to select, with discretion, from a particular information spectrum (a range of previously created distinctions) which is both broader and richer than those open to lay people (non-designers). Therefore, by implication of professional training and experience, designers ought to have developed a greater creative potential and worldly understanding: to the extent that they are able to devise more informed boundaries concerning any architectural problem. Readily acknowledged by society in the past, this capability for undertaking a broader and richer view of recognized information, gives architects a status which goes some way in answering the second question posed by Cooper.

However, having attributed this status to architects, society increasingly feels that on too many occasions this status is abused by the architectural professions (Prince Charles, 1984). Why?

It is our conjecture that designers place particular constraints on themselves, and have constraints placed upon them by the groups with whom they affiliate. This limits their perception of what should be regarded as relevant information. Each designer should be capable of covering a broad spectrum of information, but, as suggested previously, it will be a spectrum which is dependent on the self-conscious mind unique to each designer – the particular self-informing strategy adopted by each designer.

Both the perceived spectrum of information, and the discretion in choosing from it, may appear wholly inappropriate in the eyes of others who recognize different information as being legitimate and/or give credence to different selection principles. Our concern here is to understand why these constraints have developed and why information that others feel is relevant to design is ignored by designers. Indeed, we hoped to begin to respond to the research needs noted by Lera *et al.* (1984) in their conclusion to an investigation on this subject of designers and information:

> If the transfer of information to architects is to be improved, it will be necessary to gain some understanding of how their existing means of acquiring knowledge operate, what sort of information they feel really is informative – in the sense of enabling them to give form to an idea – and a deeper systemic understanding of designer audiences, their activities, their problems, their attitudes, and their needs.

These summary requirements are in accord with Ritter (1981), who proposes that to induce designers to select information, it must reflect their personal perceptions and be relevant to them. It is our conjecture that these requirements imply that designers' preferences for the selection of information are predominately due to the context in which it is perceived. We define context here as the quality and texture of the events that surround and give an outline to anything that purports to be design information; in other words it is the light in which information is first seen, that enables the designer to recognize particular distinctions. Unfortunately, for most designers the majority of (so-called) design information fails to illuminate in this sense. As Powell (1968) found out from his large scale survey of design guidance, notes and literature, most of it is just too diffuse to aid designers' recognition of necessary distinctions, or worse, it often appears in opposition to their predisposed view of what constitutes relevant and useful information.

5.1.3   The knowledge gap

So what is the nature of the contexts designers need to perceive before they are predisposed towards information selection, and are these contexts so entirely individual to each designer that catering for this design information tailoring would necessitate unbounded resources?

It is our contention that what is selected to be information by designers, actually reinforces and creates their perception of the world. It follows from this that in order to develop appropriate contexts for information transfer, it is important to understand designers' existing world views and, in particular, the information they already use to interpret their world. Our aim has therefore been to gain an insight into such designerly interpretations and world views and, in so doing, define any descriptors that could help improve the potential for architectural design information and technology transfer.

5.2   THEORETICAL UNDERPINNING OF THE INTERFACE

5.2.1   Approaches to knowledge structure

As previously stated, over the last decade, concern has developed due to designers' inability or unwillingness to refer to important design information. Previous research suggested to us that the restricted self-informing strategies of designers played a major role in the generation of this concern. In particular, we found a suggestion by Cross and Nathenson (1981) helpful in enabling us to establish a description of possible designers' strategies for action. For they suggested that the way designers design is closely related to the way they learn. So perhaps learning, or rather individuals' different styles of learning, might be a key issue in determining their self-informing strategy. We decided, therefore, to make a major aspect of this study an investigation of designer learning. Learning is a particular domain of research exploration that centres on two fundamental aspects of an individual's involvement with the environment. Individuals who have become conscious of their distinctness in the environment also initiate and respond to directed changes in that environment. Knowledge can be re-constructed by the individual's own efforts. However, it is more usually gained by individuals through their interactions with others. We suggest the two processes work in combination, individuals receive parts of their knowledge about the world via others but use their personal learning abilities to fill in the gaps. Of paramount importance is the individual's gathering of social knowledge through learning and interpersonal interaction.

One possible typology of understanding learning better is suggested by Kolb's (1984) studies in developing his experiential learning model. He indicates that his theories and methodology are based on the work of Pepper who in 1942 created a typology of different forms of social knowledge from his own concept of *World Hypotheses*. Pepper and Kolb's theories, which underpin the interface development, are described below in some detail. To give meaning to these stances we also suggest possible design orientations which could arise from them. Stances that could help an understanding of the world views held by designers.

5.2.2 Pepper's world hypotheses

Pepper's work has, in recent years, been gaining deserved recognition. Altman and Rogoff (1987), for example, have used Pepper's thesis to help them comprehend the multiple world views now being recognized in psychology. Pepper's (1942) work argues with adroitness that there exist only four approaches for understanding the world we inhabit that have: 'the capacity both to explain the observed phenomena accurately and to encompass sufficient knowledge about these phenomena to be termed a world hypothesis'. In other words, only four strategies can exist for any individual to permit them to operate in some viable way and thereby allow a continuous flow of sensory input about the world to be structured into their working personal world view. Pepper believes these world hypotheses are generated by a process of structural corroboration. In particular, Pepper indicates that in order to reduce the degrees of doubt and attain a refined knowledge of reality (a common sense knowledge revealed by multiplicative corroboration), individuals actively and continuously seek structural corroboration while developing their own world views. Pepper uses the example of how we gain an understanding of what a chair is to exemplify this point.

If we were to enquire into the strength of a chair, we might ask a number of people to sit on it; if it doesn't break, we would then have multiplicative corroboration for our hypothesis that the chair is sound. Structural corroboration occurs when, from an examination of the parts – the woods used, the fastening devices, the overall design – we infer that the chair is sturdy. Structural corroboration is an agreement of many different facts in the determination of one central fact and a massiveness of convergent evidence upon the same point of fact. In other words evidence to support a hypothesis is not founded upon repetition of one particular observation, but through collecting a diversity of observations that provide consistent validity.

In Pepper's opinion, there are only four self-organizing approaches open to structural corroboration which are relatively adequate in both precision (how accurately they fit the facts) and scope (the extent to which all known facts are represented); in this typology a philosophical stance develops. Figure 5.1 summarizes the central philosophical perspective each stance encourages.

Contextualism

Central to Contextualism is the question: How can you be so sure that nature is not intrinsically changing and full of novelty? The only means to gain stability with such an outlook is to rely on the consistent knowledge that there will always be change. It is the continuous awareness of the active, present event which gives the basis for understanding reality and the stance:

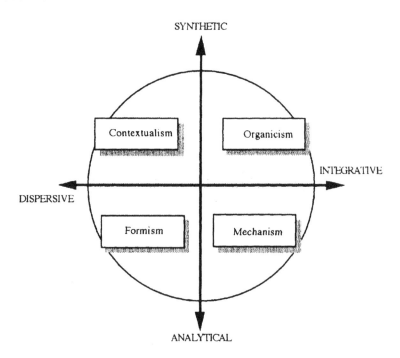

Figure 5.1 The four philosophical stances of Pepper

is very definite about the present event and the premonitions it gives of neighbouring events, but less and less definite about the wider structure of the world. It is willing to make more or less speculative wagers about the wider structures of the world.

Pepper, 1942

Contextualism is unique among the four philosophical stances in denying the existence of unchangeable structures in nature. Instead the perception of reality emphasizes acceptance of change and novelty.

Organicism

Striving for a unique wholeness, Organicism takes the view that everything is a fragment bursting with an inherent ability to coalesce. Man is assumed to be the channel through which data is gradually being transformed towards an absolute whole. The eventual goal is total synthesis, which is the outcome of a continuous circular process, described thus:

(1) fragments of experience which appear with (2) nexuses or connections or implications, which spontaneously lead as a result of the aggravation of (3) contradictions, gaps, oppositions, or counteractions to resolution in (4) an organic whole, which is found to have been (5) implicit in the fragments, and to (6) transcend the previous contradictions by means of coherent totality, which (7) economises, saves, preserves all the original fragments of experience without any loss.

Pepper, 1942

Pepper believes that these stages, (1) to (7), are repeated until total synthesis is completed. The ultimate achievement is the transformation of all the perceived fragments into an ideal whole.

Mechanism

The philosophical stance of Mechanism assumes there are structural features which underlie the experienced nature of objects. This network of abstracted qualities and laws defines what objects can be and how they can act. These descriptive properties can be organized into primary and secondary categories. The primary categories describe the field of location and qualities such as size, shape, mass/electric charge and motion. While the secondary categories indicate appearances such as colour and sound.

All the cognitive evidence for this way of perceiving the cosmic structure is derived from the secondary categories. It is from these resultant emanations that the symbolic primary categories are constructed. Interestingly, as Pepper points out:

All immediate experiences are emergents from neural configurations inside the organism, so all immediate evidence is, therefore, private to each individual organism.

The structural categories perceived in these personal realities can be formalized through social agreement with others. More than any other, this philosophical stance relies on the rigid consistency of an archetypal order.

Formism

The philosophical stance of Formism relies on correspondence, and searches for empirical uniformities. Like Mechanism the stance uses categories, but these are dispersive, e.g. character, particular and participation. Their meaning is described thus:

> This is yellow, is a sentence epitomising these three categories. This represents the uncharacterised particular; yellow, the unparticularised character, is the participation of each in the other to produce the object.
>
> *Pepper, 1942*

The relations that are determined between objects are considered as ties of similarity which are based on directly experienced evidence rather than through the implications of abstracted rules. The logic for these analogies is paleologic, a term which is defined by Cross *et al.* (1986):

> Paleologic therefore insists that it is not, as the metaphysicians claim, empirically inapprehensible, formal or essential concepts that account for our ability to group individuals into classes that have a common predicate (e.g. chalk and cheese grouped by the concept of whiteness), but on the contrary it is, as the empiricists claim, a primordial apprehension of the commonality of predicates (e.g. that chalk and cheese are the same colour) that enables us to group individuals into analogically related classes, to which a concept is then applied.

This quote is cited from a research document entitled 'Designerly Ways of Knowing: A clarification of some epistemological bases of design knowledge'. One of the main contentions of this document is that there exist a variety of valid ways of knowing and that they each have relevance to design. On the basis of Pepper's typology the intention is now to describe four possible orientations relevant to a design context.

5.2.3   Kolb's learning style theory

Kolb's work in the domain of learning had its first major reference in 1973 when, in conjunction with Goldman, Kolb undertook a study of MIT Seniors to create a typology of learning styles. The two major publications that resulted from this study and subsequent investigations are *The Learning Style Inventory: Technical Manual* (1978) and *Experiential Learning* (1984). A review by Curry (1983) of 21 models of learning style and strategies found 10 to be psychometrically acceptable. However, only three of these were, in her opinion, assessing learning in the strict sense of information processing style. One of these was Kolb's (1978) *Learning Style Inventory* (LSI), based on his *Experiential Learning* theory which had been used with much effect by Davis and Talbot (1987) in their studies of Royal designers.

It is Kolb's contention that learning is primarily experiential (following the ideas of Dewey (1934)) and centres on two structural generators. The first is the concept of prehension: how the individual gathers experience from the world; the second is that of transformation; how the individual's ephemeral representation of experience becomes frozen into understanding. Both prehension and transformation are dimensions which extend to bipolar attributes. Further, Kolb contends that an individual will be inclined towards a particular pole in each dimension (see Figure 5.2). Taking the argument a little further, prehension relies either on the tangible, felt qualities, which Kolb terms the apprehension of experience, or on the tendency towards comprehension, a conceptual, symbolic representation of experience.

The polar attributes on the orthogonal transformation axis (Figure 5.2) deal with individuals' choices for either internal reflection on the prehen-

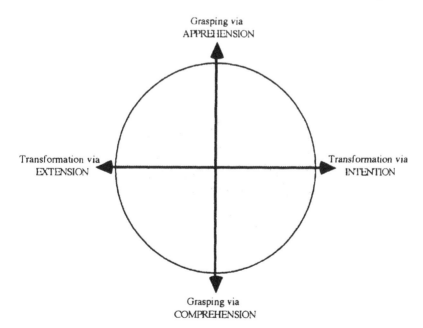

Figure 5.2 Structural dimensions underlying the process of experiential learning (Source: Kolb, 1984)

sioned experience, which Kolb calls intention, or reaction to the experience by external manipulation of the world, an act of extension. Kolb (1984) summarizes these underlying themes thus:

> The central idea here is that learning, and therefore knowing, requires both a grasp or figurative representation of experience and some transformation of that representation. Either the figurative grasp or operative transformation alone is not sufficient. The simple perception of experience is not sufficient for learning; something must be done with it. Similarly, transformation alone cannot represent learning, for there must be something to be transformed, some state of experience which is being acted upon.

To affect learning, individuals need to have been aware of an experience and register a reaction to that experience. With two options in each of the two dimensions of Kolb's model, a 2 × 2 matrix of decision is brought into existence. This matrix allows a typology of four quadrants, each representing a particular learning style (Figure 5.3).

The development of a fourfold (an integrated set of four options) learning typology is based on the previously mentioned Fourfold World view suggested by Pepper and is not dissimilar to the outcome of the work of Piaget and Inhelder (1975) on learning and cognitive development (see Figure 5.4). With this in mind we will, therefore, digress for the moment to explain Piaget's theoretical stance on learning (or in his terms, intelligent

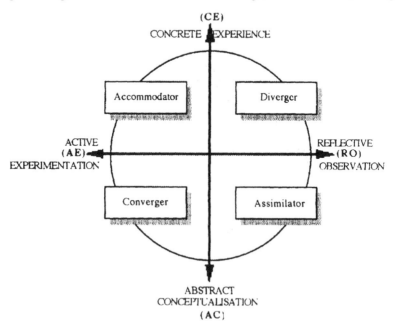

Figure 5.3   Experiential learning style model (Source: Kolb, 1978)

adaptation). A comparison between how Piaget views this typology and how it is perceived by Kolb should enrich the present understanding of Kolb's work. In the first instance, it is interesting to note that the more metaphoric descriptions of the poles of Piaget's axes are not far removed from those descriptions given to the poles by Kolb in his final and most widely adopted typology.

For Piaget the key to learning lies in the tension which is created between the desire to accommodate and the desire to assimilate. Accommodation enables individuals to mould themselves to the environment, while assimilation is the ability of individuals to project their own mould or concept onto the environment. Cognitive growth through childhood to early adulthood can be seen as a change in emphasis from accommodation to assimilation. For individuals to enhance their intelligent adaption, Piaget contends they move away from concrete phenomenalism towards abstract constructionism and employ active egocentricism and internalized reflection when necessary.

According to Piaget and Inhelder (1975) the progression of learning from childhood follows a clockwise direction around the four quadrants (see Figure 5.4). The first stage (0–2 years) is a period of enactive learning – feeling, touching and handling are uppermost. In the second stage (2–6 years) emphasis moves to iconic learning, a manipulation of observations and images. Stage 3 (7–11 years) leads onto inductive learning where

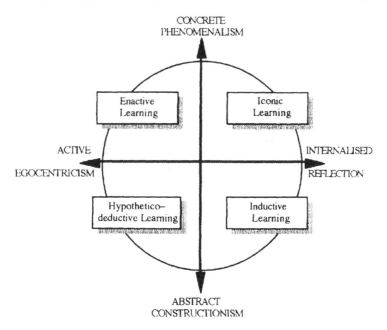

Figure 5.4   Model of learning and cognitive development (Source: Piaget and Inhelder, 1975)

things are classed and relations abstracted. The final stage (12–15 years) is hypothetico–deductive learning, the implications of constructed theories are investigated.

Kolb's interpretation of learning, is that all four quadrants in this typology remain equally valid for any stage of an individual's development, and not just in childhood as Piaget would suggest. Indeed, Kolb originally felt that complete learning would be achieved only by engaging with each polar attribute (see Figure 5.3) in a continuous circular process where concrete experience (CE), an open, feeling-based involvement, is followed by careful consideration and taking a holistic viewpoint, reflective observation (RO). Then, using abstract conceptualization (AC), a logical, analytical approach, hypotheses could be formed that are tested with active experimentation (AE); undertaking such a practical application creates new experiences and initiates a fresh learning cycle.

However, Kolb (1984) soon realized this was an ideal, and began to show that individuals adopted a restricted scenario of preferred options from each dimension. For it appeared that a complementary pair of orthogonal descriptors gave an identifiable, but personal understanding of the world to each individual. So, for instance, an individual who favours concrete experience rather than abstract conceptualization, and active experimentation rather than reflective observation, will be identified by the categorization 'accommodator'. Continuing around the circle the other categorizations would be diverger, assimilator and converger. These are terms which Cross and Nathenson (1981) and Pask and Scott (1972) have already suggested as pertinent descriptors of designers.

Quadrants labelled according to Kolb are shown in Figure 5.3. This fourfold typology of processing structures for designers' learning, are best exemplified by depicting the various options they give a designer who needs to learn about a proposed new urban building site. (The clockwise convention, initiated with the description of Piaget's model and carried into Kolb's work, should not be thought of as having any bearing on the importance of any one quadrant over another; as is explained later, all have equal validity.)

Accommodator design learners

The accommodator relies heavily on direct sensory experience and exploration (Figure 5.5). This is sought for in the immediate surroundings and so a trip to the site would be essential. Physical presence at the site would provide the opportunity for this designer learner to experiment and manipulate the actual reality. Since the prime action of accommodators is to enrich reality, they would make themselves and their ideas known to residents. It is important for them to gain a feel for the place, know what risks are possible, what challenges are present and what statement they can leave as their signature.

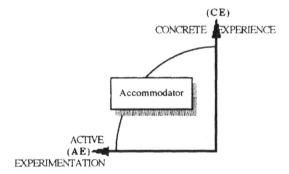

Figure 5.5 The Accommodator has preferences for active experimentation (AE) and concrete experience (CE)

Diverger design learners

The diverger (Figure 5.6) has an inherent interest in people and in gaining an awareness of values and meaning. As with the accommodator, visiting the site is important. The aim of this designer learner is to understand inherent natural relationships, and to generate ideas that are in accord with the present and immediate future concerns of the residents. Their strategy looks to adapt reality. Sketches and measurements would be used to assist in understanding the building site.

Assimilator design learners

The aim of the assimilator is the accumulation of a wide variety of observations which can be incorporated into a well reasoned, holistic conceptualization (Figure 5.7). First-hand experience of the site by this designer learner is likely to be supplemented by photographs, and climate data. They are keen to search for detailed assessments of the site's physical

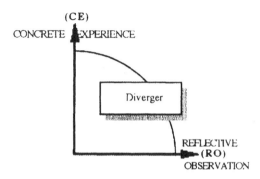

Figure 5.6 Divergers have preferences for concrete experience (CE) and reflective observation (RO)

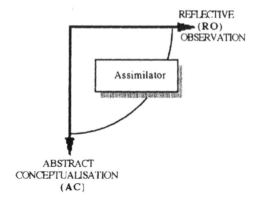

Figure 5.7 Assimilators have preferences for (RO) and abstract conceptualization (AC)

attributes. Finally they have a preoccupation for global implications of development. Assimilators attempt to absorb reality.

Converger design learners

The converger (Figure 5.8) is concerned with dealing efficiently with problems in a logical manner. This designer learner would make a rational assessment of the site, prepare formal recommendations and clarify necessary procedures. Emphasis for them is towards recreating reality by abstracting a theoretical model which can lead to experiments that imply correct action. The converger directs attention to the site's history in order to make predictions for the future:

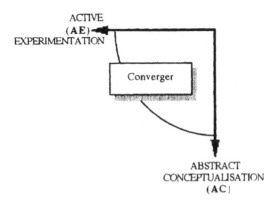

Figure 5.8 The Converger has preferences for abstract conceptualization (AC) and active experimentation (AE)

knowledge of the site's immediate condition is perceived as secondary to recorded knowledge of its past.

In Kolb's theory all these options are open to designers who could employ any combination of them. However, and counter-intuitively to most designers who feel they should be all-rounders, Kolb's empirical studies found the choice was often not as open as it first appears, being in many respects rather limited. For Kolb (1978) showed that we all bring a particular pair of tinted spectacles to bear upon the options open to us.

The concept of a possibility processing structure gives central importance to the role of individual choice in decision-making. The way we process the possibilities of each new emerging event determines the range of choices and decisions we see. The choices and decisions we make, to some extent, determine the events we live through, and these events influence our future choices. Thus, people create themselves through their choice of the actual occasions they live through. Human individuality results from the pattern or programme created by our choices and their consequences.

Kolb's studies indicate that individuality will always preferentially affect an individual's approach to learning and further, once engendered, such a preference appears to become relatively stable. Such a stability in human strategies is further supported by the work of the anthropologist researchers, Douglas (1978) and Thompson (1982a,b).

It should be clear from the above that Kolb's experiential model of learning provides a useful starting point for the development of an information system which attempts to match designers' learning needs. Since it is:

1. a robust theory in much general use;

2. well-founded, being supported by numerous validation studies;

3. encompasses other design researchers' understanding of designer learning.

5.2.4   The related processes of learning and design

To state that the learning process and the design process are synonymous is in accord with the views of Cross and Nathenson (1981), Pask and Scott (1972), Glanville (1980) and Powell (1988). For instance, Cross and Nathenson reason from observational studies that:

The design process is often likened to a learning process. This is because a designer is quite clearly learning about the design problem as he attempts to solve it. Much of his activity is concerned with attempting to clarify the problem as given, with seeking information, and with attempts to come up with an acceptable answer. As he goes through the design process he is learning more and more about the problem, its constraints, and its potential solutions. At the end of the process he usually regards himself as appreciably both older and wiser than when he began it.

Of further significance here, is Cross and Nathenson's conclusion that the experimental evidence indicates that designers exhibit preferences for learning on the basis of particular cognitive styles: divergent/convergent, impulsive/reflective, field-dependent/field-independent and serialistic/holistic. Cross and Nathenson emphasize that no-one is exclusively divergent or convergent, or exclusively any of the other polarities in these duals of cognition. However, they are clear that these duals represent the existence of preferential options for designers, which allow only a limited number of possible ways for them to understand reality.

In recent years the acknowledgement of these preferential options has produced a range of models of the designer. Most aim to create typologies that account for the different learning approaches which are adopted by design individuals.

5.2.5   A fourfold of design orientations

If Pepper's and Kolb's contentions are correct there should be four approaches to design presently distinguishable; one relating to each of the four distinct philosophical stances. For design, in our opinion, should be fundamental to the construction of all social knowledge and should therefore exhibit a distinctive role in any philosophical stance. Such descriptions move Pepper's typology and our argument back to the realm of designers. Furthermore, embedded and implicit within these descriptions are suggestions for how preferences in the learning domain will match with each of the design orientations described from our work.

Design orientation based on Pepper's Contextualism

Pepper's philosophical stance of Contextualism places emphasis on events in the active present. We suggest this necessitates designers being overtly and continuously aware of their actions and sensations. It is this constant recognition of sensing concrete experience and doing through active experimentation, which assists them in maintaining a viable reality. Within this context we see design as ephemeral for each event, but ever present in the continual flux of their experience. As Wallas suggested back in 1926, designers continually inhabit the fringe consciousness of bi-sociation (Koestler, 1978). These terms have been used by Davies and Talbot (1987) in an attempt to convey the sense of rapid superpositioning of images that successful British designers they studied, known as the Royal Designers, use to facilitate their openness to present events. These designers' creative acts seemed to arise directly from such openness being a sort of transformation state from being and knowing to being the knowing. Davies and Talbot describe this transformation as the imago. The imago represents the ephemeral design solution. In the context of designing it is the most difficult part of the process for the designer to control. For, if the flux of events during the design process is too rapid or they are inappropriate, a suitable imago can easily allude capture.

Good designers here seem to be able to sense a melting pot of potential imagos – a close relationship to the grasping of creativity. They have a confidence to remain in the primordial soup of creative fusion. However, this demands in them a perception of interpersonal interaction which is benign and tolerant to the point of submission, so that cascades of originality can be allowed to permeate their minds. Such a designerly orientation would seek dominance to provoke the required acceptance of novel ideas.

Design orientation based on Pepper's Organicism

The final aim for the philosophical stance of Organicism is total synthesis. A designerly approach to assist this aim would be one that was looking to transcend polarities. We contend that reflective observation combined with concrete experience are the most appropriate learning preferences to emphasize for this stance. Creating the right context is essential here so that these designers can ensure that they can take into account the multitude of perspectives needed to facilitate a holistic attitude.

The central ability of these designers is to fuse and go beyond polarities. It is a design orientation which has been described in some detail by Rothenburg (1976), who gives it the name Janusian thinking, after the God with two opposite facing heads. Rothenburg sees this particular design orientation commonly occurring within architecture, especially with those architects who adopt an Organic design style:

> integrated opposition and, by implication, Janusian thinking, can be seen to have an important role. In architecture, the Janus metaphor is particularly appropriate, since it is necessary for the creative architect to conceptualize the inside and outside of a building simultaneously. For example, convex outer shapes produce concave inner shapes and the architect must reconcile these contradictory spatial characteristics with the overall conception of a building to be built. Moreover, the best buildings do not convey a quality of spaciousness on the outside that is contradicted once one is inside. Since external shapes conveying spaciousness often deceptively require a great deal of internal buttressing structure and consequent cramping, it is necessary for the creative architect to overcome this. He does this by formulating designs that accomplish spaciousness in opposite spatial orientations simultaneously. Frank Lloyd Wright has described the operation of Janusian thinking on an even wider scale than this in

his description of the development of Organic Architecture, the type of architecture he created. He referred to the Organic Architecture idea as an affirmative negation, meaning that it negated the three-dimensional concept in architecture and affirmed it simultaneously.

Clearly the above-mentioned designers are attempting to unite diverse aspects – placing man and nature in reciprocal harmony. The attitude of this type of architectural designer would be extremely friendly in all interpersonal interactions.

Design orientation based on Pepper's Mechanism

Pepper's philosophical stance of Mechanism supports the notion that there is a set of underlying principles for the universe at large. In our opinion the discovery of such principles would reside primarily in some designers having the capacity for abstract conceptualization and reflective observation. Indeed, Cross *et al.* (1986) describe a design approach which emphasizes such a capacity, which they call abduction:

Abductive logic proceeds by abstracting the universal form or essence applicable to a class of individuals, and explicitly articulating the general properties of all such analogically-related individuals by which they are regarded as members of that class. Hypotheses, rules, theories and scientific laws are arrived at by abduction.

The cultural stance necessitated for such a thorough investigation would resign the architectural designer to the fact that an external system has to have authority. Such interpersonal interaction based on submission is fitting for this deterministic approach to design.

Design orientation based on Pepper's Formism

In essence the basis of Pepper's philosophical stance of Formism would see designers seeking to explore the world by fragmentation (Glynn, 1985). We see the learning preferences of active experimentation and abstract conceptualization prevalent in this desire to actively split reality into parts; parts which are firstly named and then assessed. Designers adopting this fourth design orientation use a great deal of their designerly effort in protecting the very practical things and views that they themselves and their support groups recognize as being appropriate – in the sense of best fit. In doing so they express a need to perceive other points of view as slightly hostile. These designers therefore adopt and portray a critical stance for interpersonal interaction.

Summarized in Table 5.1 are what we perceive to be the interconnections across the typologies of Kolb and Pepper, with our own title given to

Table 5.1 The interconnections between the typologies of Pepper and Kolb

Powell and Newland	Pepper (World Hypothesis)	Kolb (Learning)
Dynamic	Contextualism	Active experimentation
		Concrete experience
Focused	Organicism	Concrete experience
		Reflective observation
Contemplative	Mechanism	Reflective observation
		Abstract conceptualization
Rigorous	Formism	Abstract conceptualization
		Active experimentation

each category. The fourfold of philosophical stances described by Pepper give a framework to understand the underlying facilitators for a range of design orientations. Each of the design orientations which emerge provide stable, basic convictions to which a designer can adhere. However, if designers intend to manifest their convictions there should be a corresponding range of motivations which are also available to them. This fourfold framework has been used to develop a multimedia information system aimed at removing the barriers to communication to designers and to help them integrate their values with those of others.

5.3 INFORMATION SYSTEM DEVELOPMENT

The preceding fourfold strategic action patterns have guided our design and production of information or learning systems, which enable the acquisition of useful professional knowledge about different aspects of building design. The important means by which our systems are different from others, but thus become acceptable to busy building professionals, relates primarily to the special ways in which they handle the interaction with the learner and especially through the tone of the information portrayed to the learner. We place particular emphasis here on getting the tone of knowledge in the system right as a means of achieving resonance between any individual and the imagery, learning, interpersonal interaction preference, philosophical basis and social assumptions they naturally seek. These issues are of fundamental importance in assuming that integrated construction information systems get implemented in practice. However, of equal importance is the development of an acceptable acquisition mechanism for the information in the system. For example, the visual portrayal of a topic could be by a set of still photographs with text annotation, via a film with an audio commentary, through self-paced questions and answers, or as audio/visual sequences, etc. The mix and style of media presentations provided by any system such as film, audio and text will facilitate different self-informing strategies.

Our prototype system, using advanced 'intelligent' interactive video techniques, tries to present designers with pictures, moving video, graphics and text, in the right balance, to form a congenial and compelling information transfer environment for design. Furthermore, to help designers keep total control of their access to the different information in the system, we have also produced a simple user control mechanism – a refinement of the window–icon–mouse–pointer (WIMPS) environment of the Apple Macintosh. The combined interface, an 'ideal advisor' according to the *Sunday Times*, has already proved successful in helping some architectural designers gain a better understanding of key 'energy target' and 'fire escape' issues for building design. Very different types of architect seem to readily accept the information designed for themselves, while finding that information designed for other designer types was not so valuable.

The following descriptions indicate the general characteristics of media presentation, tone and interaction relevant for each strategic action pattern.

Dynamics

The tone of information needed for a Dynamic individual centres around the notion of 'information transfusion'. In terms of portraying a particular topic, a sense of transfusion is only invoked when a topic is made tangible. Therefore, for these individuals, any topic of information transfer should draw heavily on personal experiences or shared experience events leaving them to weave a description of the topic into their own viewpoint.

Dynamics portray their knowledge through their actions, and to be of a matching tone, an information source needs to be similarly active. It appears to us that retelling anecdotes to others is one such active expression by Dynamics and helps them personalize, retain and find meaning in information. Another acceptable alternative for Dynamics is listening to autobiographical experience from other similar individuals. Taking a topic and describing it by having individuals talk of their personal experience with the topic would encourage the Dynamics to give this information source attention. However, such communications need to be short, for if their interest is not caught then such Dynamic individuals will soon want to present their own personal experience of the topic in question instead of taking on board the alternative views of others.

Appropriate interactions with information sources would be ones that attempt to facilitate such self-discovery and allow these professional learners to engage in a series of raw and challenging experiences. Interaction, for them, needs to be novel with respect to both style and access. We have found that one useful mode of interaction for Dynamics is setting up a framework of challenges. Such challenges can give a near real life quality to involvement with an information source thereby demanding spur-of-the-moment acquisition of knowledge.

The Focused

These types choose to inform themselves by a strategy which seeks information that has a 'tone of transposition'. For this individual there should be provision for practical knowledge both at a general and specific level. For knowledge to be acceptable to these individuals it should be capable of grabbing their attention and being naturally incorporated or transposed into their own normal professional processes.

The Focused engage with the environment in order to understand why processes are as they are and theirs is essentially a 'small is beautiful' philosophy which demands information with a clear-cut, down-to-earth quality. Focused individuals are encouraged by and take an interest in information which is voicing beliefs of adaptation and equilibrium of man with nature. The necessary tone of information required to support such beliefs is only possible with directness and presentation by similar believers. In essence Focused individuals want to be assured there are others of like mind who support their views.

We suggest the best presentation for this learner is an audio/visual documentary style, where a presenter, who comes across as both friend and supporter, gives faithful, step-by-step and coherent guidance on any topic. The main emphasis in such documentary productions would be parsimonious case studies of relevant problems and their sound solutions. These individuals would appreciate an interaction which makes it possible to readily interrupt the documentary style case studies when detailed questions are raised. Such interrupts would give the Focused an opportunity to review the material presented, enquire about the general usefulness of the topics covered for their present concerns, and to select particular points in the material for which greater detail could be requested. In essence, the aim of such an interaction is to allow these individuals to navigate their own path through a tone of information which is in good faith with their beliefs and seeks to provide well-founded guidance on practical issues, which is endorsed by respected others.

The Rigorous

These types attain viability through the principle of 'transcribing' codes into practice. Individuals who find this strategy appropriate are those who seek information which reinforces the validity of the rules they already use to guide their action. Such rules are recognized as emanating from authorities. In a similar fashion, information which is seen to have been generated by these authorities will also be given credence as well. Therefore, they would be assured by information whose tone implies it is a code or standard practice. Procedural information would also be welcomed for its strength in providing straightforward, economical solutions.

An appropriate interaction for the Rigorous is with a stable, historically accumulated knowledge base, on a given topic of precedents which would emphasize the training algorithm of pretest–teach–post-test. In our view Rigorous individuals appreciate being stage managed through an introduction which gives a topic authoritative approval, then delivers a logical set of procedures and finally gives an opportunity to quantify their expertise. In essence Rigorous individuals require formalized knowledge which can be retrieved from a well-structured hierarchy of authoritative findings.

Contemplatives

Our research shows that those who choose to adopt a Contemplative strategy are interested in a broad spectrum of information. For them a comprehensive montage is necessary to adequately support their desire to accumulate a mass of knowledge which is sufficient to allow them to transcend a unifying principle. These individuals are given confidence through having access to a wide range of primary information sources. Though they are willing to listen to others' opinions and summaries of primary sources, they much prefer delving into the unedited data and drawing their own conclusions. Unlike the Focused approach, commentaries which in any way pre-empt the Contemplative arriving at their own conclusions would be a low priority option. Contemplatives would prefer to self-pace themselves through annotated high quality visual stills and be given access to comprehensive references and global views of a topic's causes and effects. In essence, these individuals are seeking the means to undertake an integrative analysis of a topic and a welcomed information source would need to divulge uncensored data and incorporate guidelines to unify the known facts.

Embracing the above scenarios has led us to investigate information technology which could facilitate educational environments which allow response on a bespoke basis to at least four different types of user. We have already developed one such information system, which presents information on energy conscious design from four perspectives and aims at engaging individuals of any strategic action pattern preference.

Recently, we have been able to refine our learning/information system using Apple Macintosh series II computers extended by the addition of transputers and real-time video digitizing boards. These hardware improvements have enabled us to make more responsive educational environments to support specific strategic action patterns.

5.4 SYSTEM TESTING

5.4.1 Demonstrable environments

The following detailed descriptions concern educational environments we have developed using this refined system. In particular, they have been developed mainly to present information to Dynamic individuals. Although the experiential environments described here are not directly aimed at construction, they do relate to areas of considerable importance in finished structures, namely, the command and control of a major (fire) problem and quality in design. The demonstrations also portray the possibilities of state-of-the-art interactive technology. As such, the techniques involved in their creation are easily adaptable to any profession which requires attainment of adaptive competency in its members.

Case study I: ICCARUS (intelligent command and control – acquisition and review using simulation)

At present in Britain, there are 5000 fire station officers. Each could find themselves having to take charge of a large fire incident (defined as one requiring the attendance of five or more pumps). In addition, within the ranks of the fire brigade there are 500 potential fire officers per year who require training before promotion and 200–300 per year who require refresher courses. The West Midland's Fire Brigade alone estimate that they spend approximately £1 million per annum on training and there are 62 such brigades in the United Kingdom. The efficient management of large fire incidents entails considerable cost benefits. The Fire Research Station estimate of the cost, for each minute that each fire rages, is about £10,000. Thus a reduction of each large fire by as little as 1 minute would produce a saving to the United Kingdom of almost £10 million per annum.

Effective command and control by an Officer-in-Charge of a large fire involves overall co-ordination of a wide range and high complexity of human and other resources. To do this, the officer has to maintain a large-scale real-time information communication network. Acquiring management skills to do this effectively and efficiently is vital for potential fire officers – to prepare them for command – and for existing officers – to keep them alert to situations they might only deal with very occasionally. However, a primary problem in the effectiveness of Fire Officers' training lies in the realism of the training task when compared with real life incidents. Present training consists of a blackboard and chalk discussion in which a training officer describes a fire incident, followed by a question and answer session. Rarely is it possible to create a practice session involving real men and hardware and, even when such role playing exercises are mounted, the situation is contrived and cannot easily be given the sort of realism and variety that would truly tax the skills of strategic command and control.

ICCARUS is designed as a multimedia environment in which realistic emotive experience is gained of attending, commanding and controlling a major fire incident. The implementation of AI techniques, supported by a Mac IIx, transputers, various optical technology and the use of a sophisticated WIMPS interface allows the participating officer, as in real life, to become an information nexus: able to request and allocate resources within an audio/textual/visual environment, creating a real-time, event-based qualitative simulation, capable of realistically modelling the behaviour of both the fire and the decision-making capabilities of the agents under the officer's control. In the main, the final simulator is a self-learning tool for those individuals of a dynamic persuasion. However, there is an intelligent tutoring package back-up to the system which accommodates the more Rigorous preference of fire training officers who can review and analyse the trainee's performance.

So far we have only undertaken a preliminary evaluation of ICCARUS being used by 15 fire officers from different parts of Britain. This has shown the restricted, but genuine, nature of the simulated environment we have created, is an emotive experience which recaptures the stress and actuality of true-to-life command and control situations. Officers using ICCARUS have commented that the iconic interface, indeterminate audio/visual interrupts and semi-intelligent actors give a suspension of disbelief not experienced in previous simulations. Such officers also point out that their private one-to-one involvement with the system allows them to investigate the outcomes of 'wrong actions', which they find difficult to initiate in traditional role-playing sessions where their colleagues and peers are present. The knowledge of how mistakes can occur is therefore gained without loss of face and reduces the necessity for this experience to be obtained through real life occurrences.

Case study II: Quality – our definition

In order to gain an understanding of different people's strategies for attaining quality design, it is important for those managing a design project to have a clear, reciprocal awareness of what each active participant involved in the design process means by quality. At present quality is a noun looking for appropriate adjectives. In the round table design discussions normally associated with most early phases of designing, each participant of a design team, and even the client, can bring their own particular and idiosyncratic view of quality to bear. In many cases they will not be sure themselves what they mean by quality. The project manager should somehow experience that quality too easily becomes a catch-all noun to represent many facets and contradictions that can be left undefined.

Through the use of a Mac IIx equipped with a video digitizing board and CD-ROM/videodisk storage sources, a one-screen mouse-controlled environment is created giving the learner access to a montage of information on designed artefacts portraying very different aspects of quality. A changing array of menus gives access to the views, thoughts and expressions on quality from architects, engineers and product and fashion designers using nine key artefacts as a suit of common parlance. Filmic vignettes of impressions of quality and a database of Design Council award products are also open to interaction. The agreements or otherwise of the participator are constantly stored and participants find themselves negotiating their own perception of quality, which is summarized for them when they leave the package. The environment can also act as database of raw information appropriate for a more Contemplative individual.

University students, industrialists, managers and designers have found their use of the package illuminating and rewarding. All find they are familiar with at least one of the key artefacts, and the range of comments accessible for each artefact is often surprising and an inducement to constructive discussion with colleagues when they have finished using the system. They leave with a clearer understanding of their own perception of quality, and many comment that the experience encourages them to seek out the opinions of others on this concept. We have also seen that participants become more aware of the nature of language and the creation of definitions as a social construct. The design process itself then benefits from this heightened understanding that negotiation of as many viewpoints as possible to achieve an agreed design objective(s), is a fundamental basis for quality design.

5.5   CONCLUSION

Research by the authors has shown how a fourfold model of design information transfer can be used to enable, rather than disable, content-rich construction decision-making. Multimedia information systems, configured according to this model, have been shown to have a far better chance of breaking through professionals' natural information rejection strategies and properly connect them to new and different understandings of the world. If professionals are to grow in understanding and work better together, they need to know the pattern underlying any data set in their own and other people's terms. This chapter demonstrates the potential of such an information system interface to help technically creative innovation. Its contribution to the overall discussion of what is meant by the integration of construction information and why it is important, is to introduce the human dimension and to clarify the need to integrate values.

REFERENCES

Altman, I. and Rogoff, B. (1987) World views in psychology: Trait, interactional, organismic, and transactional perspectives, in *Handbook of Environmental Psychology*, (eds) D. Stokols and I. Altman, John Wiley & Sons, New York.

Burnett, C. (1979a) The Architect's Access to Information: constraints on the architect's capacity to seek, obtain, translate and apply information. *National Bureau of Standards GCR 78–153*, Washington.

Burnett, C. (1979b) Making Information useful to Architects. An analysis and compendium of practical forms for the delivery of information. *National Bureau of Standards GCR 78–154*, Washington.

Cooper, I. (1988) Personal communication.

Cross, N., Cross, A. and Glynn, S. (1986) Designerly Ways of Knowing: A clarification of some epistemological bases of design knowledge. *ESRC-SERC Research Project Report*, Design Discipline, Faculty of Technology. The Open University.

Cross, N. and Nathenson, M. (1981) Design methods and learning methods, in *Design: Science: Method*, (eds) J.A. Powell and R. Jacques, Westbury House, IPC Business Press Ltd, Guildford, Surrey, pp. 281–96.

Curry, L. (1983) *An Organisation of Learning Styles: Theory and Constructs*. Paper to American Educational Research Association, Montreal, Canada.

Davies, B. and Talbot, R.J. (1987) Experiencing ideas; identity, insight and the imago. *Design Studies*, 8(1), 17–25.

de Bono, E. (1992) *Surpetition*, Harper Collins, London, p.234.

Dewey, J. (1934) *Experience and Nature*, Constable & Company Ltd, London.

Douglas, M. (1978) Cultural Bias. *Occasional Paper No. 35*, Royal Anthropological Institute of Great Britain and Ireland, London.

Glanville, G. (1980) The architecture of the computable. *Design Studies*, 1 (4), 217–25.

Goodey, J. and Matthew, K. (1971) Architects and Information. *Research Paper 1*, Institute of Advanced Architectural Studies, University of York.

Koestler, A. (1978) *Janus: A Summing Up*, Hutchinson, London.

Kolb, D.A. (1971) Individual Learning Styles and the Learning Process. *Working Paper*, MIT, Boston, MA.

Kolb, D.A. (1978) *The Learning Style Inventory: Technical Manual*, McBer and Co., Boston, MA.

Kolb, D.A. (1984) *Experiential Learning: Experience as the Source of Learning and Development*, Prentice-Hall, Englewood Cliffs, New Jersey.

Kolb, D.A. and Goldman, M.B. (1973) Towards a Typology of Learning Styles and Learning Environments: An Investigation of the Impact of Learning Styles and Discipline Demands on the Academic Performance, Social Adaptation and Career Choices of MIT Seniors. *WP-688-73*, Alfred P. Sloan School of Management, Boston.

Lera, S., Cooper, I. and Powell, J.A. (1984) Designers and Information, in *Designing for Building Utilisation*, (eds) J.A. Powell, I. Cooper, and S. Lera, E. & F.N. Spon, London.

MacKinder, M. and Marvin, H. (1982) Design Decision-making in Architectural Practice. *IAAS Research Paper 19*, University of York.

Pask, G. and Scott, B. (1972) Learning strategies and individual competence. *International Journal of Man–Machine*, 4.

Pepper, S.C. (1942) *World Hypotheses*, University of California Press, Berkeley, CA.

Piaget, J. and Inhelder, B. (1975) *The Psychology of the Child*, Basic Books, New York.

Powell, J.A. (1968) A Design Guide: Open Office Acoustics. MSc Thesis, UMIST, Manchester, Unpublished.

Powell, J.A. (1987) Is architectural design a trivial pursuit? *Design Studies*, **8** (4).

Prince Charles (1984) 150th Anniversary Gala Speech for the Royal Institute of British Architects.

Ritter, J. (1981) *Building Design: Information and Aids*, Percey Thompson Partnership

Rothenberg, A. (1976) Janusian thinking and creativity, in *The Psychoanalytic Study of Society 7*, (ed.) W. Muensterberger, Yale University Press, New Haven.

Talbot, R.J. (1981) Construing Problems, in *Design: Science: Method* (eds) J.A. Powell and R. Jacques, Westbury House, IPC Business Press Ltd, Guildford, Surrey, pp. 253–65.

Thompson, M. (1982a) A Three-Dimensional Model, in *Essays in the Sociology of Perception*, (ed.) M. Douglas, Routledge & Kegan Paul, London.

Thompson, M. (1982b) The problem of the centre: An autonomous cosmology, in *Essays in the Sociology of Perception*, (ed.) M. Douglas, Routledge & Kegan Paul, London.

Wallas, G. (1926) *The Act of Thought*, Harcourt Brace, New York.

The history of construction integration

Part Two of the book contains four papers which begin our search for some solutions to the problems whose needs have been justified and clarified in Part One. We have entitled this 'The History of Construction Integration'.

We have been most fortunate to be able to include in this part papers from three authors of the 'Scandinavian School' of pioneers in information classification. The issue of integrated construction information is one that has long troubled scholars, policy-makers and practitioners in different parts of the world, and our attempts at finding a solution first entered into a phase of concerted internationally collaborative work soon after the Second World War. Three of the leading players in that work, whose efforts resulted in the more broadly based interest in the subject that we now find, are contributors to this section.

In Chapter 6, Giertz, from Sweden, provides a broad overview of the different classification efforts that have occurred. He documents the origins of international co-operation, drawing particularly on work with SfB-related systems. He concludes by discussing the implications of all of this work and making conclusions as to where 40 years of effort has brought us. This should convince us of the non-trivial nature of the problems we are dealing with here.

The chapter by Bindslev from Denmark (Chapter 7) illustrates in more detail some of the classification work. This is done by a detailed description of the co-ordinated building communication model. The chapter contains details of the tables within the faceted classification system and has numerous examples of how the coding system defines and describes work, resources, products, activities, etc. The way these facets are integrated within the classification system is also demonstrated. Whatever form of integration technology is ultimately applied, it seems likely that some form of classification system will be necessary and as such the type of system described in this chapter may well be as relevant to future integration solutions as it has been to an understanding of how and why it is that we have reached our current position.

This part of the book is then concluded with two chapters from Karlén of Sweden. These are written much more from a philosophical standpoint.

They place our historical classification efforts, as summarized in Chapter 6, into the philosophical context of the need for ordering and formalization. The first of Karlén's papers (Chapter 8), which looks back, attempts some broad definitions of the concepts of integration, information and knowledge. Karlen also introduces the concept of 'systems' as an important basis to our previous work and future needs. The relevance of the quality and performance concepts are also related to information integration in these chapters.

The second of Karlen's papers (Chapter 9) looks towards the future. It argues the case further for systems thinking and a systems approach to guide our work in ordering and classification. The relationships of some of the philosophical concepts to emerging technologies such as product models and object-oriented database are also discussed.

Overall then, the chapters in Part Two give detail of previous classification and integration efforts but set these within the context of concepts that we are still using today in our attempts to find solutions, as documented in the later parts of this book. There is a temptation with new technologies, when using them to solve long-standing problems, to ignore previous efforts that were made. The lessons we can draw from the chapters in this second part of the book are that in doing so we would be discarding the wealth of internationally collaborative efforts that have been taken to the stage of detailed implementations and which have been rationalized into basic concepts as relevant to our current needs as they were to previous efforts.

CHAPTER 6

Integrated construction information efforts since 1945

Lars M. Giertz, Bromma, Sweden

6.1   THE BEGINNING OF INTERNATIONAL CO-OPERATION

The end of World War II in 1945 was a milestone in construction efforts. Problems were enormous. More than 45 years later the story was almost forgotten. In 1945, among the ruins, the demand for international exchange of reconstruction know-how grew. The keyword in those days was 'Documentation'.

Two names may not be forgotten, Jan van Ettinger and Hugo van Kuyck.

Jan van Ettinger was originally a mechanical engineer from Delft, specializing in industrial quality control. During the war he was part of the secret team that planned a post-war reconstruction programme for the Netherlands. The Germans had nothing against his setting up of a humble Stichting Bureau Documentatie Bouwwezen. After the war, this foundation was the beginning of the important Bouwcentrum, which grew rapidly on the most central site in devastated Rotterdam.

Hugo van Kuyck, an architect from Antwerp and also professor in theoretical physics, was the inventor of infrared photography from above the clouds. A sailor-boy, knowing the Normandy coast by heart, he helped the Americans adjust their wrongly developed photographic maps before invasion in the final stage of the War. After the War Hugo van Kuyck was the enthusiastic initiator of international co-operation – by improved documentation – between those who were going to be the reconstructors. He went around Europe to collect those who could start the co-operation he envisaged.

6.1.1   CIDB is created

After a series of preparatory meetings, the first International Conference on Building Documentation was held in Paris in 1947. The results

Integrated Construction Information. Edited by Peter Brandon and Martin Betts. Published in 1995 by E & FN Spon, 2–6 Boundary Row, London SE1 8HN.
ISBN: 0 419 20370 2

achieved were confirmed more officially by an ECE (UN Economic Commission for Europe) Expert Conference in Geneva in 1949. ECE initiated the International Council for Building Documentation, CIDB, which was formally established in Paris in 1950 with the programme outlined by the ECE Conference (ECE, 1949).

6.1.2   CIDB develops into CIB

In the meantime, ECE had called for another conference of building research experts. They recommended the establishment of an International Council for Building Research. The ECE then initiated CIB (International Council for Building Research, Studies and Documentation); CIDB was changed into CIB in 1953. In its first stage, CIB had three sections: Documentation, Studies and Research. The Documentation Section, CIB W1, focused on IBCC (International Building Classification Committee) jointly working for CIB and FID (International Federation for Documentation), which since the beginning of the century administered the widespread UDC (Universal Decimal Classification) for library arrangement.

The CIB arranges congresses every third year. In 1959 the Congress was held in Bouwcentrum, Rotterdam, under the presidency of Jan van Ettinger. The main report of IBCC with its recommendations (IBCC, 1959) was confirmed by CIB at the Congress. The interest in building documentation at that time may be illustrated by the fact that 25 information officers from 17 countries (Australia, Belgium, Canada, Czechoslovakia, Denmark (Secretariate), Finland, France, Italy, Japan, The Netherlands, Norway, Portugal, Spain, Sweden (Chairman), UK, USSR, Yugoslavia) participated in the work of IBCC over five years arriving at a unanimous agreement. This agreement was the starting point for 30 years of struggle for co-ordination of supply and demand of information in the construction industry.

6.2   THE CONTENT OF INITIAL INTERNATIONAL AGREEMENTS

6.2.1   Basic recommendations

The main recommendations made by the 1947 Conference, and further confirmed at the ECE expert meeting (1949), CIDB programme declaration (1950) and CIB constituting congress (1953), were as follows (ECE, 1949):

1. Abstract service

    (a) Each country presents its (internationally) important publications, (special emphasis to be put on research reports and articles in periodicals) on card size 74 × 105 mm (or double for folding). Abstracts in the original language on one side and in English on the other side (French and Russian also accepted).

  (b) Exchange of abstract cards through appointed information centres. Preferably one centre in each country (preferably a member of CIB) to be the focal point for international co-operation.
  (c) Abstract cards to be numbered by UDC classification numbers for filing.

2. Data sheets
  (a) All data sheets in A4 size.
  (b) A filing system for data sheets to be developed (in combination with UDC for libraries).
  (c) An international building classification committee (IBCC) to advise on:
  (i) Co-ordinated use of UDC in different countries.
  (ii) Filing system for data sheets.

6.2.2   IBCC 1954–59

When IBCC took over its responsibility in 1953, UDC was practically the only library arrangement known and used in building information centres all over Europe including USSR, but applications differed widely. After five years of work, however, IBCC managed to make the FID agree to an extract of numbers called *ABC* (Abridged Building Classification for Architects, Builders and Civil Engineers). The ABC was translated and published in 17 languages, most of them identically laid out to serve also as a multinational dictionary for the construction industry.

The secretary of IBCC, Rasmus Mölgaard-Hansen (of Denmark), later Chairman of the FID Research Committe, studied and evaluated filing systems related to construction, which were used in 19 countries. He came to the conclusion that UDC could be used for filing of data sheets except for data on construction (building) materials and works. Mölgaard's studies of filing systems also led to the conclusion that the Swedish SfB System should be used for this purpose.

Consequently, IBCC recommended the composite classification UDC + SfB for classification and filing as follows. Basic library classification, to be used also for filing of data sheets, should be UDC (in the field of construction according to ABC) except for information related to materials and works. The Swedish SfB System should be used for this purpose. The scope of SfB, thus, was information related to building materials and works, complementary to UDC. The main use of the SfB codes was assumed to be similar to their use in Sweden.

CIB confirmed the IBCC recommendation. The CIB Report No. 6, 'Building Classification Practices' (1966), in which UDC was presented by Agard Evans (UK) and SfB by Egil Nicklin (Finland), was published after some further national testing.

| S f B code |
| UDC number |

Figure 6.1

The recommended indicator to be placed in the upper right-hand corner is shown in Figure 6.1. The SfB code was the one developed in Sweden for the arrangement of Work Sections, Trade Catalogues and related information. The UDC number had to be selected from the ABC manual. This method of arranging, called 'Principle of Complementarity' (Mölgaard-Hansen), could also be used for libraries, as published by the UN Economic Commission for Africa in 1968 (Figure 6.2).

6.3   ORIGIN OF THE SFB SYSTEM

The SfB system was developed between 1946–49 in Sweden by a co-ordinating committee (37 member organizations) for the construction industry, SfB (Samarbetskommittén för Byggnadsfrågor). The system consists of two sets of main groups, one for Materials and one for Works (Work Sections). The number of such groups is 25 in each set. Each group is indicated by a capital letter. There is an intimate relationship between a group

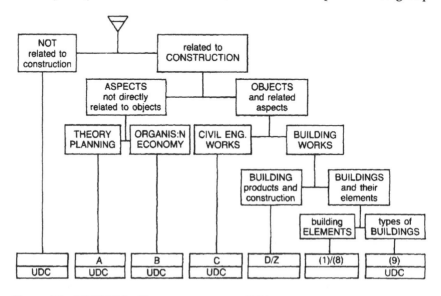

Figure 6.2   SfB/UDC for library arrangement (ECA, 1968)

of materials and the work section coded with the same capital letter. For instance, F=bricks and blocks (e.g. for trade catalogues) and F=brick and block work (for specifications and bills of quantities). All kinds of attributes (prices, etc.) may be related to objects thus grouped. Capitals A, B, C and Y, Z were left 'free' for specific (non-related) chapters in specifications, bills, trade information (materials and components), etc.

Two complementary facets were developed. One listed Substances of which materials are made; 107 substances were specified ('substances' here includes 'formless materials'). They were coded with a lower case letter followed by a one-digit number. For instance, g=burnt clay; g1, g2, g3 and so on for different qualities of burnt clay. The combination of a capital letter and a lower case letter – with or without a number – gave a reasonably detailed grouping of materials, e.g. Fg=clay bricks and blocks, Ig=clay pipes, Ng=clay roof tiles, etc.; a, b, c and x, y, z were 'free' to be used for other specifications.

The second facet contained a list of Building Elements indicated by two-digit numbers within brackets, e.g. (22)=Partition Walls, (31)=Windows, (32)=Doors. Even if factory-made parts of buildings were not common in Sweden in the 1940s two capital letters (G and X) indicated prefabricated units with the names of building elements: G for larger units, G(22)=prefabricated partition units, further specified as G(22)i=prefabricated partition units, (mainly) wooden. Similarly, X(31)i=wooden windows (marketed). In the Specification of Works, installation of prefabricated windows had also the indicator X(31) but in a list of parts in the finished building (product model) windows were numbered (31) without the X referring to building materials and/or works.

Contemporary publications in 1950 demonstrated the co-ordinating effect of the SfB system:

1. *Svensk Byggkatalog*, an annually renewed collection of general quality rules and trade catalogues for building materials and components (Publisher: AB Svensk Byggtjänst).

2. *Bygg AMA*, the Swedish National Specification, with advice on how to arrange building specifications, relating them to the National Specification clauses without repeating the wording (Publisher: SfB Committee).

3. *Aktuella Byggpriser*, current prices presented on loose leaves for updating; aimed at architects for preliminary cost estimates on different levels of detailing (Publisher: SAR Centralkontoret (Central Office of the National Association of Swedish Architects, SAR).

The meaning of the capitals when used for classification of building materials and components and the two complementary facets of the SfB system may be summarized as shown in Figure 6.3. Codes used in the two facets were chosen for easy retrieval of materials according to attributes

Codes for main groups

Sundry materials	D
Materials for concrete works	E
Building bricks and blocks	F
Structural units (prefabricated)	G

Long units
– profiles, bars, etc.	H
– pipes, tubes	I
– thread, wires, mesh, etc.	J

Insulating materials
– heat, sound, vibration	K
– water, wind, (incl. asphalt)	L

Covering materials
plane flexible (metal) sheets	M
corrugated sheets, 'roof' tiles	N
– for rendering, plastering	P
plane rigid sheets, panes	R
'wall and floor' tiles	S
– for flooring (special)	T
– for decoration (sheets)	U
– for painting (generally)	V
– for varnishing, polishing, etc.	W
Prefabricated units (non-structural)	X

A, B, C and Y, Z had different meanings on different applications.

Codes for subgroups
facet: substances and formless materials

Substances of which formed materials and components are made
metal	d (d1–d8)

stone
– natural	e (e1–e8)
– artificial, concrete, etc.	f (f1–f8)
burnt clay	g (g1–g8)
gypsum, asbestos cement, etc.	h (h1–h8)
wood	i (i1–i8)
fibre (board), paper	j (j1–j8)
cork, chips (board), etc.	k (k1–k8)
felt, wool, etc.	m (m1–m8)
asphalt, linoleum, rubber, plast.	n (n1–n8)
glass (and similar)	o (o1–o8)

Formless materials
fills	p (p1–p8)

mortars
– with cement	q (q1–q8)
– with other binders	r (r1–r8)
asphalt, tar	s (s1–s8)
adhesives (incl. soldering, etc.)	t (t1–t8)
corrosion-, rot-protection, etc.	u (u1–u8)
oils	v (v1–v8)
chemicals	w (w1–w8)
Other criteria for subgrouping	x,y,z

(a could be used for 'general'
b could be used for 'tools'
c could be used for 'work instructions')

Codes for subgroups
facet: made for specific element

general accessories, ironmongery	(0); (01)–(08)

component for
element of substructure — (1); (11)–(18)
element of superstructure
– primary element	(2); (21)–(28)
– complementary element	(3); (31)–(38)
– surface element	(4); (41)–(48)
element of utility service*	(5); (51)–(58)
– mainly piped, ducted	(6); (61)–(68)
– mainly electrical (incl. transport)	

fixture, room equipment element
– general	(7); (71)–(78)
– special	(8); (81)–(88)

* not complicated in the 1940s when the SfB research was accomplished

Figure 6.3 Structure of the Swedish SfB System, 1950. Conceptual coding for (information on) materials and components. The Work Section arrangement was based on 'work with...'

other than those used for the definition of the capitals. Computers were not used in the 1940s when the SfB system was created.

6.4   DEVELOPMENT IN THE 1960S AND ITS CONSEQUENCES

Long before World War II UDC had been the generally accepted library classification. Therefore, the Abstract Service on UDC-numbered cards rapidly developed after the ECE recommendations in 1949.

In Sweden, *Teknisk Tidskrift* (periodical for technical development) was printed in A4 size with all important articles starting on a fresh right-hand page and numbered UDC for cutting (or copying) and systematic filing according to subject. After 1950, architects replaced some UDC tables by SfB codes for library filing of information on materials and components (trade catalogues), and information on the work on sites with them. Thus, SfB/UDC was the Swedish solution to architects' library filing and retrieval problems. It is still in use.

In the United Kingdom, in the 1960s, the RIBA Services accepted the CIB recommendations and published the first version of their *Building Filing Manual* introducing SfB/UDC. So far so good; but now the difficulties started. After 1959 both the Chairman and the Secretary of IBCC had left, since its main task had been fulfilled. IBCC split into two subcommittees, one for further development of UDC (later Working Commission UDC 69+ under FID) and the other one for further development of SfB (later SfB development group, SfB DG).

CIB W1 lost its status as one-third of CIB. The CIB was more and more dominated by national building research institutes. A few working commissions were created for specific information problems. Documentation was not a popular title any more. The computer era had been born.

The further development of the old-fashioned UDC library classification is left aside in this chapter. The fate of the SfB complementary method for arranging information on materials and works, however, may be of interest to anybody trying to contribute to the problems of how to co-ordinate supply and demand of information in the complexity of the construction industry.

The SfB system has been changed. There were two main reasons:

1. Its restricted scope as complementary to UDC was not accepted.

2. The different use of capitals – both as indicators of work sections in specifications and bills, and as indicators of groups of materials and components (for trade catalogues, etc.) – was not accepted. In other words: it was agreed that the capitals must represent a single facet.

Two developments emerged, both trying to create a 'logical system' out of SfB. They were CI/SfB (RIBA (Royal Institute of British Architects)

Services, UK) and CBC/SfB (Doctor Björn Bindslev, Architect and Building Economist, Denmark). Both CI/SfB and CBC/SfB claimed that SfB must be a system of three tables with well-defined faceted contents. Each code must be known to have a constant meaning. But the two competitors could never agree. *CIB Report No. 22* appearing in 1972, published 'SfB Basic Tables' as a compromise: Table 1 Elements, Table 2 Constructions, Table 3 Materials and other resources.

The words used to identify 'classes' in each of these tables (facets) were similar to those of the original SfB, but the whole character of the original system was gone and different users interpreted the meaning of the words as they found them suitable for their purposes. Publishers of trade information on building materials and components used SfB for coding in a way which was partly reminiscent of the original intentions, but which nevertheless tried to follow the wording presented in *CIB Report No. 22*. The situation ended up in a continuous fight over interpretations.

6.5   SFB-RELATED SYSTEMS AND REACTIONS

6.5.1   The CI/SfB and CPI common arrangement

The RIBA Services soon (early 1960s) felt that the double coding UDC + SfB was unnecessarily cumbersome and therefore developed CI/SfB by replacing all relevant (for architects) UDC tables by two more tables added to the three of SfB. They were called Table 0 and Table 4. Librarians were proud of the five-faceted library system for architects, of which three facets could be used for arranging trade catalogues as well as work sections in specifications and bills.

The editing of a national specification similar to the Swedish AMA was attempted. It was, however, vigorously resisted by those who for many years, had relied on the *Standard Method of Measurements* (SMM), published by RICS (Royal Institute of Chartered Surveyors). The SfB Work Section arrangement did not rely on the traditional subcontracting habits of the British construction industry, but on the inherent attributes of the materials (mainly shape and substance) and of the components (mainly elemental characteristics).

In order to come to a conclusion, different parties of the British construction industry set up the Co-ordinating Committee for Project Information (late 1970s), the work of which resulted in the *CPI Common Arrangement of Work Sections for Building Works* (finally published in 1987). The arrangement is well known. The following factors were observed for the grouping of works:

1. Responsibility for design and performance

2. Methods of working, related to subcontracting practice

However, special sections were created for:

3. Location; where complex operation interrelationship merit such grouping

4–5. Avoiding repetition and specifying reference paragraphs

6. Separation of design alternatives

7. Grouping to avoid overfragmentation

8. Creating homes for things

The first two factors (attributes for the grouping of objects) relate to constructions according to SfB, but they are different from the attributes used by SfB for main definition of the objects (the constructions).

Other 'Work Section' definitions (points 3–8) use different attributes for editorial convenience in project documentation (specifications and bills). The strict definition and conceptual separation of substances and elements from 'Capitals' (grouping Materials and Works), which characterizes the original SfB editorial and retrieval method, has now been abolished. The reason seems to be lack of interest in the apparent relationship between materials and works, and between the works and the elements which they materialize. With the return to 'who is doing the job' as a main characteristic of a construction (work) and no other attribute for retrieval, the links between materials, works and results have been abolished.

It may be observed, however, that the 'who is doing the job?' question prevails for arranging work sections in most countries. This, of course, simplifies editing of specifications and bills of quantities in a nationally stable subcontracting practice, but it has the disadvantage of depending on changing habits. It neglects the designer's need for easy retrieval of alternative materials and components for a specific work (construction). It promotes national isolationism and retards development.

6.5.2   The CBC/SfB

Many papers have been written by the inventor of CBC (Co-ordinated Building Communication), Dr Björn Bindslev, starting in the 1950s when his method of planning and follow-up of big construction projects was first tested. His most recent views of the subject form Chapter 7 of this book. Most influential (even for the development of CI/SfB) were his articles in 1964.

The CBC uses three tables

1. Space functions, whether Open Spaces or Elements (Elements are functions (spaces) to be filled by constructions)

2. Combinations of resources (called Constructions) to fill (materialize) the elements ('Construction' in CBC means both the work (activity) and its result)

3. Resources, not only materials and components but also administration, labour and tools.

Doctor Bindslev has been the most influential theoretician over the years of international co-operation for SfB development. He used the three SfB tables. He gave the free symbols a definition and changed the meaning of the words used. Thus the CBC is a logical system of construction in three steps:

1. Finding resources   (Table 3)

2. Combining them into constructions   (Table 2)

3. Which materialize the elements   (Table 1)

When designing the building the process is the opposite:

1. Which are the elements of the building?   (Table 1)

2. How are they to be constructed?   (Table 2)

3. Which are the resources required?   (Table 3)

Any building consists of open spaces and 'elemental constructions'. Each elemental construction – according to CBC – has to be coded by a bracketed number for element and a capital letter for construction (SfB coding).

CBC/SfB is an ingenious system of construction analysis; but it differs from the original SfB, and is in permanent conflict with the defenders of CI/SfB and the defenders of the principle of complementarity (UDC + SfB). The conflict has resulted in the fact that 'SfB Basic Tables' can only be understood as a 'logical system' in one of its different applications (e.g. CBC) in spite of efforts by the SfB DG to explain SfB as 'The System' by publishing 'An Introductory Guide to the Use of SfB' (*CIB Report No. 40*) in 1977.

6.6   FURTHER DEVELOPMENTS

It may be of interest to follow the further development within CIB through the 1970s and 1980s in the field of information. CIB works through working commissions. Since the 'death' of W1 for Documentation, there has been a series of working commissions with different names (and numbers) for solving problems related to the hopes for easy availability of answers to questions generally and for efficient 'flow of information' within a project.

After a series of different working commissions, two main commissions emerged.

1. W57 for co-operation between information centres with big libraries, which cover the total field of environmental development.

2. W74 for discussions on the problem of how to make the flow of information within – and related to – a project more efficient.

Large information centres, including members of CIB W57 are of course, fully computerized. Most of them work with keyword retrieval method(s) combined with some classification system(s). When a question is raised a centre may find a number of documents by which the question may be answered. The centre will present these documents. Members of CIB W57 seem to have no interest in relating their retrieval methods to the co-ordination needed for efficient flow of information in the construction process (W74 problems). W57 and W74 have little in common; perhaps because W74, so far, could not present any proposal for a 'System'.

The inability of W74 (as well as of its predecessors) to come to any conclusion relates to the uncertainty, over the period in and after the 1970s, concerning the consequences of computer and telecommunication development. W74, therefore lately, co-operates with W78 dealing with integrated computer applications. An interest in related problems is nowadays growing within ISO too, but, generally speaking, nations go on acting without any internationally recommended co-ordinating system. Will the EC have an influence? Here, however, the present takes over from the history.

6.7 CONSEQUENCES

6.7.1 Building classification

Since 1960, when the computerization process started, there has been a split between the old documentalists and the new computer enthusiasts. For more than 20 years the computer enthusiasts argued that classification was meaningless. Those who still believed that classification is relevant – and necessary – for sensible computerization, were left aside. Gradually, however, attitudes have changed. Classification is the base for 'structuring'.

The classification needed for the database structuring of today meets the same problems as classification needed 40 years ago. The fact that options have been multiplied – complexity is growing – does not change the basic principles. The basic classification principle, adopted in 1959 (after 10 years of international co-operation), was called the principle of complementarity (Mölgaard-Hansen). This implied acceptance of one 'universal' system and separate (more or less) related 'special' systems for different 'fields of activity'. The universal system then in use, was UDC. 'Building' was accepted as a field of activity including anything related to any single building (not civil engineering works). For this 'field' the SfB system was accepted.

The UDC/SfB combination may not be relevant any more. UDC is out of date and SfB is too restricted if UDC tables are not used for 'universal' concepts. Three different ways to widen the scope of SfB (CBC/SfB in Denmark, CI/SfB in England and BSAB in Sweden) confuses the situation, It will be necessary to find 'universal' classification tables (facets) to replace some of the UDC tables which are of interest to 'our' field of activity.

The 'field of activity' may have to be redefined. Should its name be 'Building' or 'Construction Industry'? Within this book we have adopted the word 'construction' for the process and industry, and 'building' for the product. Which agreements are internationally desirable and how are international agreements reached today?

6.7.2 'Construction' in the 'universe'

Since the UDC tables are apparently not wanted any more for 'universal' concepts related to the specific concepts belonging to 'Construction', it may be useful to consider the relation between building, other fields of activity and 'universality'. In Figure 6.4 three conceptual levels are recognized:

1. The lowest level (3) contains all concepts (and words) for things and activities 'on earth', experienced directly with our five senses.

2. Next level (2) contains all concepts (and words) for scientific analyses of the things and happenings on earth and in its surroundings.

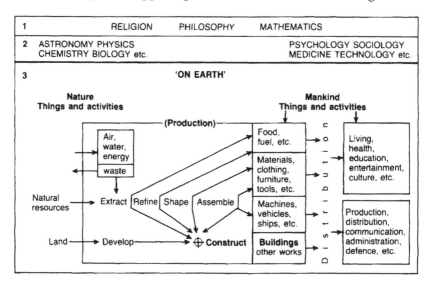

Figure 6.4 'Construction industry' in a wider context. ⊕ = 'building materials'

3. The upper level (1) contains spiritual concepts and reasonings 'above' what we experience with our five senses.

Construction is part of the production class of activities in human society. Each product is a thing for use (consumption) in one (or more) of the human activities. The thing has been produced for this purpose. After production, things are distributed into their place of use. The products have become artefacts. Construction works remain on place. Distribution, then, implies a change over from 'product' to 'artefact'.

'Expert systems' rely on concepts belonging to Level 2, whereas the flow of information in building projects relies on concepts belonging to Level 3. The 'coloncombination' (UDC method of combining concepts from different classes) needed will be found when systematically arranged computer programs for design support (expert systems) are being worked out.

6.7.3   The flow of information in the construction industry

The SfB system was originally created to improve relations between owners (mostly represented by architects), constructors and suppliers of materials by rationalizing the flow of important information in order to avoid mistakes and loss of time. The SfB thus created an editing and retrieval system for trade catalogues, for building specifications (in Sweden, 1946, separate bills of quantities did not exist) and for related drawings. The idea of 'product model' was dormant.

The method used for systematizing information retrieval may be of interest. After cutting 98 building specifications to pieces for comparison, SfB found that all headings used relied on 'single concepts'. These single concepts could be arranged and named by symbols in such a way that one symbol – or (normally) two or (seldom) three symbols combined – presented any relevant heading (work section) in a building specification and in the related group of materials. This is the reason for the use of capitals, lower case letters and numbers for the symbols in SfB.

The way in which the CBC later used the three 'SfB Basic Tables' shows that the artificial language combining single concepts from no more than three tables: Elements, Constructions and Materials ('other resources' added by CBC) is still relevant.

Today a Product Model is recognized to be the backbone of information on buildings. Later sections of this book describe the most recent innovations in product modelling in detail. The Elements are logically grouped into Systems (and subsystems). Agreement on a product model defining elements and systems of elements is urgently needed. After this has been done, different Technical Solutions to 'fill' the elements may be discussed. Each technical solution may be analysed as a number of Constructions (or sometimes only one).

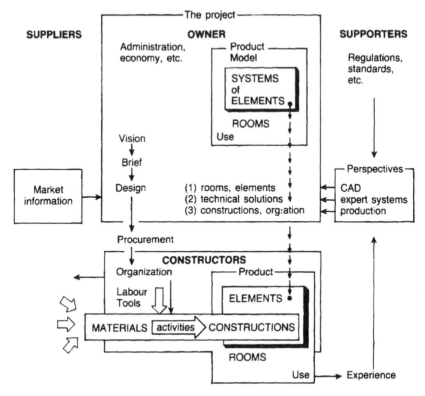

Figure 6.5 Pattern of information flow in the construction process. Note: In practice, of course, there is no fixed sequence between design and procurement. Much design continues during the production phase. Nevertheless, design remains design – the preparations and decisions for production – regardless of who is the actor. When design is performed by constructors or suppliers, they may be considered to be 'owner's agents' for that purpose

Obviously a separate product model should be created for each individual project. This product model should exist through all phases of the process development of that project, from first design to the end of the use phase. The product model should be structured to allow for development from 'vision' to 'reality'. This is why the pattern of the information flow in the building process is so important (Figure 6.5). The product model develops in the stages of design (owner's decisions) and it must cope with changes in the period of production.

When speaking of different perspectives, we may be wrong in focusing on the actors who are engaged in the development of the project in different capacities. The base for the 'perspectives' to be recognized is the logical sequence of the maturing process of the product model. Interface problems are secondary. Primary problems relate to the element/construction relationship found by SfB and highlighted by CBC.

When we speak in general terms of a 'product model' applicable for the structuring of support programmes for designers, this 'product model' is not a model really but a type model for the structuring of models. There may be several type models since there are several different types of buildings.

6.8   CONCLUSION AFTER 40 YEARS OF CONCEPTUAL DEVELOPMENT AND TESTING

Whatever the future development may be, the method used by SfB to code 'single' concepts, found by the analysis of titles in relevant documents, may help. Especially important seems to be SfB's finding that different 'perspectives' relate to conceptually definable properties of the objects looked upon. The differences observed in different spectators' views is secondary – and even uninteresting – when the objects have been correctly classified and information on them has been stored for retrieval according to 'perspective' (e.g. the CBC computer application, PROXIMA).

Another major experience from the fate of the SfB system is the danger of introducing artificial languages to building practitioners. SfB coding is, of course, an artificial language. If the publication of the SfB system had been restricted to the demonstration of its co-ordinating effect in the use of the three basic documents published in 1950 (see Section 6.3) the method of coordination by standardized chapter headings might have been accepted. The mistake was the effort to explain the meaning of each single symbol. People started to discuss 'faceted classification' with the aim of giving each symbol in the artificial language an independent definition by simple wording. This turned out to be impossible. Each symbol, of course, had its distinct conceptual base, but in practice, a single symbol – as well as the combination of two or three symbols – could only be translated into 'every-day language' (irrespective of country or profession) when used for a title of a chapter in a document dealing *either* with parts of buildings (e.g. product model) *or* with activities (work sections) *or* with materials (e.g. collection of trade catalogues).

Maybe, in the future, the conceptual analysis of descriptors for information retrieval should only be applied, not explained, to the users. The SfB system – both the method of conceptual analysis, its result (in 1950) and the reasons for the rejection of it – may be interesting as an historical background to the structuring of databases today.

The classification of things should be independent of interfaces. Things may be looked at from different perspectives, but it is misleading to relate characteristics of the thing – necessary for its classification – with interface problems due to language habits of the receiver of information related to the thing. The criteria according to which classes are defined must be understood as conceptual rather than verbal in order to avoid the fallacies of semantics.

REFERENCES

CIB (1966) Building Classification Practices. *CIB Report No. 6.*

CIB (1973) The SfB System, Authorized Building Classification System for Use in Project Information and Related General Information. *CIB Report No. 22*, Rotterdam.

CIB (1977) An Introductory Guide to the Use of SfB, *CIB Report No. 40.*

ECE (1949) *Proceedings of the Conference on Building Documentation*, United Nations Economic Commission for Europe, Geneva, November. Reprinted by CIB Working Commission W57, January 1978.

Giertz, L.M. (1982) *SfB and its Development 1950–1980*, CIB/SfB International Bureau.

IBCC (1959) Recent Developments in Building Classification. *IBCC Report 1959*. (At that time available from IBCC Secretariat or Bouwcentrum, Rotterdam.)

CHAPTER 7

Logical structure of classification systems*

Björn Bindslev, Architect, Det Kongelige Danske Kunstakademi, Denmark

The work model described in this chapter is a result of ongoing research since the early sixties with the development of a comprehensive classification and coding system for Computer Integrated Construction (CIC). The work model is a sub-model within a much broader framework called 'Co-ordinated Building Communication' (CBC), first mentioned in a series of articles in the *Architects' Journal* 1964/65 [01]. The SfB-version named CBC/SfB was based on the international SfB system, CIB/SfB.

This chapter gives a description of a specific classification system for use in integrating construction information. The CBC/SfB system consists of four *general* classification tables and certain *specific* codes. For a *building project* the following is true:

$$\Sigma \text{ buildings} = \Sigma \text{ elements} = \Sigma \text{ activities} = \Sigma \text{ resources}$$

Classification symbols alone or in conjunction classify text items in databases or files for the preparation of project documents like specifications, bills of quantities, time schedules, etc. Items are *identified* by primary keys (reference numbers). The tables are described briefly below.

Table 0

A classification or *calculus* of form elements of the **town**, including *environmental* elements in the **country**. **Society** is divided in main **building** classes denoted by numbers from 0–9. Table 0 is further divided so that each building is denoted by symbols from 00–99. The universe class is denoted by ~ (tilde). Subclasses are created by addition of further symbols. Symbols stand for the *geometry* of **town and country** and are used to denote **macroeconomic** *cost places* in the project.

* This chapter is an elaboration of the work presented in an earlier paper (Bindslev, 1993).

Integrated Construction Information. Edited by Peter Brandon and Martin Betts. Published in 1995 by E & FN Spon, 2–6 Boundary Row, London SE1 8HN.
ISBN: 0 419 20370 2

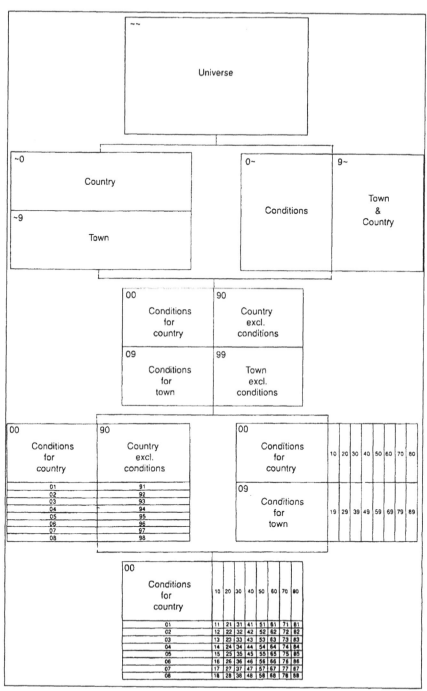

Structure of Table 0

```
0 Conditions (environment). . . . . . . . . . . . . . . . . . . . . . . . . . £ xx

  1 Administration . . . . . . . . . . . . . . . . . . . . . . . £ xx
  2 Industry . . . . . . . . . . . . . . . . . . . . . . . . . . . . £ xx
  3 Commerce . . . . . . . . . . . . . . . . . . . . . . . . . . £ xx
  4 Health  . . . . . . . . . . . . . . . . . . . . . . . . . . . . £ xx
  5 Recreation  . . . . . . . . . . . . . . . . . . . . . . . . . £ xx
  6 Church . . . . . . . . . . . . . . . . . . . . . . . . . . . . £ xx
  7 Education  . . . . . . . . . . . . . . . . . . . . . . . . . . £ xx
  8 Housing . . . . . . . . . . . . . . . . . . . . . . . . . . . . £ xx

9 Town & country (excl. environment) . . . . . . . . . . . . . . . . . . £ xx

~ Total project cost  . . . . . . . . . . . . . . . . . . . . . . . . . . . . . . . . . £ xx
```

Table 1

A classification or *calculus* of form **elements** of the **building**, including *environmental* elements on the **site**. **Building and site** are divided in main element classes denoted from (0)–(9). The table is divided further into elements from (00)–(99). The universe class is denoted by ~ (tilde). Subclasses are created by the addition of further symbols. Symbols stand for the *geometry* of the **building and site**, and are used to denote **microeconomic** *cost places* in the project.

```
(0) Conditions (environment)  . . . . . . . . . . . . . . . . . . . . . . . . . . . . . £ xx

   (1) Foundations . . . . . . . . . . . . . . . . . . . . . . . . £ xx
   (2) Superstructure . . . . . . . . . . . . . . . . . . . . . . £ xx
   (3) Completion . . . . . . . . . . . . . . . . . . . . . . . . . £ xx
   (4) Finishes . . . . . . . . . . . . . . . . . . . . . . . . . . . £ xx
   (5) Engineering services . . . . . . . . . . . . . . . . . . £ xx
   (6) Electrical services  . . . . . . . . . . . . . . . . . . . £ xx
   (7) Fittings  . . . . . . . . . . . . . . . . . . . . . . . . . . . £ xx
(8) Furniture  . . . . . . . . . . . . . . . . . . . . . . . . . . . . . £ xx

(9) Building & site (excl. environment) . . . . . . . . . . . . . . . . . . £ xx

~ Total project cost  . . . . . . . . . . . . . . . . . . . . . . . . . . . . . . . . . £ xx
```

Table 2

A classification or *calculus* of human **activities**, comprising those related to *production* in factory and on-site as well as those related to *use* of the building. Human activity is divided in classes A–Z. Y denotes a union class of activities from E–X. The universe class is designated by the symbol ~ (tilde). Subclassification is made by addition of numbers. Symbols are used to denote microeconomic *cost bearers* in the project.

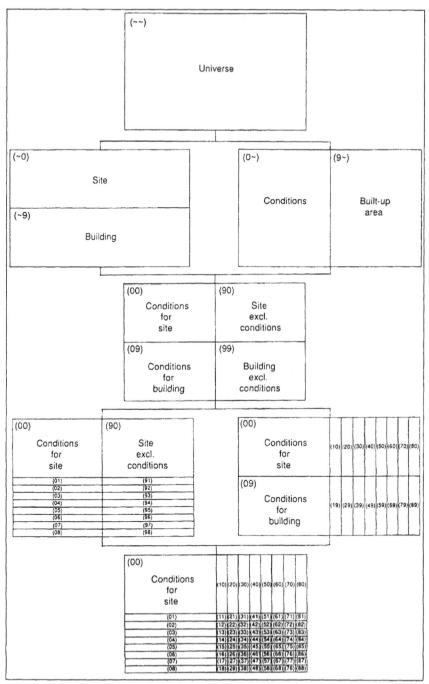

Structure of Table 1

A Production conditions (use) .... £ xx

B Demolition works .... £ xx
C Earthworks .... £ xx

D Construction conditions (prefab.) ...... £ xx
Y Construction works .... £ xx

Z Production works (excl. use) .... £ xx

~ Total project cost .... £ xx

Table 3

A classification or *calculus* of basic factors of human activity, i.e. **resources**. Each located activity may be divided into basic resources according to the table. Classes are designated by *lower case letters* from a to z with the addition of a number from 0 to 9. The class y is a union class of materials from e–x. The universe class is designated by ~ (tilde). Subclasses may be made by addition of further numbers. The symbols are used to denote *cost classes* in the project. By the term 'cost' is meant resources measured in **money**.

a Administration .... £ xx

b Labour .... £ xx
c Tools .... £ xx

d Operations .... £ xx
y Materials .... £ xx

z Work .... £ xx

~ Total project cost .... £ xx

7.1   THE RESOURCE MODEL

The resource table in CBC/SfB, shown as Table 3 above, was developed according to the principles of symbolic logic: the class operations of *universe class*, *union class* and *complement class* were applied to the original SfB table, so that the new classes of z 'work', d 'operations', and y 'materials' could be accommodated into the system. Class a 'administration' was treated as a complement class to z 'work', and was not *directly* related to the resources of labour, tools and materials, but to their *co-ordination*, i.e. in the summarized concept of *work* (Figure 7.1).

Coding by means of symbols in Table 3 takes place in the *detail design stage* (a3) in conjunction with activity type codes that have been established during the scheme design stage. It also takes place in the *detail records stage* (a4) and in the *detail control stage* (a7). Resource codes on

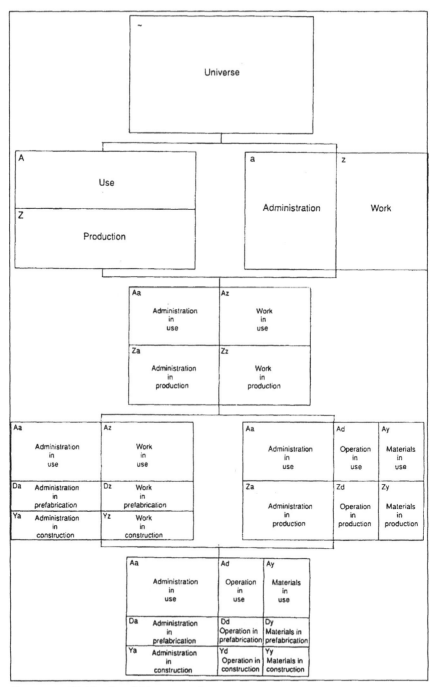

Structure of Tables 2 and 3 intersected

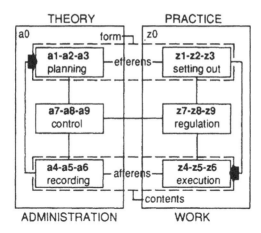

Figure 7.1 Common structure of administration and work

drawings refer to the same codes in the *project database* and vice versa. In the practical use of Table 3 for project management, i.e. for preparation of documents in the form of drawings, specifications, bills of quantities, drawing lists, time schedules, etc., there was need for a more *detailed* classification of administration (a) and work (z) including related operations (d) and materials (y). The resulting model was based on the precondition that there is an *isomorphic* connection between manual work (z) and those operations of the mind (a), which administrate the work, i.e. they have the same *structure*.

The idea of the common structure of intellectual and manual work constituted the basis for the development of the a/z model shown in Figure 7.1. In the model, planning (a1–a2–a3) is followed by setting out (z1–z2–z3) of the dimensions shown on the drawings, cf. ISO 4463 (1979). Setting out establishes the precondition for actual work performance (z4–z5–z6), which again forms the precondition for recording of work execution (a4–a5–a6). In order to secure the quality of the product, it is being controlled if executed work performance corresponds to agreed specifications via comparison between records and plans (a7–a8–a9). In case of deviations, necessary regulations are executed (z7–z8–z9). The complete administration model is shown in Figure 7.2, whereas the complete work model is shown in Figure 7.3.

By isomorphism between work and administration of work is meant the conformity in pairs between classes in the two submodels, by which a work process is being *realized* by an administrative process, or an administrative process is being *realized* by a work process (Ashby, 1965). Similarly, there is isomorphism in the correspondence between the control and regulation processes of the model. In this way, the isomorphism results from the fact that the class division in the *subject*'s activity (administration and

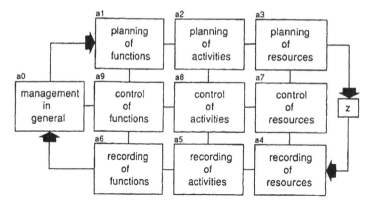

Figure 7.2   The administration model

work processes) is based on the division of the *object* in elements, activities and resources. This means that the structure of the subject is, so to speak, 'merged' with the structure of the object, so that the human activity in a Gödelian way appears as its own precondition (Nagel and Newman, 1958).

7.2   THE WORK MODEL

By 'work' (z) we will understand *handling of materials* (in the widest sense) by means of human labour power, tools or machines. Work and sub-processes of work (z) differ from work *operations* because they *include* labour power (b), tools (c) and materials (y). A work *operation* (d) includes

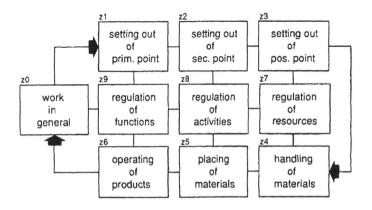

Figure 7.3   The work model

consumption of labour power (b) and tools (c), while materials (y) are described *separately*.

In the earlier stages of SfB development the *materials* (e–x) were used for text items describing work performance. In the Swedish ByggAMA the code Fg2 was used partly for classification of the *commodity* 'bricks of burnt clay', partly for the *work* 'brickwork with materials of burnt clay'. In other words, formed materials of burnt clay was regarded as the *primary characteristic* of the *work process* and the description of this process was assigned a *commodity* code, not an independent work code. This implied that material descriptions, e.g. in brochures, ended up carrying *material codes* that were identical with the *process codes* in work specifications: equality signs were put between building materials and building works, a principle which appeared to be untenable in computer-aided project management.

In CBC/SfB the starting point is not in the world of objects outside the human being, but in human *activity*. This is seen as consisting of an *outer* work process (z), whereby a person enters into physical contact with the material world (y) by using their own labour power (b) and the available tools (c), and an *inner* administration process (a) which is a *reflection* of the work process in consciousness. This conditions the collection of experience and determines the objectives of the subject. It therefore *steers* the outer work process. So the concrete work process is the primary concept. It is only through analysis thereof that we realize work as a *reciprocal process* between operations and materials, and realize operation as a *reciprocal process* between tools and human labour power.

The definition of *work* that comes closest to the work classification mentioned above is contained in the rules for the definition of *unit rates* agreed by Construction Economics European Committee (CEEC):

Unit rates include the labour, materials and plant content of the specified work item together with a reasonable allowance for contractor's profit. Unless otherwise specifically stated the following shall be deemed to be included with all items:- labour and all costs in connection therewith; materials, goods and all costs in connection therewith; fitting and fixing materials and goods in position; plant and all costs in connection therewith; waste of materials; square cutting; overhead charges and profit. Rates should not include any allowance for site establishment, site huts, supervision, setting out etc., which will vary from one site to another, and will normally be allowed for and priced separately.

Where *work* (z) does not imply consumption of material, e.g. in cases where materials are delivered by the owner, or where materials for other reasons are required to be described separately, the tasks are considered (work) *operations* (d). The subclasses of d are isomorphic with the subclasses of z. The same is valid as regards b, c and y.

Figure 7.4 Difference between product and material

The following is a brief demonstration of the use of the work classification in Bills of Quantities and specifications without quantities. The classification is regarded as a *subdivision* of the activity classification, e.g. (21)F, (23)G, etc., that shows the logical relation between *products* and *materials*. The code (23)G on Figure 7.4 stands for a *site product*, whereas the code (23)D stands for the corresponding *prefabricated product*. In CBC/SfB the term 'product' is always used for the *output* of a production, while the term 'material' is used for the *input* to a production.

7.2.1   z0 Work general

'Work general' is work which *cannot* logically be referred to any of the *concrete* work processes z1 to z9. General work processes are processes which are *conditions* for proper work processes. They are generally concerned with site establishment *not* related to proper working places in the *building*, e.g.:

1. Interim roads, bridges, crossings, etc.

2. Interim sheds, canteens, sanitary services, etc.

3. Interim offices, installations, etc.

4. Interim telephones, radio, TV, etc.

5. Interim screens, fences, catwalks, rails, etc.

6. Interim scaffolding, etc.

7. Interim signboards, name plates, etc.

8. Interim transport, cranes, vehicles, lifts, etc.

9. Interim winter arrangements, protection, etc.

Stage z0 may be regarded as the *external* regulation stage, i.e. containing arrangements against external disturbances in the work process. In this way, class z0 is complementary to the external control stage a0. The disturbances may follow from alterations in the weather, errors or shortcomings in project documentation, changes in the client's decisions, clamp-downs from authorities, strike, lock-out, lack of labour and plant, delays in material supplies, etc. The stage therefore includes a number of *insurances* which

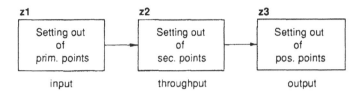

Figure 7.5 The setting out stages

may protect against the effects of human mistakes, fire, storm, illness, accidents, malicious damage, etc. For protection against excess of budget, items for *contingencies* should be allowed for. Similarly, *buffers* should be inserted into time schedules in order to protect against unforeseeable fluctuations in productivity, etc.

Costs in relation to interim installations and services, etc., are normally referred to the *general* activity of (0)A and are regarded as belonging to the production conditions. Similar costs relating to *prefabrication* should be referred to (0)D. If it is agreed that there ought to be separate items for work operations and materials, (0)Az0 is divided into separate classes under (0)Ad0 (work operations) and (0)Ay0 (materials). As required, materials may be divided in separate items for materials under e to x.

7.3   SETTING OUT

By 'setting out' is understood transformation of the dimensions of *drawings* to *true* scale in work places. Setting out *primary points* of building elements, stage $z1$, constitutes the condition for setting out *secondary points* of constructions, stage $z2$. These constitute again conditions for placing of single materials by means of *position points*, stage $z3$.

The setting out stages $z1$, $z2$, $z3$, shown in Figure 7.5, are *isomorphic* with the planning stages $a1$, $a2$, $a3$, i.e. they have the same structure. The three SfB tables are in both cases good instruments for structuring the necessary information. The three stages correspond to international standard ISO 4463. Setting out in $z1$ may be regarded as the *input* of the process, $z2$ may be seen as *the throughput*, while setting out in $z3$ may be seen as the *output*, i.e. the dimensional result, without which the contractor cannot place the components of the construction in position. Every placing depends in principle on the transfer of dimensions from the drawings in true scale.

The model in Figure 7.5 should not be regarded as a 'time-plan', but as a *logical* model. However, setting out is, in practice, tied to the activities of the project and is therefore connected to the time-plan that is again based on the mutual relations of the activities set out in the *net plan* and the related *logic file*.

Setting out takes place by means of instruments, including levelling instruments (theodolites, etc.), measuring rods, steel tape measures, chalk string, templates, etc., besides a number of aids in the form of sheet piling, concrete formwork, centering, etc., that are necessary to establish concrete, dimensioned rooms or places for the proper *work execution*, z4, z5 and z6.

7.3.1   z1 Setting out of primary points

The SfB symbol z1 stands for setting out of primary points, including labour and aids in the form of tools, machines, instruments, etc., together with materials, if any. The requirements are related to the dimensions of buildings and building elements, cf. Table 1. They are coded as *operations* under d in cases where no materials are involved. Where setting out requires delivery of materials, requirements should be described under z1. Primary points are normally the responsibility of the *Client*. They may generally be coded under (0)Az1. If a specification according to building elements is required, setting out is referred to elements of Table 1, e.g. (19)Az1, (29)Az1, etc.

7.3.2   z2 Setting out of secondary points

The SfB symbol z2 stands for setting out of secondary points, including labour, aids in the form of tools, machines, etc., together with materials, if any. Setting out may include proper *formwork* such as sheet piling, concrete forms, centering, etc., that are *left in* the construction. For example:

(21)	**EXTERNAL WALLS**
(21)E	**WORKS WITH *IN SITU* CONCRETE**
(21)Ej8	**Formed materials of wood–wool cement**
	Wood–wool cement sheets 0000 mm × 0000 mm type XYZ to be used as left in formwork, external walls.

If forms are left in the construction, they are not regarded as aids but as *materials* similar to all other materials in the building. They could be given in *separate* items as above.

(21)	**EXTERNAL WALLS**
(21)E	**WORKS WITH *IN SITU* CONCRETE**
(21)Ez2	**Setting out of secondary points**
	Formwork to 200 mm thick vertical concrete walls (one side only, forms left in). Form materials described separately, cf., **(21)Ej8**.

In most cases forms would be *interim*. They should therefore be described as work operations under d2. For example:

(21)   **EXTERNAL WALLS**
(21)E   **WORKS WITH *IN SITU* CONCRETE**
(21)Ed2   **Setting out of secondary points**
                  Formwork to 200 mm thick vertical concrete walls
                  (one side only).

In this case the form is regarded as an *aid* to be taken down after use and to be reused a number of times in other jobs. Forms left in the construction could be described under d2, if *separate* items are given for the form materials concerned.

7.3.3   z3 Setting out of position points

The SfB symbol z3 stands for setting out of position points, including labour, tools and materials (if required). In other words, setting out of *details* may include supply of *forms*, e.g. lists, blocks, bushings, etc., for execution of *holes* or for use in fixing of components, e.g. as windows, doors, bolts, etc.

(23)   **FLOORS**
(23)E   **WORKS WITH *IN SITU* CONCRETE**
(23)Ed3   **Setting out of position points**
                  Setting out of holes in 250 mm concrete floors for
                  drainage pipes, 000 × 000 × 000 mm.

(24)   **STAIRS.**
(24)E   **WORKS WITH *IN SITU* CONCRETE**
(24)Ed3   **Setting out of position points**
                  Setting out of holes in concrete in situ stairs for later
                  fixing of bannister, 000 × 000 × 000 mm.

In these examples the items of setting out have been regarded as *operations*. They have, therefore, been coded (23)Ed3 and (24)Ed3. If form materials should be calculated as left in, the codes would have been (23)Ez3 and (24)Ez3, respectively.

(21)   **EXTERNAL WALLS**
(21)F   **WORKS WITH BRICKS AND BLOCKS**
(21)Fd3   **Setting out of position points**
                  Setting out of holes in 350 mm brickwork for later
                  fixing of bolts, 000 × 000 × 000 mm.

In this example setting out is clearly an *operation*, without any supply of materials. The code is therefore (21)Fd3. If it were to include the bolts, etc., the code would be (21)Fz3.

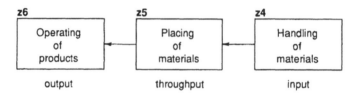

Figure 7.6   The execution stages

7.4   EXECUTION

While work in connection with setting out has the aim of securing the intended *form* of the product, it is the aim of work execution to secure the concrete *contents* of this form. The essential part of work execution is the *placing in position* of materials or goods, specified in the design stages. To this are added certain *preparatory* works (input) and *final* works (output). Work execution, therefore, has been divided in the following three stages in the work model, cf., Figure 7.6 (from right to left).

7.4.1   z4 Handling of materials

The SfB symbol z4 stands for handling of materials *prior to placing*, including labour power, tools and materials. Handling work assumes that all materials have been delivered to the site. It is defined as reception, transport and handling from delivery to placing of materials in position. Handling includes preparation of materials (e.g. mixing mortar, bending steel), necessary cutting and shaping of materials prior to placing in position. Such preparatory work should not be mistaken for proper *prefabrication* in factory or on site. Handling may include transport of materials from site store to lift and from lift to positions for placing the materials, i.e. delivery from *local* store to *local* working place in the building. Handling comprises any form of *local* loading or unloading.

Preparatory cutting, shaping, etc., may comprise excavation of earth, cutting, grinding, etc., of bricks, pipes, sheets, tiles, etc., i.e. materials delivered in commercial formats and which require shaping to a certain degree before placing. Handling does not require consumption of materials, unless *waste, breakage* (y4) should be regarded as such. Where this is not the case, handling should be treated as *operations* and coded d4, and maintenance, z7. Examples are:

(11)Cd4	Excavating trenches of external wall foundations
(21)Fd4	Cutting bricks or blocks for external walls
(42)Rd4	Cutting sheets for internal wall finishes
(43)Sd4	Cutting tiles for floor finishes

It depends on agreed *rules of measurement*, whether handling, including transport, cutting, etc., should be given in separate text items for pricing in

Bills of Quantities. According to English rules of measurement, *local* handling and transport should be understood as included in the placing of items (z5).

7.4.2   z5 Placing in position

The symbol z5 stands for placing of materials in position, such as filling, casting, laying, mounting, fixing, fitting, painting, etc., including labour, tools and materials. It is estimated in the United Kingdom that 50–80% of work items belong to z5. Examples are:

(10)Cz5	Earth filling on site
(11)Ez5	Casting concrete in foundations external walls
(21)Fz5	Laying bricks in external walls
(43)Sz5	Laying tiles in floor finishes
(45)Vz5	Painting of ceilings

If operations and materials are required to be given in *separate* text items, the code d5 should be used and separate items for materials should be added. For maintenance, see z8. Examples are:

(21)	**EXTERNAL WALLS**
(21)E	**WORKS WITH CONCRETE** *IN SITU*
(21)E**d5**	Casting of concrete
(21)E**q4**	Delivery of concrete
(21)F	**WORKS WITH BRICKS AND BLOCKS**
(21)F**d5**	Laying of bricks
(21)F**g2**	Delivery of bricks

By giving operations and materials in separate items, it becomes possible to *select* material items from the project file for direct ordering from the suppliers. In connection with CAD, material codes may be used as cross references on drawings. In quality control, material items in the plan file can be checked directly against items in the record file.

Requirements regarding separate pricing of operations and materials force the building economist to formulate separate items *within each activity* (product). Operational bills give many advantages in project management and quality control. The introduction of operational bills of quantities means that separate classification schemes must be available for *products* as well as for *materials*, cf., (43)S and n6 in Figure 7.7.

7.4.3   z6 Operating of products

The symbol z6 stands for operating of finished products, including *testing* before handing over. Operating of finished products includes *caretaking* in the form of cleaning, supply of electricity, water, gas, oil; watching, guarding, insurance, etc., including labour, tools and materials, if any. Examples are:

DATABASE STRUCTURE for Building Works						
Sample Specification Items for Bill of Quantities						

(43) SO51						30 MAR 92
DRAWING	SfB 1	2	3	ACT KEY	RES. KEY	TEXT
(43) Sn6 (43) Sa3	(43) (43) (43)	S S		051		FLOOR FINISHES WORKS WITH TILES Construction of floor finish with vinyl tiles, 300 × 300 × 20 mm thick.
	(43)	S	a2	051	0677	ADMINISTRATION Scheme design drawing no. 0677 scale 1:50. Date: 30 JAN 92.
	(43)	S	a3	051	0683	Detail design drawing no. 0683 scale 1:10. Date: 30 Jan 92.
	(43)	S	d4	051	2097	OPERATIONS Cutting and fitting around pipes or the like; not exceeding 0.3 m girth.
	(43)	S	d4	051	2098	Cutting and fitting around pipes or the like; 0.3–1 m girth.
	(43)	S	d5	051	2279	300 × 300 × 20 mm vinyl tiles to floors on power floated concrete base; level or to falls; over 300 mm wide; butt joints.
NOTE primary key activity primary key resource	(43)	S	d5	051	2280	Ditto in compartments not exceeding 4 m2 on plan.
	(43)	S	n6	051	2366	MATERIALS Vinyl asbestos tiles to be Armstrong 'Acoflex' or equal approved BS 3260; to regular pattern; selected colour; 300 × 300 × 20 mm thick.
	(43)	S	t3	051	2480	Adhesives to be according to manufacturer's instruction.

Figure 7.7 Operational bill

(43)Az6	Care of floor finishes (cleaning, vacuum-cleaning)
(51)Az6	Care of refuse disposal services (collection)
(54)Az6	Care of gas services (supply of gas)
(56)Az6	Care of heating services (supply of oil, etc.)
(63)Az6	Care of lighting services (supply of electricity)

If *separate* items for operations and materials are required the code d6 is used and a separate item for the materials concerned should be given, cf., above under z5. Examples are:

(53)Ad6	Care of water services
(53)Aw4	Supply of water from waterworks
(53)Aw7	Supply of cleaning materials

(54)Ad6	Care of gas services
(54)Aw3	Supply of gas from gasworks
(54)Aw7	Supply of cleaning materials

(56)Ad6	Care of heating services
(56)Aw3	Supply of fuel (oil, gas, coal)
(56)Aw4	Supply of hot water from thermal power station
(56)Aw7	Supply of cleaning materials

If a division of operations in *separate* items for labour power (wages) and tools and machines is required, the following items may be relevant, e.g.:

(56)	**HEATING SERVICES**
(56)**A**	PRODUCTION CONDITIONS
(56)**Ab6**	Wages, care of heating services
(56)**Ac6**	Tools, care of heating services etc.

In addition items for *purchase* of fuel (oil, gas, coal), cleaning materials, etc., should be given, see above.

7.5   REGULATION

In this presentation we have chosen to regard 'management' as a synonym for 'administration', which again includes organization, planning (design), recording and control (*Construction Industry Thesaurus*, 1972). *Control* in this context designates the *relation* between planned and actually recorded production, which constitutes the condition that the process can be carried out.

However, we have maintained that administration of work cannot be regarded as separated from work performance and that there exists a connection between *control* and *regulation*, which results from the deviations that are demonstrated and which are again caused by outside disturbances of the process. This connection is an expression of the *feedback* of information necessary for the execution of the regulation.

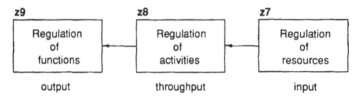

Figure 7.8 The regulation stages

It has been mentioned under *control*, stages a7, a8 and a9, that by comparing actually executed work recorded and existing plans, stages a1, a2 and a3, there may often be found *deviations* regarding quality, quantity, cost and time. Deviations of this kind will normally require some *regulation* or other in order that work may be carried on according to plan. In other words, by the term 'regulation' we will understand, correction, repair, making good, etc.; in brief, *maintenance* in order to establish accordance or equilibrium between planning and execution. In the model, regulation has been divided into the following three stages shown in Figure 7.8.

7.5.1   z7 Regulation of resources

The symbol z7 stands for resource regulation in the sense of *maintenance* of labour power, tools and materials *prior* to proper work execution, including labour, tools and materials necessary for carrying out the maintenance. The work comprises repair of shutters, scaffolding, machines, etc., making good of damages on delivered components, etc., *before* placing in position. Maintenance may result from deviations between actually delivered tools and materials and already agreed requirements, found by resource control, stage a7. Examples are:

(23)Ez7   Repair of forms for concrete floors
(31)Dz7   Repair of window components (in factory)
(31)Xz7   Repair of window components (on site) etc.

As mentioned above, stage z7 may also include regulation of the labour force connected to the project.

7.5.2   z8 Regulation of activities

The symbol z8 stands for regulation of activities in the sense of *maintenance* in order to maintain equilibrium of activities, i.e. the intended structure of constructions *during* work execution, until and at handing over, including labour power, tools and materials. Maintenance includes tasks in the form of making good, repair, down-rigging, etc.

Items may be coded under z8 in the project database, e.g. as in the following examples:

(31)Xz8	Repair of windows and external doors
(41)Pz8	Repair of plaster external wall finishes
(42)Sz8	Repair of tile work internal wall finishes
(42)Vz8	Repair of painting work internal wall finishes
(43)Sz8	Repair of tile work floor finishes

Regulation of activities include maintenance until handing over and at handing over; compare with the general conditions of contract according to which the contractor shall *maintain* the works.

7.5.3 z9 Regulation of functions

Symbol z9 stands for regulation of functions in the sense of *maintenance* after handing over, i.e. in the *use period* of the building in order to uphold the function and mode of operation (performance) of the finished product(s). Maintenance includes work in the form of repair, making good, etc., including labour, tools and materials.

Regulation of functions depends on functional control, stage a9, according to which there are certain discrepancies between actual function and stated requirements in the contract, stage a1. In Denmark, rules for rectifying defects *after* handing over are given in the General Conditions of Contract. Regulation of functions (maintenance) in the use period should normally be specified under A 'Costs in use', for example:

(31)Az9	Maintenance of windows, external doors
(40)Az9	Maintenance of site finishes
(41)Az9	Maintenance of external wall finishes
(43)Az9	Maintenance of floor finishes
(47)Az9	Maintenance of roof finishes
(56)Az9	Maintenance of heating services

Division in *operations* and *materials* may take place as shown above under z6, for example:

(47)	**ROOF FINISHES**
(47)A	**PRODUCTION CONDITIONS**
(47)A**d9**	Maintenance of roof finishes
(47)A**g2**	Delivery of roof tiles
(47)A**i1**	Delivery of lathes, boards
(47)A**n2**	Delivery of felt
(47)A**t7**	Delivery of nails, screws, etc.

For convenience all materials may be classified using y9 of Table 3, as follows:

(47)	**ROOF FINISHES**
(47)A	**PRODUCTION CONDITIONS**
(47)A**d9**	Maintenance of roof finishes
(47)A**y9**	Delivery of materials for maintenance

If required, the operation of maintenance may be distributed on *labour* and *tools*, e.g. as follows:

(47)	**ROOF FINISHES**
(47)A	**PRODUCTION CONDITIONS**
(47)Ab**9**	Labour in maintenance (wages)
(47)Ac**9**	Tools, machines, etc., in maintenance

If required, maintenance may be distributed on *activity types*, e.g. as follows:

(31)Xz9	Maintenance of windows, external doors
(42)Vz9	Maintenance of painting works, internal walls
(43)Sz9	Maintenance of tile works, floor finishes
(45)Rz9	Maintenance of sheet works, ceiling finishes
(47)Lz9	Maintenance of felt works, roof finishes
(56)Iz9	Maintenance of pipes, heating services

In case of proper *work alterations*, coding should take place according to the rules for new work.

7.6   CONCLUSION

The CBC/SfB classification system, part of which has been described above, has been used since 1963 for the structuring of databases on building and civil engineering projects in many countries. Projects include houses, schools, universities, libraries, office buildings, shops, hospitals, roads, bridges, off shore installations, etc. Theory and practice relating to the system has been described in more detail in a recent publication (Bindslev, 1994).

REFERENCES

Ashby, W.R. (1965) *An Introduction to Cybernetics*, University Paperbacks, Methuen & Co. Ltd, London.

Bindslev, B. (1993) Problems in the development of a work model. *Building Research and Information*, 21(6), 325–38.

Bindslev, B. (1994) Paradigma. A Project Data Model. Published by CBC Systems, 2950 Vedbæk, Denmark. (500 p., 300 ill.).

Burgess, D. *et al.* (1964–68) Co-ordinated Building Communication (CBC). 20 articles in *Architects' Journal* from 25 March 1964 to 31 January 1968.

CIT (1972) *Construction Industry Thesaurus*. Property Services Agency, Department of the Environment, Croydon, February, G0311.

ISO 4463 (1979) Measurement Methods for Building. Setting out and Measurement. Permissible Measurement Deviations.

Nagel, E. and Newman, J. (1958) *Gödel's Proof*, New York University Press.

CHAPTER 8

Construction integration – from the past to the present

Ingvar Karlén, Unit of Informatics and Systems Science, Royal Institute of Technology, Fiskartorpsvägen 15A, Stockholm, Sweden.

8.1 FROM THE 'BEGINNING' TO THE PRESENT STAGE

When we discuss construction information we have to consider both semantic information (information as meaning)[1] and signal information (as known from the mathematical communication theory). We should therefore consider both 'meaning' and 'signal'.

The complexity of construction (both as a process and as a physical object) already required in the 1950s improved organization and improved integration, harmonization and co-ordination of regulations, specifications, information about products for construction, as well as information and knowledge of reports and other documents from research and development. Co-ordination is supported by classification and coding of interest for construction and construction research. In this book, integration is the focus of interest. Co-ordination of projects and processes and of related documents was applied in the 1950s. Harmonization was, and still is, important for the international work with construction regulations. All these activities require co-operation between the actors concerned.

Contacts and meetings between actors in research and in practice were arranged more frequently since the information flow in the building process came into the focus of interest. Two CIB (International Council for Building Research, Studies and Documentation) symposia were arranged by CIB S47 in Oslo (1968) and Rotterdam (1970) with the themes 'Information flow in the building process, classification and coding for computer use' and 'Some problems of information flow in the building process' respectively. Work for the improvement of building information

---

[1] I have used an information concept which is close to the information concept in Dretske (1981).

Integrated Construction Information. Edited by Peter Brandon and Martin Betts. Published in 1995 by E & FN Spon, 2–6 Boundary Row, London SE1 8HN.
ISBN: 0 419 20370 2

started in CIB in around 1950. The first 30 years gave several basic results, which are described in *CIB Publication 65* (1982). References to these contributions and also to some other contributions may give some ideas about the development of construction information with roots in the 1950s.

Thus, from the beginning of these CIB activities, those involved in work on the improvement of construction information were, in principle, aware of the breadth and importance of information and knowledge. They were also aware of information resources, processes and results, as well as of the necessity for co-ordination, harmonization and integration of information and for structuring of information and knowledge (e.g. with the help of classification). A lot of this work discipline was hitherto unknown within information technology (IT).

One of the reasons for construction research bodies dealing with construction information projects was the need for improved handling of competence and the need for an improvement in the communication between research and practice. Maybe buildings are objects which require variety and a personal touch. However, technology as support for construction information has proven to be an important factor for development. When we are discussing the present stage of development we therefore must also regard technology-supported information systems as 'systems in the focus' of our interest.

8.2   THE PRESENT STAGE

It is important for us to consider that we are dealing not only with information but also with knowledge, and that knowledge is both a wide concept and a very determinative concept for decisions and actions. We must be aware of the present situation, which means that several actors in projects concerning information technology (IT) have limited knowledge of the current development of the handling of information and knowledge.

It is important for us to know which are the barriers that make it so difficult for construction practitioners to discuss data, information and knowledge. What can be a feasible level for discussions about construction research and informatics (both theory and technology), and for construction practice? This is an important and critical question! We need methods and means which can support co-operation so that we can use words expressing those precepts and concepts we need.

The interest for improvements in construction practice and research seems today to focus on production. This is evident in several regional and international development activities. In such a situation, one must be aware of the manifold aspects of construction and construction processes, which create complexity. Improvement of the present situation requires steps to be taken from harmonization to co-ordination and further to integration. Design, production, maintenance and operation of the building often require different perspectives and different aspects of the studies required.

The levels of a construction (relevant system levels of the construction as a physical object (system)) should be co-ordinated and also harmonized, for example, with regard to statements concerning functions and malfunctions of the relevant construction as a 'system in focus'. One should also consider the relations between relevant expressions for 'stimulus–response' and the analogue relations between 'cause–effect' (c.f. Ackoff and Emery, 1972). We should strive towards co-ordination and harmonization which may facilitate the communication of information as well as the decision-making.

There exists an uncertainty about what belongs to the subject field of construction informatics, for example, theory and practice of construction information – both technology and human behaviour; both natural language and formalized communication; both 'old fashioned' and electronic exchange of data. We are sometimes uncertain whether ordering (for example, with the help of classification and coding) should be based on existing professional languages or on theories and concepts. In Sweden, we have been aware of a tendency, e.g. in the application of the BSAB system[2], that the influence of language on the structure of the provisions in Master Specifications is stronger in the application of the BSAB system than in earlier SfB-based versions of the Master Specification with its faceted structure. In order to avoid unnecessary complexity, no combinations of codes are applied in the BSAB system which you, however, have to 'pay' for, because that means that you cannot benefit from code combinations, which was possible with the earlier SfB system. As far as I know, very few studies have been made since the development of SfB, CI/SfB and CBC/SfB with the intention of finding or creating effective rules for code combinations.

When formalization of information is increasing, as seems to be the case today, classification and coding systems seem to be regarded as less important. Further, a parallel change has been made for the code combinations applied to products for building in the Swedish BSAB approach. The two codes for 'building elements' and for 'works' are used as notations for products for building but the codes for materials or material resources are not used for products for building. This means that the BSAB system (1972) seems to be neither systemic nor an example of applications of set theory.

Which is the best way to go? Can we find methods for such studies and for evaluating such studies? Some considerations can be found in a report to the system committee of the Swedish Building Centre by Bindslev and Karlén (1993). The report can give ideas about what is required from 'informatic systems' serving design, production, operation and maintenance of buildings.

From studies I can roughly recognize that the original CI/SfB system (CIB Report No. 22, 1973) with its relations to system thinking can help us

---

[2] The BSAB system was developed by BSAB mainly for the classification and coding of the Swedish Master Specification Bygg AMA 1972.

to deal with 'organic' wholes, and that CBC/SfB (Bindslev 1991) with its relation to set theory can help us to deal with 'additive' wholes. The two terms 'organic' and 'additive', respectively, I have borrowed from hermeneutics. In *CIB Report 22* (1973) these differences between organic and additive respectively are considered.

The discussions about sets, systems, classes and codes should not tempt us to disregard the contents and the formulation of the messages given in the provisions, e.g. in a master specification. On the contrary, it seems to be a step forward to consider the well-expressed meanings of information markers when one is designing classification and coding systems and other support systems for communication of information.

The Swedish Master Specification 'Bygg AMA' (with Giertz as the main responsible expert and author for the first edition 1950) was a pioneer work with its prescriptive unambiguous 'direct' language. Bygg AMA 1950 was structured with the help of a classification and coding system, the SfB system, which about 10 years later was accepted as an international system, CI/SfB, which together with UDC could meet the requirements for a combination of a general information system and a project information system.

We often meet different definitions of concepts from various fields of application methods and from rules for decisions and actions. Some lists of definitions should, when possible, guide the relevant practice. They can give us a good help within the relevant field of activity as steps on a way to a 'common' language, see the British development work for the improvement of construction practice, which, for instance, produced the CIT thesaurus (1976). Such efforts can give us conventions, which often get their authority with the help of circulations of proposals to selected persons for criticism and advice. Such scrutiny requires some kind of background knowledge and ideas about future trends.

We will get into trouble if we do not solve problems caused by 'word conflicts' (one word has different meanings). Also the doubts, whether an actor belongs to an information system or not, are critical. In the text of ISO/OTR 9007 (1985), the actor does not, for example, belong to the information system concerned. Such a situation will certainly lead to misunderstandings.

CIB has during the decades worked with construction information. I think it is very important, particularly today, that research and development of construction information is undertaken, not least because 'information technology' is expected to give new possibilities for formalization. We should, however, not forget global co-operation not only in the earlier mentioned *CIB Publication 65* but also as an example in a UN/ESA/CHBP report (Karlén, 1976).

8.3 TECHNOLOGY-SUPPORTED INFORMATION AND INFORMATION SYSTEMS, AS WELL AS INFORMATION TECHNOLOGY, SHOULD BE CONSIDERED TOGETHER

Technology-supported information and information systems are, in the many ways in which they are described in this book, not the same as information technology. A lot of work, including international work today, hastens to use 'words with power' which makes both understanding and co-ordination difficult. We must therefore study the concept 'information technology' carefully.

According to Webster's Dictionary, 'information' means

1. Knowledge communicated or received concerning a particular fact or circumstance; news.

2. Any knowledge gained through communication, research, instruction, etc.

3. Computer technology concerns any data that can be coded for processing by computer or similar device.

   Similarly, according to Webster's Dictionary 'technology' means

1. The branch of knowledge which deals with industrial arts, applied science, engineering, etc.

2. The terminology of an art, science, etc: technical nomenclature.

3. A technological process, invention, method, or the like.

4. The sum of the ways in which a social group provides themselves with the material objects of their civilization.

Information technology is, according to Danielson (1990), 'knowledge and methods and the technology tied to information handling (IT). The construction process is, from idea to the management of existing buildings, an information process'. Some definitions of 'information technology' (IT) state that a criterion for IT is that it should include a computer. In ISO/OTR 9007; Concepts and Terminology for the Conceptual Scheme and the Information Base (1985), it is stated that an information system and subsystems only comprise an information processor and information base.

As an extension of the 'IT' development of today, we can study some lines from the research programme for 'IT-building centres' at the technical universities in Stockholm, Gothenburg and Lund. The IT Bygg Center at KTH, Stockholm describes, according to Bröchner (1990), its programme through three themes:

1. Integration: a successive growth and 'construction' of knowledge in an integrated construction and management process for existing buildings.

2. Man-system: human being–system communication in the construction industry.

3. Construction: structures of information for construction technique.

The goal for the centre at the Royal Institute of Technology in Stockholm, Sweden (KTH) can be summarized as follows:

With the utilization of information technology (IT) we can create conditions for improved quality, value for the users, and effectiveness in construction and management of existing buildings.

The programme should give possibilities for a long-term growth and development of knowledge, parallel to effective and at the same time short-sighted cooperating projects. The programmes are developed together with the actors in the construction branch, and with exchange of knowledge between a university (as one part) and industry and administration (as another part).

The idea is that the problems which should be treated should be initiated from practice.

The text of Bröchner tells us that 'the program is structured in three themes. Most of the projects can be referred to one or more of these themes. The projects bridge over traditional borders between different fields of disciplines and presuppose active support from knowledge in data science.' Thus no other sciences of a general character seem yet to be of any strong interest for the development of IT.

This is the perspective from the Swedish construction IT centres.

8.4   SYSTEMS RESEARCH

The two words 'research' and 'systems' express broad concepts. The concepts are related to theories, to accepted methods, to ordering methods, etc. Research often requires the help of scientific methods, which in their turn require experiences as well as fundamentals of ordinary theories, which can also create relevant contexts for meanings and for relevant scientific theories.

Systems comprise subsystems and components and relations (e.g. connections between entities, components, subsystems, etc.), which can together create a recognizable 'whole' interacting with the relevant environment. Some of the concepts discussed can be ordered, e.g. with the help of classification. Ordering can, for example, be used for the reduction of complexity.

We may, for instance, need knowledge about Living Systems which are important for our studies of human and user's requirements including demands for healthy buildings, and for studies of ecological considerations, etc.

In several fields of activities, information can be regarded as a forgotten resource which is 'parallel' to matter and energy. It can help us to serve as an input to planning and design, production as well as operating and maintenance processes in construction, and also as important parts of construction processes.

Maybe we will want to come back to the British Meta System Approach from around 1970 if we want to develop a Meta System. This may be a basis for relevant 'world pictures' as an aid for management which, for instance, can handle the kinds of problems mentioned in this book and can serve as a context both for the present understanding and for further development.

In construction the long-life perspective for a building has come more and more into the focus of interest. I think that the aspects of time (the past, present and future) and meta-level are important, not least because they can help us to:

1. take care of existing thoughts, plans, implementations and experiences for the future and

2. discuss, with the help of a possible meta-language, possible ways for future harmonization, integration and co-ordination of new efforts for the development of information in building.

Experiences from already developed and implemented paradigms, systems and methods, could have a decisive influence on our future activities and on results of activities.

Within the Systems Research Group in CIB W 74 some further reports have been drafted since 1987, see Karlén and Bindslev (1987) and Karlén (with Bindslev) (1990, 1992). We can also follow the work of CIB S 47. (Around 1970, CIB S 47 arranged two large symposia which treated the information flow in the building process (Oslo, 1968 and Rotterdam, 1970). The reports can still give us ideas for solving problems in our work for the improvement of information flow in the building process, and we also could have a lot to add.) Important is also the work of the performance commission CIB W 60 (see CIB, 1973; Karlén, 1992a and b).

8.5 PHYSICAL OBJECTS, THEIR PROPERTIES AND PERFORMANCE

For a long time, at least since the 1950s, we have in building research activities, been very interested not only in the physical objects and their structure, but also in the activities of the human beings and their requirements, and in the behaviour of a building in use (its performance). CIB W 60 'The performance concept in building' is the CIB Working Commission which works in this field of research. The increasing interest for CIB Master Lists of properties, etc. (now within CIB W 74), is positive. We need such 'tools' as support to decision-making in the construction process.

The CIB commission W 60 has pubished a report on examples of the application of the performance concept as contributions from the commission members. In my contribution to that report, I bring attention to two items (Karlén, 1992a,b).

We can regard the building in a reductionistic way, and we can apply system thinking and try to regard the building as 'a system' or 'a whole'. We are interested in the properties of the building and of the subsystems, components, etc., the relations (connections) between them, the values of the properties and the time factor. We are interested in the structure, which makes the building ready to accomplish its functions. We can regard the structure as comprising a number of levels (system levels) (Karlén, 1979).

The word 'property' seems to be more distinct than the word 'attribute'. In IT development 'attribute' is related to 'Minsky's frame', which we meet in object-oriented databases. We 'give' attributes to an object. Properties characterize the object per se and 'define' the object. Objects are often possible to describe as systems.

We often use scientific theories in our strivings for reliable technical structures and technological processes. Bunge (1980) states that 'a theory is scientific (as a General Systems Theory) if (1) it is compatible with the bulk of scientific knowledge and (2) jointly with subsidiary hypotheses and empirical data it entails empirically testable consequences'. This means, for example, that we can find or develop feasible and correct methods for the handling of systems theories and of development activities based on these theories.

We must be aware that we should always have well-defined relations to the 'real world', until we may be able to create an 'artificial ontology' (Ganascia, 1991).

In the present state of the art and affairs, we should always be aware of the risks for disintegration and disorganization of applied technological rules and methods. These disadvantages are often caused by insufficient knowledge.

8.6 APPLICATION OF THE QUALITY CONCEPT

The quality concept is used more and more in construction. Such applications are, for example, supported by ISO documents in the ISO 9000 series. Quality in building is also treated in CIB and in other international documents. The quality concept helps us to bridge between the more precise technical data of a reductionistic character and it thus serves judgement, and classification of functional achievements; it as a tool for harmonization and integration (Karlén, 1988; 1994).

The CIB list of properties (1964, 1972, etc.) can be regarded as a tool in the use and promotion of the quality concept. The CIB lists are overlapping lists from which relevant properties and performance requirements can be selected for use mainly in the establishment and use of project documents and in the publication of documents which can guide the activities in production, operation and maintenance of buildings.

I have studied the quality concept in a research project concerning the need for extension of the application of this concept (Karlén, 1988). I have tried to deal with quality, risks and failures within the same context, and with the help of a set of selected concepts expressed by relevant terms. This approach has the capability of 'bridging' between human requirements and considerations, and descriptions of relevant technical solutions to design problems, the connections between units, their properties and use of relevant performance. Such an approach seems to be capable of reducing the complexity of construction as a process and of a construction as a physical object, the building, and the building parts, components, etc. We must consider the users of buildings and their requirements. This can be done most often in connection with our considerations about quality.

I will also like to mention a presentation by Lundequist (1992) of the two-dimensional quality concept, namely, the 'technical–functional quality' concept and the 'ethical–aesthetical quality' concept; while Sörbom (1974) discusses art, value and value solidarity. We are considering values with the help of structuring of values, according to their mutual relationships or kinds of connections. We very seldom meet documents which deal with both technical and aesthetical background knowledge. Albarn and Smith (1977) is an example of bridging between these two kinds of knowledge.

Ethics and quality are thus related to one another. Both quality and ethics are related to the concept of value. Both are dependent on knowledge. The various actors concerned need access to knowledge, not only well known and stabilized knowledge, but also 'new' knowledge and ideas, in order to be able to handle exploratory thinking, corroboration and confirmation, as well as art. This kind of thinking seems to be relevant today when so many architects, consultants, contractors, researchers, etc., are searching for a 'new' construction process and also for renewal of our thinking about the industrial production of products for construction. These considerations should be regarded as a concern for the complete construction process and for its development.

8.7   CONTRIBUTIONS TO THE IMPROVEMENT OF BUILDING INFORMATION.

The increasing research and development on construction involves responsibility for the growth and development of knowledge. The Swedish Council for Building Research (BFR) and its Scientific Board (BVN) are working with programmes for studies of the development of the exchange of information between research and practice. The existing knowledge, and increasing research activities require organization, for example, with the help of categorization, facet formation and classification. Some of these efforts require systems thinking and/or set theory as well as knowledge about topology and taxonomy.

We need to know more and more about cognition: how we think. In logic, we study the principles for thinking correctly. The use of logic in construction is thus much more than a tool for formalization as a condition for computerization. Many ideas and examples of solutions are available in the preparatory work in CIB W 74 within its Systems Research Group for recommendations concerning 'building information' (see Karlén (with Bindslev), 1990, 1992).

We know that we can, with the help of set theory, topology and logics, deal with many of the problems which concern formalization as an important part of mathematics, psychology, linguistics, etc. Also system theories and methods can give us help by formalizing our strivings for improved planning, design, production, maintenance and occupancy of construction works. This formalization may support bridges to computerization and information technology (IT).

We know from our experiences and from several kinds of theoretical considerations, that we often meet and work with two basic kinds of classification (at least in biology): logical division and empirical grouping (Mayr, 1982). Earlier we could, with the help of the character of a document (e.g. a master specification, etc.), know the kind of information concerned. You could have a shorter code, e.g. for provisions, than today when you use computers and cannot get any guidance from the characters of the documents concerned. It seems important that the present knowledge about construction information and about its integration should be known, and that theories developed in these processes should be given opportunities for further development.

REFERENCES

Ackoff, R.L. and Emery, F.E. (1972) *On Purposeful Systems*, Aldine-Atherton, Chicago.
Albarn, K. and Smith, J.M. (1977) *The Instrument of Thought*, Thames and Hudson, London.

Bindslev, B. (1991) Introduction to Construction Administration, in *Paradigma* (in Danish and in English).
Bindslev, B. and Karlén, I. (1993) Goals, criteria and requirements concerning informatic systems in construction and in the management of existing buildings – a basis for system development. Svensk Byggtjänst (in Swedish). Draft.
Bröchner, J. (1990) Paper presented at the Conference Build with IT Information Technology in the Building Process (in Swedish). Translated by Karlén, I.
CIB (1972) CIB Master Lists for Structuring Documents Relating to Buildings, Building Elements, Components, Materials and Services. *CIB Report No. 18*, Rotterdam. Later editions have been printed.
CIB (1973) The SfB System, Authorized Building Classification System for Use in Project Information and Related General Information. *CIB Report No. 22*, Rotterdam.
CIB (1982) The Information Concept in Building. Report of the CIB Information Study Group. *CIB Publication 65*, CIB, Sweden.
CIT (1976) *Construction Industry Thesaurus*. Property Services Agency, Department of the Environment, Croydon.
Danielson, U. (1990) IT Bygg. Byggarnas informationsteknologi på 90-talet (IT Build. Information technology for actors in construction in the 1990s), Byggförlaget (in Swedish).
Dretske, F.I. (1981) Information is what yields knowledge, in *Knowledge and the Flow of Information*, Basil Blackwell, Oxford.
Ganascia, J.G. (1991) L'âme-machine. Les enjeux de l'intelligence artificielle.
Giertz, L.M. (1950) Presentation of the SfB System, in *The Swedish Bygg AMA*, 1st edn, Appendix to Bygg AMA.
ISO (1985) Concepts and Terminology for the Conceptual Schema and the Information Base. *ISO/OTR 9007*.
Karlén, I. (1976) *Information Exchange on Human Settlements Needs, Problems and Actions*, UN/ESA/CHBP. With an annex containing reports mainly from members of the 'UN family'.
Karlén, I. (1979) Informatics for Design, Production, Maintenance and Occupancy of Construction Works – A Search for Simplicity. *BFR Report D13*.
Karlén, I. (1988) Aktiv Kvalitet (Active Quality). *BFR Report No. 8* (in Swedish).
Karlén, I. (1992a) Background Knowledge for the Application of the Performance Concept in Building Processes. Paper presented to the Meeting of CIB W60, Athens, October 1992.
Karlén, I. (1992b) Research and Development of the Handling Properties, Performance and Quality in Building Processes. Paper presented to the Meeting of CIB W60, Athens, October 1992.
Karlén, I. (1994) Kvalitetsproblem ibyggande och för valtuing. (Quality problems in design, production and maintenance of buildings). BFR Report No. R16.
Karlén, I. and Bindslev, B. (1987) Building Informatics and the Approach to a Paradigm. Paper presented to the Systems Research Group of CIB W74, Lund, 120 pp.
Karlén, I. (with Bindslev, B.) (1990, 1992) Recommendations for Building Information. Paper presented to the CIB W74 meetings, Eindhoven, 1990 and Montreal, 1992.
Lundequist, J. (1992) Kvalitetsbegreppets två dimensioner (The two dimensions of the quality concept), in *Arkitekturmuseets Årsbok* (in Swedish).
Mayr, E. (1982) *The Growth of Biological Thought*, Harvard University Press.
Sörbom, G. (1974) *Konst, Värde, Värdegemenskap (Art, Value, Solidarity of Value. Views of Aesthetic Theory of Value)*, Almqvist & Wiksell, Uppsala (in Swedish).

CHAPTER 9

Construction integration – from the present to the future

Ingvar Karlén, Unit of Informatics and Systems Science, Royal Institute of Technology, Fiskartorpsvägen 15A, Stockholm, Sweden

9.1   A SET OF COMMON CONSIDERATIONS

This chapter contains a set of considerations which may be of interest for the future work. Parts of the texts in this chapter are based on extensions to two papers by the author prepared for a publication within CIB W 60 (Karlén, 1992a,b).

We must be aware that experts and craftsmen in various professional fields constitute the foundations of knowledge for construction practice. The actors in the construction processes should be considered to have their own capacity as 'reflective' practitioners, who do not only organize their activities and get help from their own silent (implicit) knowledge but also listen to colleagues and other actors who give reliable information. Reflective practitioners thus do not only rely upon 'objective knowledge' but also upon their own 'subjective knowledge'.

We must consider the need for technological and scientific knowledge which often has the character of 'expert knowledge' which is sometimes integrated in 'expert systems'. Information and knowledge can be transformed from expressions in professional and trade languages to formalized expressions in order to fit the requirements for formalized information. These are the kinds of languages we meet in the use of computers and in 'information technology'. Wedberg (1984) discusses the concept of formalization, a concept which is important for the handling of information, e.g. human information and information technology (IT).

If we try to co-ordinate or integrate models, ordered structures, etc., we can make our statements and intentions clearer with the help of 'world pictures' which can give the required common contexts to the actors involved in a topical communication of information. We must also consider that the knowledge which we need is dependent on the perspectives and the aspects which are used or which can be used, as well as on the

Integrated Construction Information. Edited by Peter Brandon and Martin Betts. Published in 1995 by E & FN Spon, 2–6 Boundary Row, London SE1 8HN.
ISBN: 0 419 20370 2

objects which we are studying and using. Furthermore, our thoughts and actions are dependent on knowledge gained in education.

In technical developments, including technological progress, for the handling of information and knowledge, we should be aware of the importance of attempts which give good foundations and, when possible and feasible, necessary flexibility. The work for rationalization of activities in the construction industry comprises technology – also information technology. Chapter 8 argues in favour of a broad interpretation of the term 'information technology' (IT). Integration is also required here, including the use of information and knowledge about 'building information' in the broadest sense. Other interpretations may state that all information technologies (IT) have in common is that computer application is used in any form – it could be a complete computerized information system or memories in microchips in any form and size. Several seminars, conferences and books help us today to understand and develop IT further (see Wikforss, 1993). However, as stated earlier, we must know how to deal with uncertainty with regard to several concepts and terms expressing the concepts we will need in the future.

In all kinds of information processing applied in construction, we should always in one way or the other meet the final responsible receiver of directives, instructions, specifications, etc. This person is the user of our efforts. Dialogues should therefore be opened between 'producers' and 'users' of information. Such dialogues are necessary conditions for the integration of information and actions.

The importance of the context or background for our way of thinking and for our way of using different kinds of professional languages should also be considered. We can use systems approaches, classification and other supports for perception and conceptualization, in order to help. We can get help from systems approaches, e.g. through the application of the concept of level, the bottom–up and top–down approaches, for considering processes, construction, function and performance. Certain help is already available because decades ago system thinking already influenced national and international R & D work, for example, in the Working Commissions of CIB (see Karlén, 1979).

Our strivings for co-ordination and integration can be regarded as an indication telling us that we are aware of the fact that we need theories and methods and technological rules, based on theory and fit for coordination and harmonization as well as for integration. The co-ordination can be 'external' as well as 'internal'. We also need good methods for the communication of information and for the interaction between building research and building practice. We need methods and tools for information and for the communication of information, including information technology (IT). All these methods and tools have to serve the human mind. The change from now to the future thus works at a broad front.

The human being as well as other animals have contact with their environment through perception and through communication of information. We try to find and/or develop 'representations (images) of reality' (in the real world) which we meet as physical objects, with their properties (and property values), and their relations (connections). We are also interested in the activities of the actors concerned in other kinds of processes which lead to change of the physical objects and of the human beings involved.

With the help of technology we try to create better artefacts and a better environment for our work with these artefacts in the relevant environment. Our ways to co-exist with our artefacts, including our buildings, and also to use the artefacts are creating the technology as part of our culture.

From different studies, I have learnt that several of the problems we meet in the context concerned have the character of system problems (see Karlén, 1979; Karlén with Bindslev, 1990, 1992). Some of these problems are also related to properties, performances and functions of a building, of building elements, constructions, etc., and also of changes (for example, by ageing). We thus seem to be forced to consider the buildings in work and functions as well as required or achieved performance. We cannot isolate this to concentrate our interests on the production phase of the total construction process.

In system theories we meet matter, energy and information as the basic inputs to a system, which inputs can be transformed to planned outputs. Matter and energy have been studied during a long period of time. The study of information as another important input or output of processes, respectively, helps us to take a relevant step forward in construction practice and research. We cannot concentrate our interest only in information. We need also to deal with knowledge, and in that respect also with 'new knowledge' based on research (scientific knowledge).

A systems theory can, according to Bunge (1980), be regarded as a scientific theory if and only if:

1. it is compatible with the bulk of scientific knowledge and

2. jointly with subsidiary hypotheses and empirical data, it entails empirically testable consequences.

A statement from Klir (1985) gives us a similar message, namely, 'systems science is not directly comparable with other sciences; instead it is more appropriate to view it as a new dimension in sciences'.

Experts, and often also laymen, seem to begin to understand that systems approaches are not so rigid as they earlier might have thought. Further research of AI and of human intelligence seems to require closer contact and co-operation with systemic approaches.

'System' means, in a way, order and ordering, and 'information' also means, in another way, order and ordering. The study of informatics and systems science will therefore certainly be able to deal with both, together or separately, thus contributing to our common strivings for new knowledge which should be well structured and easy to find and use.

We meet several of the discussions mentioned above in our strivings for the handling of the quality concept, which now is in the focus of interest. Quality is related to objective knowledge (also scientific knowledge), to users' needs and to value as regarded by the actor(s) concerned, including the users. Further, quality is related to the relevant 'whole'. We can apply the quality concept as a way to deal with value (both in a philosophical and in an ecological/economical meaning).

The structure with the connections between the components of a building, a building element, etc., make it possible for a building, etc., to function so that the building satisfies functional requirements (for instance, users' requirements). This way of thinking is relevant within both the Systems Approach and the Performance Approach, which have relations to structures, processes and functions – all factors of major importance for decision-making processes. Thus, it seems possible that we should be able to find methods for descriptions of and discussions about processes and physical objects in construction, which are strong enough to help us formalize our statements, provisions and product descriptions, statements etc. which can include or be part of 'product modelling', and also to help us to make, use and discuss relevant 'packages' of information and knowledge.

9.2   ORDERING AND CLASSIFICATION

In the 1950s the interest for ordering and classification in the building process was concentrated on the structuring of documentation of products for building, and also the documentation of their properties, including databases for these products, including properties and connections between entities concerned. References to specifications, e.g. in Construction Master Specifications for construction belong to these tasks.

The development in the 1960s had to consider how to get human and machine information processing closer to one another. This concerns both the methods for improving meaning, and the methods for the signal technology, e.g. the methods for the mechanical and the electronic support of formalistic transmission and transformation of data and information.

We must consider that some of the theories and methods discussed in this paper, for example logic and set theory, systems theory, classification theory (e.g. taxonomy), 'performance theory' are often misunderstood not only by ordinary people and journalists but also by building practitioners. Sometimes misunderstandings depend on a lack of basic knowledge. To avoid such mistakes the practitioners, information intermediaries and researchers should have knowledge about relevant basic theories and about relevant methods and technological rules for actions, and methods.

In literature, see for example, Roberts (1992), the relations between categories, classes, sets and systems are discussed. A fact which widens the discussions about these kinds of concepts. Some considerations are also given in Karlén (1979).

Neelameghan (1988) discusses systems thinking in the study of the attributes of the 'universe of subjects', and refers to Ranganathan. 'System' is here defined as 'a set of interrelated elements, each of which is related directly or indirectly to every other element, and no subset of which is unrelated to any other subset. A purposeful system has a goal or objective'. It certainly will be of interest to study further the relations between systems within Ranganathan's and Neelameghan's theories and systems and General Systems Theories (GSTs) such as General Information Systems Theory (GIST) and General Living Systems Theory (GLST).

As we in this context seem to have possibilities of finding basic theories which are overlapping and not contradictory, we can create 'bridges' which, in a correct way, can combine relevant theories. Thus we can establish a relevant integration.

9.3   INFORMATION AND INFORMATION TECHNOLOGY (IT)

We can find various headings for what is written in the different chapters of this book about problems and required background knowledge at the present 'state of the knowledge' and with considerations about the possibilities for the 'next future'. We can now see the contours of an integrated development of general development of building information and building knowledge, and information technology, although the borderlines between the two approaches are blurred. The concepts and conceptual relations should, however, still be regarded as preliminary.

We must also at this state of development try to be very open-minded and not keep guarding our 'esoteric boxes'. For future organization and management, we may need meta system approaches and discussions about possible paradigms (see work of CIB W 74 within its Systems Research Group). We certainly can learn a lot from ongoing research and development in the complex fields of 'integration of construction information'.

Information technology (IT) in construction serves various kinds of processes such as production supports, for instance, management of production (including management of costs), material administration, and trade of products for building. The development hitherto has focused very much on production, on CAD, on 'product modelling' and on maintenance. 'Product modelling' has, in principle, a counterpart in the form of some chapters of CIB Master Lists (1972).

A lot of work has been done earlier in order to improve the descriptions and requirements of building elements etc. (expressed, for example, as specifications, as requirements for properties/performance and as expressions of functional requirements). Examples can be found in reports and other documents from CIB W 60, CIB W 74 (earlier CIB W 52), W78, W57, etc., and from seminars and workshops arranged by universities.

The so-called performance approach can be co-ordinated with the systems approach. The performance approach can also serve as an objective method for our considerations about quality and determination of quality.

We must, according to Bunge (1967) and to our experiences, be aware of the fact that we cannot treat only properties and property values of objects. We must also consider the relations (connections) between the objects concerned (see also CIB Master Lists 1964 and 1972 (1982)). We are therefore interested in the kinds of objects concerned, their properties (and qualities) and property values, their relations (or connections), and their conditions and changes. We are also interested in the handling of the physical objects.

In construction we have to deal with functions, and also malfunctions. It seems possible to find theories and methods which can be co-ordinated. Such an approach is discussed in Karlén (1988) and in Karlén (1993).

A later section of this book is concerned with 'product modelling'. In the work with 'product modelling' we meet 'frames' with their 'facets' and 'attributes'. Frames are used in object-oriented systems. Hughes (1991) states that 'frames are passive data structures which may be manipulated by processes which are external to their structure. These procedures must be retrieved and involved by some agent other than the frames.'

A frame is, according to Minsky (1982), a generic physical object or concept (Hughes). Hughes brings attention to 'an important difference between frames and the objects in object-oriented systems, which encapsulate both state and behaviour, and execute operations in direct response to the messages received.' Object-oriented database systems have, according to Hughes, been developed to meet 'complex, large scale, data intensive programs', such as those found in the areas of CAD and computer integrated manufacturing.

In informatics we meet 'elementary messages' (Langefors, 1973; Langefors and Samuelson, 1976). We also meet 'frames' as technologies used in object-oriented languages, programs and databases. Elementary messages and frames contain data and information about objects, their properties (or attributes) and property values (or attribute values), and last but not least, their relations (connections).

The striving for 'product models' in IT takes schemes of concepts which can serve the 'frames' with attributes used in object-oriented systems. We must be aware of the risks of 'word conflicts' because of the differences between words expressing concepts related to theories (applied in reports from research projects also in an international context), and conceptual schemes as bases for STEP and other means for international co-ordination of EDP and IT standards.

Parallel to studies of how systems science can be applied as a tool for harmonization and co-ordination of theories and technological rules for both the building process and for the buildings as physical objects, we can also study how researchers and other experts working with the development of information technology in building try to develop and introduce object-oriented databases.

9.4   FURTHER CONSIDERATIONS

In this chapter, I have dealt with various approaches, e.g. approaches which are acquainted with 'bridging theories' in science (see Hempel, 1966). Such approaches can help us use both the knowledge from our 'old' information technology without IT facilities and new approaches, e.g. with IT facilities, in the context of object-oriented solutions, conceptual modelling, product modelling and other advanced computer support. Further, we must study whether and how earlier structures and new structures, e.g. structures in 'line' with formal rules in IT, can be integrated. This also means that we must study well-defined IT approaches and their capability of being adaptable for use in practice.

The whole discussion in this chapter is an example of integration of entities and of systems and subsystems at the levels which are of most interest to us. A lot of work is in progress. It is therefore very important to bring those persons together who are working with these kinds of problems, which have been put forward in this book. We may be able to find 'right' levels for integration. Certainly we need help from categorization, classification, set theory, topology and systems theories. We have as yet no firm ideas about the various roles of the different approaches.

We must also find a balance between practice and research and we must improve the exchange of knowledge about the different kinds of technologies which we have to deal with. This means, for example, that we must know a lot about cognition, information and about creativity as well as about ethics. Ethics is one of the important fundamentals for the development of information and knowledge. All this is as supplement to other wishes expressed in this chapter.

Information technology (IT) and Informatics and Systems Science as they are presented and applied today constitute an important potential source of common knowledge. The present incomplete situation of construction informatics is often described as missing required harmonization, co-ordination and integration as well as missing required feedback loops. These situations lead to shortcomings for the construction processes and for their results, the buildings. We know examples which tell us that the situation can be improved by construction management based upon systems theories, informatics, classification, coding, cognitive sciences and information technology as well as economics. Economics seems to be familiar with system theories and methods. We are thus starting to be aware of both 'organic' and 'additive' wholes.

In construction, we are interested in both the construction processes and in the buildings as physical objects. This creates a need for some kind of duality, for example, between action and physical object, which can be applied both in theory and in practice. During the latest decades, we have also been aware of the importance of cybernetics with its feedback loopings. We need for our daily work, statements about

1. levels and technological rules for description and prescription of physical objects and for processes, and

2. technological rules for the handling of planning/design, accomplishment (e.g. production) and management.

We are in construction as well as in informatics, aware of the importance of the human and users' needs. They can be studied and expressed in terms of functions, performance and quality. We meet more general concepts than earlier in construction research and development, for example, concepts expressed by order, set, collection, function etc. This will certainly lead to the situation where construction sciences will come closer to other sciences, making scientific and technological co-operation easier. We must in situations created by our strivings for harmonization, co-ordination and integration, be very clever to handle intra- and extra-scientific problems and to define and handle statements and concepts, which we want to regard as important in our strivings for integrations.

To conclude this chapter, I quote from a Norwegian writer Nils Collet Vogt the following challenge.

> Do not give up! Remember: the best people were created
> by life in action, not for life in bending
> in numb discouragement at all that they have lost

Translation: Ingvar Karlén

REFERENCES

Bunge, M. (1967) *Scientific Research I and II*, Springer, New York.

Bunge, M. (1980) The GST challenge to the classical philosophies of science. *International Journal on General Systems*, **4**(1).

CIB (1972) CIB Master Lists for Structuring Documents Relating to Buildings, Building Elements, Components, Materials and Services. *CIB Report No. 18*, Rotterdam. Later editions have been printed.

Hempel, C. (1966) *Philosophy of Natural Science*, Prentice Hall, New Jersey.

Hughes, J.G. (1991) *Object-oriented Databases*, Prentice Hall, London.

Karlén, I. (1979) Informatics for Design, Production, Maintenance and Occupancy of Construction Works – A Search for Simplicity. *BFR Report D13*.

Karlén, I. (1988) Aktiv Kvalitet (Active Quality). *BFR Report No. 8* (in Swedish).

Karlén, I. (1992a) Background Knowledge for the Application of the Performance Concept in Building Processes. Paper presented to the Meeting of CIB W60, Athens, October 1992.

Karlén, I. (1992b) Research and Development of the Handling Properties, Performance and Quality in Building Processes. Paper presented to the Meeting of CIB W60, Athens, October 1992.

Karlén, I. (1993) Byggorganisation och byggprocesser. (Construction, Organization and Processes). A report in preparation for the Scientific Board of BFR (in Swedish).

Karlén, I. (with Bindslev, B.) (1990, 1992) Recommendations for Building Information, paper presented to the CIB W74 meetings, Eindhoven, 1990 and Montreal, 1992.

Klir, G. (1985) *Architecture of Systems Problem Solving*. Plenum Press, New York.

Langefors, B. (1973) *Theoretical Analysis of Information Systems*, Auerbach, Philadelphia.

Langefors, B. and Samuelson, K. (1976) *Information and Data in Systems*, Petrocelli/Charter, New York.

Neelameghan, A. (1988) Systems thinking in the study of the attributes of the universe of subjects, in *Debons Information Science*.

Roberts, J. (1992) *The Logic of Reflection. German Philosophy in the Twentieth Century*, Yale University Press, London.

Wedberg, A. (1984) *A History of Philosophy, Vol. 3 From Bolzano to Wittgenstein*, Clarendon Press, Oxford.

Wikforss, Ö. (1993) *Informationsteknologi tvärs igenom byggsverige (Information Technology as Support to Construction in Sweden)*, Svensk Byggtjänst (in Swedish).

Product modelling

Part Three is the first of two parts which begin to examine the future of integrated construction information by examining potential technical solutions. The two technologies are product modelling in this part and process and information modelling in Part Four. These two modelling approaches represent the primary areas of technological research being conducted into integrated construction information at present.

This part on product modelling contains four chapters from different parts of the world. The first is by Watson from Leeds, UK. His group is one of the leaders in product modelling technology in the United Kingdom. His chapter gives a broad introduction to the place and nature of product modelling technology. It then summarizes current international product modelling efforts in construction before speculating where this technology will take us and what lies immediately beyond it.

Chapter 11 is by Eastman from UCLA in the United States. He has been conducting product modelling research in construction for some 20 years and the chapter contained here represents some of the latest technological thinking from his EDM group.

Chapter 12 is by Augenbroe from Delft in the Netherlands. He leads the large, EC-funded COMBINE project which represents the leading European product modelling project in construction. His chapter makes a number of overall reflections on the state of product modelling research. Taken together, these three chapters give a good insight into both the details and the overall state of development of this type of technology and where the future directions will lie.

Chapter 13 is from Poyet and Dubois in France who illustrate many of the details of how product modelling is being applied to the COMBINE and other EC collaborative projects. It is clear from this part that product models will form one important part of the technological solution of the integrated construction information problem.

CHAPTER 10

Product models and beyond

Alastair Watson, Computer-Aided Engineering Group, Civil Engineering Department, The University of Leeds, Leeds, UK

10.1 INTRODUCTION

Product models are concerned with facilitating the unambiguous transfer of engineering information between computer systems, and thus between the users of those systems. Standards for realizing product model-based information exchange and information sharing are being developed progressively within the framework provided by ISO/STEP (Mason, 1992). To date the construction industry has provided little input into STEP. Consequently, the particular needs of construction are not (as yet) well addressed by these standards. The aim of the standards is, however, to enable engineering information to be passed between incompatible computer systems, between different engineering disciplines and across divides imposed by geography or time. As such, STEP offers the construction industry the basis of a strategy to evolve into an industry that is centred on digital information.

Information, in the form of drawings, specifications, schedules, etc., is the life-blood of the construction industry. Substantial volumes of information flow between the various players during the design and construction of a particular project. Transfers of information between interested parties continue, at a lesser rate, throughout the life cycle of that product. The bulk of these information flows are still in the form of traditional paper documents.

This chapter reflects briefly on the progressive shift in the role of computers within the industry from a numerical tool to a production tool. This has been parallelled by a growing interest in computer integration and recognition of the wider potential offered by information technology. However, the need for a coherent information strategy is evident. The thesis behind this chapter is that product model-based techniques appear to offer the industry a key component of such a strategy. The chapter considers the current state of the STEP standards and the enabling technology and identifies some of the problem areas for construction. It also

Integrated Construction Information. Edited by Peter Brandon and Martin Betts. Published in 1995 by E & FN Spon, 2–6 Boundary Row, London SE1 8HN.
ISBN: 0 419 20370 2

attempts to predict where research in this area may be leading. Finally, some of the practical difficulties in realizing an information-centred industry are highlighted.

10.2   FROM COMPUTATION TO PRODUCTION INFORMATION

The rise of automation within the construction industry has been gradual. Most construction sites are mechanized to some extent, but the degree of automation which is commonly found in many factories does not exist. Thus, computerization of the design and planning stages has not been driven by demands for production information to be supplied in digital form.

The goal of early computer applications was to solve complex numerical problems such as difficult structural analysis or large-scale project planning problems. As computers became more accessible, such use expanded and general business applications grew. However, it was the appearance of the microcomputer which marked the start of a continuing shift in technical computing towards the preparation of production information (rather than just solving computational problems). This process has been assisted by the introduction of graphical user interfaces and by the availability of improved applications software.

Today it is expected that computers be used to solve numerical problems. Their use in the creation of production information is also substantial but does vary significantly between companies. Design software is widely used, although much still operates at the level of individual elements and single professional disciplines. Most companies have computer drafting facilities, and the proportion of drawings produced this way is growing. The leading companies have reached the stage that major projects are now set up to be CAD based. However, the use of full 3D CAD is still relatively unusual (although ground modelling systems are commonly used in civil engineering projects). The main exception being in the area of visualization, which is now an important marketing tool and, to an increasing degree, a valued design tool.

Thus, there is a growing volume of software which is deployed primarily as a tool to aid the preparation of production information. Its main justification is that it is cheaper and/or quicker than doing the task by hand. Other applications address tasks which are not realistic to tackle by hand. While the overall picture is fairly *ad hoc*, there is clear evidence that many companies are actively seeking a coherent integrated approach. This commitment is illustrated by the continuing involvement of construction companies in major information technology-based research projects.

10.3 INFORMATION AND INTEGRATION

Historically the UK construction industry has been reluctant to take the initiative regarding moves towards integration of construction information. Consequently, many different applications programs are in use and there has been relatively little provision for transferring information between them. Software vendors are increasingly providing links between the applications within their own portfolios. In this way, sophisticated environments have been created, for example, to analyse a structure, to design its members and to design their connections. While these links add value to the software concerned, the motivation may also include a desire to exclude competing software products. This illustrates some of the strategic concepts being applied to IT as described by Betts, Fischer and Koskela in Chapter 1.

There are some examples of the industry taking the initiative, the most notable relates to the DXF format for graphical data exchange. Initially intended to transfer drawings between AutoCAD systems, this published specification has been used by other software vendors to exchange data with AutoCAD and other systems. The use of DXF, in the short term, was endorsed by the NEDO Committee on CAD Data Exchange in Building (NEDO, 1989) for the exchange of unstructured 2D drawings. This, together with associated measures such as the development and publication of guidelines on the layering of drawings (BSI, 1990), reinforced the growing use of DXF within the UK construction industry. DXF now has an existence beyond AutoCAD; it is the basis of a significant volume of information distribution within the industry (for example, the libraries of standard details distributed by RIBACAD and others). It also forms the basis of many of the practical data exchanges which have been established by leading companies in the context of particular construction projects.

Another example of the construction industry taking the initiative is the EDICON Consortium, which has been developing agreed electronic data interchange (EDI) message formats (EDICON, 1991; Neuteboom, 1993). EDICON initially addressed simple messages such as orders and invoices, but it is now expanding into areas such as bills of quantities. The support for EDICON underlines the fact that the industry recognizes the potential of paperless trading and the implied need for agreed standards. It also underlines the fact that the telecommunications infrastructure necessary to send files between offices and sites is now available. New services such as ISDN enable large volumes of information to be transmitted relatively economically. This, together with the fact that a majority of companies have plans to network their computers (Dobson, 1993), means that the basic infrastructure for integration will soon be in place.

It is evident that we are already in the early stages of organic evolution towards integrated construction information. Clearly, at a company level, the primary motive must be to improve efficiency and thus profitability.

However, the potential exists for the industry as a whole to make gains in a multiplicity of directions such as:

1. Reductions in timescales arising from the rapid flow of information and more systematic (and thus more predictable) methods of working.

2. Improved decision-making due to better availability of information from all sources and integrated decision support tools.

3. Improved co-ordination due to effective dissemination and change control ensuring the presentation of accurate information.

4. Better technical solutions due to effective feedback from the latter stages of the process.

5. More appropriate solutions due to enhanced interactions with the client and other parties.

6. Re-design possible at a late stage to overcome unexpected problems or accommodate changes in requirements.

7. Enhanced quality control and management due to improved procedures and monitoring.

The relative priorities assigned to these possibilities will ultimately be determined by the industry and its clients, not by the enabling technology (full realization will probably require informatics techniques additional to those which form the focus of this chapter). It can be concluded that practical realization will require considerable interworking between the systems used by the many players in the industry. This implies either a high degree of standardization on the same system(s), or a commitment to support open standards. The latter option seems the most practical.

10.4 PRODUCT MODELS

10.4.1 The concept

The role of a product model is to standardize the way engineering information, relating to a *type* of product, is held to facilitate the unambiguous transfer of such information between applications software. It takes the form of an agreed information model which defines how information relating to a *particular* product should be coherently structured while in transit between applications.

Successful information exchange requires that all concerned agree on the product model to be used. This model only exists at a logical level. Software tools are used to generate corresponding physical schemas, within which actual information is held, and much of the associated access logic. Assuming the product model is appropriate, information transfer will (should) be successful providing both applications have translators which accurately map the information between their internal data structures and that of the agreed model.

10.4.2 ISO/STEP (ISO 10303)

Existing graphical-based data exchange standards like DXF and IGES employ a neutral file format to transfer what is essentially graphical information. In contrast, STEP seeks to transfer the engineering intent which underlies such graphical representations. STEP comprises a large set of interlocking standards (known as Parts) which should eventually span all sectors of engineering. Work on these standards started in 1983. The expected scope of the first version, which will be published as a draft international standard in 1993, is illustrated in Table 10.1. As can be inferred, the STEP standards will continue to grow and evolve for many years yet.

The aim of STEP is to create *open* standards for exchanging and sharing engineering information, which are based on agreed product models, and which can be implemented using STEP compliant software tools available from competing sources. Two complementary physical manifestations of STEP are defined: Part 21 which specifies how a product model determines the format of the exchange file used for a corresponding STEP data transfer, and Part 22 which specifies the STEP Data Access Interface (SDAI) (Fowler *et al.*, 1992). Although the specification of the SDAI is not yet complete, it is potentially very significant. It defines an application program interface via which application software can access product model compliant data held in a STEP repository. A repository can have any physical form (for example, a STEP exchange file, data structures in

Table 10.1 STEP Parts to be included in version one. More Parts will be included in later versions of STEP

- Overview (Part 1)
- EXPRESS (Part 11)
- Exchange file format (Part 21)
- Conformance testing concepts (Part 31)
- Six integrated reources:
  - Generic: Fundamentals of Product Data (Part 41)
  - Generic: Geometry and topology (Part 42)
  - Generic: Representational Structures (Part 43)
  - Generic: Product Structure Configuration (Part 44)
  - Generic: Visual Presentation (Part 46)
  - Application: Drafting (Part 101)
- Two Application Protocols:
  - Explicit Drafting (Part 201)
  - Configuration Controlled Design (Part 203)

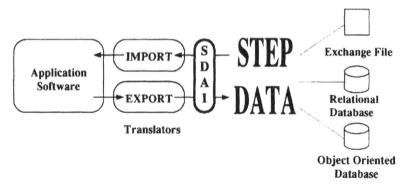

Figure 10.1   STEP implementation independence via SDAI

memory, a relational database or an object-oriented database), STEP requires only that it supports SDAI. Thus, applications software with an SDAI interface (see Figure 10.1) will be independent of the physical form of their external data.

The application of STEP within a given engineering context is governed by a corresponding application protocol (AP). Eventually STEP will have many APs, each will be published as a separate Part and thus will be a standard in its own right. The key component of an AP is an appropriate agreed product model formally defined using the data definition language called EXPRESS (Part 11) (Spiby *et al.*, 1992).

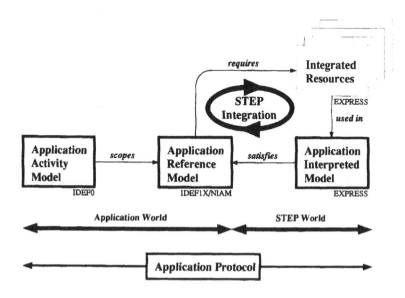

Figure 10.2   The STEP modelling process

10.4.3 Product modelling

The task of developing agreed product models is a long one, the implied need in STEP for them to be mutually compatible makes the process still more difficult. In an attempt to reduce this difficulty, the concept of APs (Parmer, 1992) was introduced into STEP during 1989 (an AP can be regarded as an operational subset of STEP). The use of the term 'product model' was (formally) dropped within STEP as part of a change in terminology which accompanied these revisions. The overall modelling process by which an AP is established is best outlined with reference to Figure 10.2.

First an application activity model (AAM) is developed using a graphical activity modelling methodology such as IDEF0. The AAM represents the activities that use product data within the area of interest. It is used to help scope the AP and thus the required product model. Then, still using the terminology of the application area, an application reference model (ARM) is developed to model the product information identified by the AAM. Again a graphical representation is employed, the data modelling methodologies IDEF1X or NIAM are frequently used. The graphically presented AAM and ARM are reviewed by experts not directly involved in the modelling.

At this stage the modelling process shifts into the domain of ISO/STEP. The STEP integration procedures are followed during which a new STEP compliant version of the product model is created. Called the application interpreted model (AIM), this model, which should be equivalent to the ARM, is constructed using existing generic entities selected from the STEP integrated resources (Lazo, 1991). The AIM is defined in EXPRESS language.

The end product of the overall modelling process, an agreed AP, is a substantial document which includes the AAM, the ARM, the AIM and other components. When creating implementations of a given AP, it is the AIM which is parsed, using EXPRESS-based software tools, to create compliant schemas, etc.

10.4.4 Enabling technology

A wide range of prototype software tools (Wilson, 1992; Boyle, 1993) have evolved alongside STEP. Based on EXPRESS, these either assist in the development of product models or enable the implementation of STEP information transfer. As the STEP standards have become more stable, the available tools have become more STEP-compliant. Tools of a commercial quality are now becoming available. Given a suitable product model defined in EXPRESS, it is now relatively easy to implement information transfer between two applications via a STEP exchange file. The importing of such a file into an application using the Caesar Systems/Femsys STEP toolkit (which is based on the emerging SDAI standard) is illustrated in

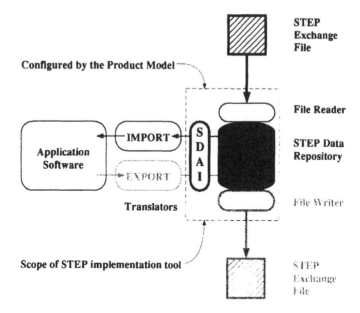

Figure 10.3 Importing STEP data using an SDAI-based tool

Figure 10.3. Such technology allows the creation of a repository which is a database; thus information can potentially be shared more directly between applications. While SDAI allows an application simultaneous access to multiple data repositories, it does not (thus far at least) support **concurrent** access by multiple applications to a single data repository.

Other STEP tools support the direct development of product models in EXPRESS. For example, DECexpress from CADDDETC which guides the user through the complex syntax of EXPRESS, ExWare from Cononial Systems, which allows EXPRESS to be specified using tables, and Extool from GOSET, which provides the STEP resource models in library form. Despite this, to date most ARMs have been developed graphically using the IDEF1X or NIAM for which good software tools exist (this also applies to AAMs and IDEF0). These tools, and the underlying methodologies, were created to aid the design of database schemas, not STEP models. However, the use of EXPRESS_G, a graphical subset of EXPRESS, for product model development is increasing.

10.4.5   Some problem areas

The aims of STEP are very ambitious. This is illustrated by STEP's long and difficult gestation period and by the fact that only two APs (drafting and configuration controlled design) will be in the first version. When viewed from a construction perspective, STEP has several problem areas.

These include:

1. the weakness of the STEP AEC group;

2. the lack of an overall modelling framework for AEC;

3. the absence of integrated resources or APs for construction;

4. the difficulty of funding the development of integrated resource models;

5. the scale of the effort required to establish a STEP AP;

6. concerns regarding the applicability of APs;

7. the absence of an EXPRESS style guide and product modelling methodology.

Many of these difficulties are interrelated, and several reflect broader problems within STEP. Construction is an application context significantly different from those which have driven STEP to date. The underlying concern is that to fully accommodate such a context, further modifications to the application of STEP may be needed. For example, the concept of APs appears to assume either that the existing applications are well defined (such as drafting), or that that new applications software will quickly aggregate about an agreed set of rational APs. Neither appears likely in the construction sector where applications software tends to be heterogeneous and long lasting.

A significant volume of construction-related product modelling research has taken place. For example, the RATAS project (Björk, 1989), which is probably the best known in the area of buildings. However, such good work has tended to remain on the periphery of STEP. This may soon change. Inputs into the AEC sector of STEP are anticipated from a number of major industrially led research programmes such as COMBINE (energy-conscious building design), CIMSTEEL (constructional steelwork), Process Base (process plants), CAESAR Offshore (offshore projects), and ATLAS (large scale engineering). A major problem facing the modellers engaged in this work is the absence of an overall STEP modelling framework for construction (to span the existing void and impose some conceptual similarity on the models to be created). The General AEC Reference Model (GARM) (Gielingh, 1988) was a notable attempt to satisfy this need. While the GARM failed to gain universal acceptance, its main achievement has probably been to raise awareness of the problems of model integration within STEP. This has been continued with the publication of IRMA as an Information Reference Model for AEC and a recent E-mail conference during which IRMA was discussed.

As Figure 10.2 implies, suitable integrated resources must exist if an AIM is to be created. However, funding for construction-related modelling tends to be directed at the creation of specific ARMs and not the development of (the more general) integrated resources.

Irrespective of these difficulties, considerable time and effort is required to develop a small suite of APs to the status of formal STEP standards. This begs the question: How many APs will be required? Currently, the answer appears to be too many. In order to be compliant with a given AP, an application needs to be aware of (i.e. be able to input and export) *all* the entities in the AIM. Given the mix of applications used in construction, this implies either the need for a large number of APs, which are narrow in scope, or a lesser number of much broader APs. The problem with the latter is that (strictly) even a simple application would need to be able to deal with all the entities in the AIM. For example, an application for designing connections would have to be aware of the complete AP for designing structural steelwork.

While the AP problem has been raised from a construction viewpoint and the basis of a solution proposed (Watson and Boyle, 1993), it is evident that similar concerns exist elsewhere in STEP. Figure 10.4 represents, in greater detail, the current procedures by which AIMs are created such that resource constructs are, as far as possible, shared by APs. The integrated resources, which are divided into 'generic' and 'application', provide a source from which entities are 'interpreted' (that is, specialized in accordance with strict rules) to provide suitable entities with which to build the AIM. The objective of the overall STEP integration process is to identify constructs in an AIM, which can be placed in the application interpreted construct (AIC) library, and reused (unchanged) in other AIMs. An AIC, which may contain other AICs, may also represent a particular unit of functionality (UofF). A UofF, which may also contain other UofFs, is some part of the ARM which conveys a well-defined concept.

Formal proposals have been made (Mohrmann *et al.*, 1993), from an automotive design context, to make it valid for a STEP AP to have a number of implementable subsets. Known as Functional Groups, each would be defined on the ARM by a coherent group of UofFs which corresponded to one or more dataflows on the AAM.

Although EXPRESS is an appropriate product modelling language, it has no formal methodology associated with it. This is one reason for the popularity of IDEF1X and NIAM. Both have well understood data modelling methodologies but, as the goal is a product model defined in EXPRESS, the methodology must be adjusted to suit. The inherent flexibility of EXPRESS is also a problem since there is no widely accepted style guide (Boyle and Williams, 1992) governing how it should be applied.

10.4.6 The future

In 1989, the NEDO Committee on Data Exchange in Building recommended, in the medium to long term, that STEP be used. Irrespective of the problems discussed above, this advice remains valid. If information transfer and sharing is to be based on open standards then STEP cannot

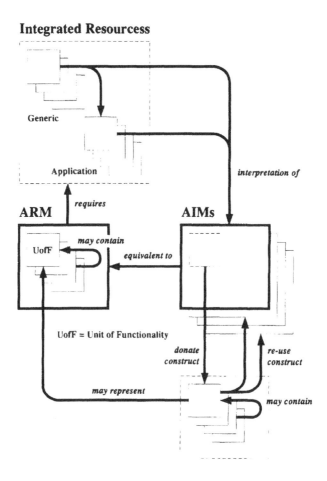

Figure 10.4 AP integration

be ignored. The scale of the investment already made dwarfs anything that construction could fund alone and the major CAD vendors are committed to implementing STEP.

Ironically, the implementation of data exchange based on product models that have not been formally agreed within STEP, will be encouraged by the availability of increasingly powerful EXPRESS-based implementation technology. Further shifts in the STEP integration procedures towards a more flexible and pragmatic AP concept seems likely. This may be complemented by the emergence of a new class of EXPRESS-based software tools to facilitate the implementation of protocol converters. These would map product data between overlapping, but incompatible APs, and thus allow interworking.

The EXPRESS language itself also provides pointers to the future. For example, it includes powerful DERIVE and WHERE rule constructs (which are not fully supported by most available implementation tools). These constructs are intended to support the inclusion of validation checks within a product model. However, they might also be used to add value to an agreed AIM to allow the transparent computation of information not explicitly stored in a repository. In this way the boundary between applications software (as the calculator) and the STEP repository (as the information store) could be shifted. Already EXPRESS is quite close to being an object-oriented language. Proposals for a fully object-oriented version two of EXPRESS could lead to more radical shifts in the architecture of future engineering systems. It is possible to imagine information objects being automatically assigned (additional) local functionality as they enter an engineering company, and being stripped of this functionality prior to being passed on to others. Equally, considerable scope appears to exist for linking knowledge bases to the standardized description provided by a product model.

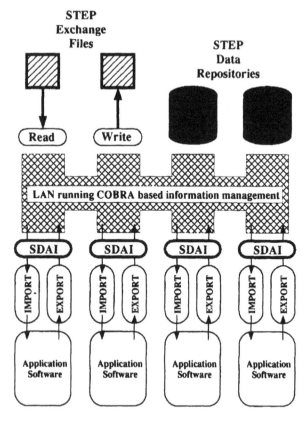

Figure 10.5 Proposed integration infrastructure

In the shorter term, the practical need to manage flows of information will become a significant implementation and deployment consideration. Additionally, attempts to generalize the existing operational possibilities of STEP will make data integrity considerations more critical. One approach to this problem is based on the evident synergy between the SDAI and COBRA standards, the latter being the Common Object Request Broker Architecture (OMG, 1992) developed by the Object Management Group. The new PISA (Platforms for Information Sharing by engineering Applications) and EIME (Engineering Information Management Executive) projects share the concept of using COBRA to insert a level of direction between applications and the networked STEP data repositories with which they interact (see Figure 10.5). This provides an infrastructure which can be used to add value to SDAI by providing information management and integrity controls.

10.5 SUMMARY

Demands for digital product information are already beginning to emerge at various points in the product life cycle. For example, elements of automation are being introduced into the off-site fabrication of steelwork and other components. Similarly, clients are increasingly introducing facilities management software. Such trends can only increase.

Agreed product models offer the construction industry a viable basis for the digital transfer of information. Critical to this viability are the STEP standards which provide construction with a framework for establishing open standards, and a context for onward evolution into an information-centred industry. Practical difficulties do exist relating both to STEP and how to bring about the change in industry.

Although this paper has highlighted a number of problems with the application of STEP, these can be overcome. Some are broadly based and will be resolved by the STEP community in due course. Others are more specific to construction and may only be addressed when the industry becomes more actively involved. An investment of resources plus a pragmatic commitment to STEP would ensure that construction was well served by these standards.

The greater practical difficulty is the need to change the industry, particularly as the technology has the potential to alter the structure and organization of that industry. Concerted action will be required to establish open standards, ensure software vendor support and achieve substantial industrial deployment. The necessary pressure for change will only be created if industrial awareness and understanding is developed. Compliance with STEP will help, as will industrial-scale demonstrators. Ultimately, the industry and its clients will need to take the initiative.

ACKNOWLEDGEMENTS

The author wishes to acknowledge the inputs to this paper – both direct and indirect – from members of the CAE Group (particularly Alan Boyle, Andrew Crowley and Gareth Knowles), and from other collaborators, and the funding of relevant research projects within the Group. The latter include the Eureka CIMSTEEL Project (Computer Integrated Manufacture of Construction Steelwork) (DTI) and the CI–PM Project (the creation and the use of product models within the construction industry) (SERC).

REFERENCES

Björk, B.-C. (1989) Basic structures of a building product model, *Computer Aided Design*, **21**(2) March, 71–8.

Boyle, A.D. (1993) STEP Tools Review Phase 2, Department of Civil Engineering, University of Leeds, *CIPM/TP/LU/7*.

Boyle, A.D. and Williams, A. (1992) EXPRESS Usage Guide, *ISO TC184/SC4/WG5*.

BSI (1990) *BS1192 Part 5: Construction Drawing Practice*. Guide for the structuring of computer graphic information, British Standards Institution, Milton Keynes.

Dobson, R. (1993) Results of the IT Questionnaire, Internal Report, CIM-STEEL Project.

EDICON (1991) In the process of construction – EDI, EDICON, PO Box 111, Aldershot, Hampshire, GU11 1YW, UK.

Fowler, J. *et al.* (1992) ISO 10303-22 STEP Data Access Interface Specification, *ISO TC184/SC4/WG7 N262*.

Gielingh, W. (1988) General AEC Reference Model (GARM). *ISO TC184/SC4* Document 3.2.2.1 (Draft), October, TNO-IBBC.

Lazo, P. (1991) Guidelines for the Development of STEP Resource Parts, *ISO TC184/SC4/WG4*.

Mason, H. (1992) ISO 10303-1 Overview and fundamental principles, *ISO TC184/SC4/PMAG*.

Mohrmann, J., Polly, A., Speck, H.-J. *et al.* (1993) Proposed extension to the AP development methodology, *ISO TC184/SC4/WG3 N126 (T19)*.

NEDO (1989) *Information Transfer in Building*, NEDC Publications.

Neuteboom, J. (1993) *Bridging the Gap between EDI and PDI*. Proceedings of the joint EDICON/EDIBUILD Conference, National Motorcycle Museum, Birmingham, UK, 27 April 1993.

OMG (1992) Object Services Roadmap. *TC Document 92.8.2*, Object Management Group.

Parmer, M. (1992) Guidelines for the Development and Approval of STEP Application Protocols, Version 1.0, *ISO TC184/SC4/WG4 N35 (P5)*.

Spiby, P., *et al.* (1992) ISO 10303-11 Description Methods: The EXPRESS Language Reference Manual, *ISO TC184/SC4/WG5 N35*.

Watson, A. and Boyle, A.D. (1993) Product Models and Application Protocols, under preparation for CIVIL-COMP 93, Edinburgh, UK.

Wilson, P.R. (1992) Processing Tools for Express, Express User Group Meeting, Dallas, October 1992.

CHAPTER 11

Integration of design and construction knowledge*

Charles M. Eastman, Graduate School of Architecture and Urban Planning, University of California, Los Angeles, USA

11.1   WHY BUILDING MODELLING IS DIFFICULT

Effective integrated CAD systems have been developed in areas where the following conditions have been satisfied:

1. the elements being composed are known;

2. the performance dimensions to be achieved are well defined; and

3. the rules of composition are well understood.

   For example, high performance CAD systems are probably furthest developed in building construction in the areas of prefabricated construction and piping. Early work in systems building in England led to the development of such systems as CEDAR and OXSYS systems and powerful CAD systems were developed to support them (for a survey, see Eastman, 1992a). With the abandonment of industrialized construction of hospitals, these CAD systems became irrelevant (though some were adapted to become object-based CAD systems, where the components and rules had to be entered by the user). Integrated piping design packages are available from such firms as Computervision and Intergraph that support design at various levels of abstraction, integrate various analyses packages, and develop bills of materials for construction. The PDMS back-end database for piping design also has its adherents. In both of these areas, the elements to be composed could be defined, the performances and rules of composition were fixed. In areas outside of construction, similar conditions exist; electronic design has proceeded far because the three

*This chapter is an elaboration of the work presented in an earlier paper (Eastman *et al.*, 1993).

Integrated Construction Information. Edited by Peter Brandon and Martin Betts. Published in 1995 by E & FN Spon, 2–6 Boundary Row, London SE1 8HN.
ISBN: 0 419 20370 2

conditions could be fixed. Manufacturing has proceeded more slowly because the above structure was more difficult to define.

I call the general framework involving a set of elements to be composed, the performances and functions by which compositions are to be evaluated, and the rules for creating compositions, *'kit-of-parts' design*. All successful CAD applications up until now have relied on a kit-of-parts approach, whether they support interactive design or are automatic design programs.

Integration of information in most areas of building design has not proceeded very far, I suggest, because the conditions for kit-of-parts design, regarding well-defined elements, functionality and composition rules, do not easily apply. A wide range of construction technologies exist for different parts of the building: for structures, for mechanical systems, for cladding, foundations, plus many others. In the performance area, different building types have fundamentally different design rules. Hospitals have one set, schools another, concert halls another, housing another and so forth. Within these building types, different performance conditions arise. It is the combinatorial explosion of these different kinds of technologies, building types and activities, and the associated knowledge that characterizes and defines them that makes the development of CAD for architecture so difficult.

Those successful efforts at architectural CAD thus far have been based on a kit-of-parts approach. There are excellent proprietary programs for parking garage design and construction, precisely because these facilities can be formulated as a kit-of-parts problem, starting with parking spaces, access aisles, ramps, then moving to slabs, beams and columns. Kitchen design programs exist in many American kitchen design centres, again, because this part of a building can be easily defined in a kit-of-parts manner and rough performance rules exist. Even room layout programs support this view, with a well-defined but abstract part called a 'space', and well-defined performances and rules of composition. While each of these individual tools does some small useful task in a building design model, the combinatorial issues of integrating them together to support realistic building design is simply overwhelming. (An important limitation of kit-of-parts design has been eloquently expressed by my colleague, G. Stiny, who has pointed to the role of emergent patterns or conditions in creative design (Stiny, 1975).)

Together, the different kinds of design knowledge used in buildings can be considered as defining a large matrix, shown in Figure 11.1 as three-dimensional, but actually of higher dimensionality. There are many alternatives in each dimension of the matrix and the number of these alternatives is not fixed; it grows continuously with the advance of knowledge and technology. The number of dimensions itself is not fixed and may vary in different projects. Every combination (i.e. cell in the matrix) represents a very specific alternative space of designs, for example, a steel girder-framed hospital, with cold air and hot water reheat, using concrete

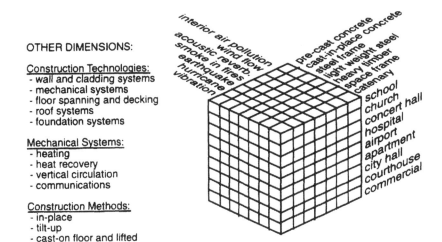

OTHER DIMENSIONS:

Construction Technologies:
- wall and cladding systems
- mechanical systems
- floor spanning and decking
- roof systems
- foundation systems

Mechanical Systems:
- heating
- heat recovery
- vertical circulation
- communications

Construction Methods:
- in-place
- tilt-up
- cast-on floor and lifted
- pre-fabricated

Figure 11.1 A matrix of the multiple dimensions of knowledge that is incorporated into a building model

exterior panels, using a defined set of analysis programs. The frequency of applying different cells in different design projects vary, and some of the cells may never be used. Many buildings use multiple technologies and are designed to support unique mixes of activities. Thus some buildings define a unique combination of cells.

Considering the problem from a knowledge-base standpoint, each application defines a kit-of-parts which corresponds to a universe of discourse, that is, some combinatorial space of designs (possibly infinite). These abstract spaces, for example, might correspond to all standard light-weight steel structures, or an abstract space of cafeteria designs. The complete building design is the union of the abstract spaces defined by the different space uses, construction and mechanical systems used.

Some aspects of this space of designs can be defined at the outset of a project: for example, those associated with building type. However, many other decisions affecting the structure of knowledge used in the design are only made as design proceeds. The program of the building changes; a new facade material is proposed that requires different jointing and structural attachment. An active HVAC system is augmented by a passive solar system. In these ways, each building design problem is defined incrementally as the project proceeds, by incrementally defining the space of designs to be considered.

Our interest is in providing a design environment that supports such a design process. We must both be able to capture the structure of design information and do so in a way that supports the process by which large design problems are incrementally structured and resolved.

11.2   DESIGN AS MODEL DEFINITION AS WELL AS MODEL INSTANCING

It is apparent, I believe to most people, that the appropriate response to this combinatorial issue is the development of a set of co-operating applications, defined as distinct modules. Modules are added to the design, as design proceeds. A variety of authors have promoted this direction of future CAD development (Gielingh, 1988; Augenbroe, 1991; Eastman *et al.*, 1992; Yamazaki, 1992).

At first glance, it may appear that this can be accomplished by a large suite of CAD applications, as is being attempted by some CAD companies (Intergraph) or by third party vendors (ASG). However, careful consideration indicates that the problem is highly complex and not easily implemented within the current CAD environments. The spaces of designs defined by each application interact, that is, one space of designs, constrains or delimits the range of another space of designs. Such interactions are not easily supported today. The combinatorial problem of all possible applications, interacting with all others, is overwhelming, if addressed a case at a time.

In order to develop the system environment that supports modular applications, we need to be able to define how the various modules interact. For example, we need to identify how one module may use information in other modules (load definitions for structures and electrical systems), how some applications constrain other applications (how an auditorium activity will constrain structural systems and construction technologies will constrain the routeing of mechanical systems), and how some applications cut across many applications (energy evaluation, spatial interference checks). We need to be able to answer these questions not only for existing application areas, but also for future applications that will emerge as the structure of building models becomes standard. Thus we must anticipate how building code checking, habitability studies, weatherability, public safety and other new applications will be modularly interfaced to a set of existing building design applications.

There have been continuous efforts to develop building models and integrated applications. Beside the issue of model complexity, already characterized, these efforts do not support the process of design. That is, building models have been developed for a specific set of applications. Most of these assume a fixed sequence of applications, each generating data from which information is derived (in a known order) to supply to other applications. Thus they fix a linear process of design. Partial updates require iteration of the whole linear process succeeding the change. Update from diverse sources not following the linear ordering are not supported.

Rather than continuing to develop building models based on some intuitively selected computer language or database technology, we chose to explore the development of new tools that would allow better conceptualization of these integration problems, prior to implementation. Our notions for such tools came from both the conceptual modelling concepts in database theory and expert systems (Brachman and Levesque, 1985) and also tools for software engineering (Meyer, 1988). Our goal was to identify a set of conceptual structures that would allow us to model any type of design information or knowledge. With such a set of structures in hand, we could then define various modules, explore methods of integration and explore much more quickly various issues of module definition. Through such exercises, we expected to define and quickly refine our criteria for future system implementations.

A strategy of incrementally adding modules to the structure of a design knowledge base, as design proceeds, does much more than respond to the combinatorial problem. It opens up a closed environment to new building uses and technologies, for example, developed by the designer. At high levels of abstraction, where walls and spaces are combined into higher level units, different architects and design styles rely on highly varied structures of a building. The strategy should support those different abstract structures, which correspond to what architects call their design vocabularies.

The main issue is the development of a system architecture that will allow the dynamic integration of information modules. Such a system architecture probably contains at its core a module defining a generic building, independent of building type (Eastman and Siabiris, 1992; Björk, 1992). To this core is added various building-type knowledge, various construction technologies and various abstract design strategies. This modular approach aims at ultimate flexibility in configuring a CAD system for a particular project and providing the designer with the ability to add new knowledge to the system during the design process, as the need arises. It views design as a dynamic process that should take advantage of continuously emerging construction technologies and design knowledge.

11.3 RESEARCH STRATEGY

We began by carefully considering the capabilities needed to represent design information and concepts, with special consideration of the semantic conditions over time. Our initial criteria were:

1. Explicit encoding of all semantic relations that are useful in managing the state of a design;

2. Allowing these relations to be testable, that is, true or false, and support management of those relations, conceptualized as partial integrity;

3. Dynamic extensibility, as design proceeds;

4. Representation of function as well as form;

5. Representation of information at multiple levels of abstraction.

We first reviewed current work in conceptual and information model-ling, and the related work in knowledge structures, to identify those con-cepts and structures from database theory and AI that were relevant for design information. Most of the work in knowledge modelling relies on a structure of entities and the relations between them. The relations empha-sized are those that are applicable to all domains, such as *aggregation* (the grouping of data into sets because it all refers to the same thing (Smith and Smith, 1977a)), *generalization/specialization* (the abstraction of a set of objects by extracting their common definitions (generalization) and the elaboration of a general object by adding to its description (specialization) (Smith and Smith, 1977b)), *composition* (a particular type of aggregation, composition focuses on the relation of the composite object to its part objects and defines the rules and relations involved (Bond *et al.*, 1992)), and *classification* (the definition of classes of objects and instances of them (Brachman and Levesque, 1985)).

We also studied and roughly categorized the semantic conditions in different areas of design and abstracted them in order to explore how they can be expressed computationally:

1. *Definitions of measurements*: watts = amps × volts; density = mass/volume

2. *Physical laws*: F = ma; single occupancy space–time

3. *Compositional rules*: all pipes must be properly connected at both ends, as one example of many different types of connection relation, e.g. structural, electrical connections

4. *Geometric descriptions*, including 2D and solid geometry

5. *Dimensional fit*: connected pipes must consist of a male and female fitting of the same size and thread type; bolt diameter must be less than the hole diameter

6. *Analyses of performance*: defining structural, energy, fluid flow or speed as properties of a composite object

7. *Activity-based information*: the size of spaces needed to support some activity, adjacency relations to support activities that relate multiple spaces; environmental conditions that support specific activities

8. *Trade practices*: standard doors are 6[ft] 8[in]; stud walls shall have blocking between all adjacent pairs of studs at less than 5 ft intervals;

9. *Aesthetic criteria*: all window headers shall have the same height; set of columns shall be equally spaced; utilization of regular grids.

In order to support open-ended relations, we have provided a general means to define relations in the form of constraints. Within a design domain, the challenge is to specify a set of constraints that economically capture the relations of importance. At the system level, the goal is to provide powerful predicate management capabilities (integrity management).

11.4   ENGINEERING DATA MODEL (EDM)

EDM has been developed by a team of researchers from the Architecture program and the Computer Science Department at UCLA. The EDM work involves the data model itself and a data model language called DML (Design Modelling Language) to support the manipulation of the product or building models within EDM. Here, an overview of EDM is presented, that supports the later presentation regarding extensibility. A full description of EDM and DML can be found in the references at the end of this chapter.

EDM is based on a small number of structures to capture the semantics of design and engineering information. We chose sets and first order logic as a neutral but powerful meta-language for defining these structures. From this, we defined three types of primitives:

1. *Domains* are sets of values, corresponding to a simple type.

2. *Aggregations* are sets of named domains, i.e. sets of variables.

3. *Constraints* are general relations, stored as procedures, i.e. they are named and called by reference.

EDM is unique in that constraints may be treated as being variant or invariant; that is, constraints may be defined as predicates that are guaranteed to be always true or they may be defined so their evaluation may be true or false. Constraints are not directional, but are used to define relations that are to be satisfied. Defined in this way, they can be propagated in a variety of directions.

As EDM is set oriented, all variables start with a set of values corresponding to their domain and can be delimited by constraints to any subset of the domain. These primitives are composed into three high level predefined forms.

The first form is the *functional entity* (FE). It is the primary data object within EDM, consisting of an aggregation and constraints. It also contains a set of FEs that it specializes, i.e. whose variables and constraints it inherits. The syntax of an FE specification is:

FE(Fename1, Aname, {Fename2}, {Cname})

Each FE is named, denoted *Fename1*, and composed of a set of FEs ({*Fename2*}) which this one specializes, an aggregation (*Aname1*) and a set of constraints (*Cname*) over the variables in the aggregation and set of specialized FEs. The concept of specialization is represented within the FE form by inheritance of additional FEs, constraints and/or variables, thus supporting a variety of inheritance models (including multiple inheritance, in effect creating an object lattice consisting of FEs). Specialization–generalization has been formally defined within EDM (Bond *et al.*, 1992).

Specialization–generalization also applies to the assignment of variable values. Making variable assignments part of specialization becomes important if it is to be available to users, in that some FEs with shared variables may be used to define design intent, e.g. that a set of windows have the same head height or alternatively the same dimensions. Such techniques are commonly used in current CAD systems by defining shared symbols. A consequence of this definition of specialization is that there is no distinction between class definitions and instances; any object may be further specialized. This accords with our understanding of design, where there is no boundary to a design specification; it may apply in ever greater detail, to the atomic level.

An important elaboration is that value specialization is not an invariant relation, but variant. That is, values propagate to all specialized FEs, if no value is assigned (the whole type domain is the assumed value). However, if the specialization propagation or an explicit value assignment encounters a different value that has been specialized, then the value is unassigned, but a constraint violation is set, notifying the system of the conflict. We will see later why value specialization is both important and should be a variant relation.

In EDM, FEs have strong scoping, in that their constraints may not refer to other, non-inherited FEs. Objects must also be consistent with regard to self-reference. That is, 'part-of' is not a specialization. All relations that can be characterized within a specialization lattice of FEs can be represented in the FE constraints.

Other relations in engineering data involve multiple objects and cross any object-oriented partial ordering. These typically involve relations of function within composite objects. Like other efforts to define composite objects, we have a structure for *compositions*:

composition(Compname, Fename1, {Fename2})

A composition is a named structure between one FE (*Fename1*), called the target and a set of parts, ({*Fename2*}) that composes it. The set {*Fename2*} can be considered the replacement to *Fename1*, defining it in more detail.

EDM does not treat composition as a single relation, but rather as an aggregation of more detailed relations, defined in various ways according to the functions and intentions that the composition must support. Each function and intention is defined as an *accumulation*. An accumulation is a one-to-many relation between the properties or structure of a composite object and the properties or structure of its parts. It defines performance relations among properties as well as well-formedness rules that the composition must satisfy to achieve that performance. Accumulations take the form:

accumulation(Accname, Compname, {Cname1}, {Cname2})

An accumulation has a name *Accname*, and reference to the composition it is part of. The relations are defined by a set of constraints ({*Cname2*}) that apply over the properties in the composition it refers to and which accumulate into the composition's target. In addition, an accumulation defines the well-formedness preconditions that the composition parts must satisfy, defined in the constraint set ({*Cname1*}). Multiple accumulations provide the detailed definition of a composition.

An accumulation is thus defined as a one-to-many relation over a subset of the composition relation. It incorporates two kinds of semantics:

1. Design rules regarding relations between the parts that are needed to achieve the function or intention, qualitatively. ({*Cname2*}).

2. Property relations, where the properties of the composite object are some function of the part objects ({*Cname1*}).

In general, the design rules are preconditions that must hold if the relations among properties are to hold. With this finer level of definition, the semantic detail of design information is greatly increased. This new definition of composition allows explicit definition of design rules, tolerances and the performance relations derived in simulations and other forms of analysis.

These three high level forms have been used to depict information ranging from abstract conceptual design information (captured by the FE hierarchy) to low level production-based information (captured by the composition structure, which describes functional relations). Constraint relations in design can be highly complex. A major issue in the definition of EDM was to organize them in a manageable way. Thus our effort has been to package constraints into forms that makes their behaviour both apparent and tractable.

EDM forms can be represented both textually, as presented here, and graphically. An example is shown in Figure 11.2, showing a small set of geometric entities. Each of the primitive and high level forms have been formally defined and the structure of EDM models axiomatized (Bond *et al.*, 1992).

EDM incorporates a variety of criteria relevant to object-oriented systems and suggests guidelines for evaluating different object semantics for representing design knowledge (Fereshetian and Eastman, 1991). It provides significantly extended semantic coverage, in comparison with the other efforts in product modelling. This includes continuous refinement, with no distinction between the concepts of class and instance, and allowing values to be defined as sets of values. It also addresses issues regarding

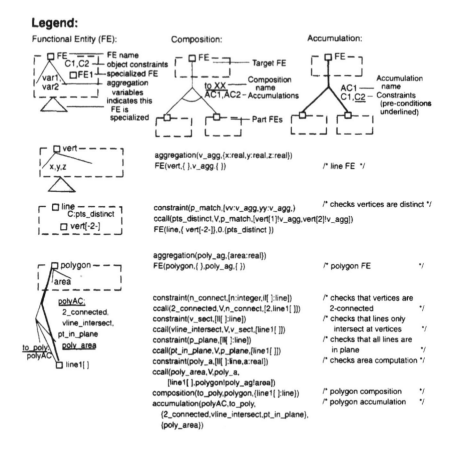

Figure 11.2 A small set of EDM geometric definitions. Constraint bodies are not included

partial integrity and the evolutionary, trial and error nature of design activities and information.

The following published work illustrates the range of design domains which have been modelled using the EDM formalism:

1. Core and panel walls, including geometric, thermal and acoustical properties (Eastman *et al.*, 1991b).

2. Composite windows (Bond *et al.*, 1992).

3. Chair design (Eastman, 1991).

4. Concrete horizontal structural systems (Assal *et al.*, 1991).

5. The basic structure of a generic building (Eastman and Siabiris, 1992).

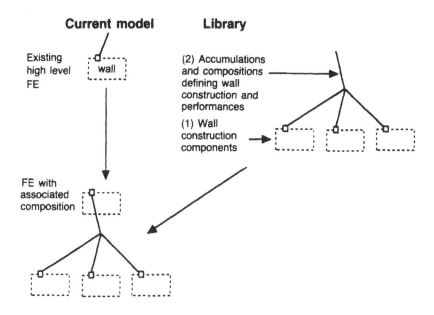

Figure 11.3 Modular extensibility, adding a new technology as a composition of an existing FE

In addition, design domains such as kitchens, parking, camping tents and Renaissance gardens have been also represented using EDM, in graduate courses in building modelling offered at UCLA.

11.5 MODULAR INTEGRATION OF CAD KNOWLEDGE

We have begun to explore the issues of modularity and extensibility (Eastman, 1992b). Taking areas of building construction that include both a variety of construction technologies and an open-ended set of performances (acoustics transmission, thermal transmission, lighting, structure, vibration), these initial studies have shown, in information modelling terms, what is required to interface a new technology into a knowledge-based design package. We have also shown in information modelling terms, what is involved in interfacing a new performance analysis into existing CAD package.

We assume that libraries exist of construction technologies and analysis packages that have EDM-based interfaces. To summarize our analysis, a new fabrication technology may be inserted into an existing design structure as a composition of some higher level FE, i.e. object. The high level FE may be a structural system, or a lower level FE, such as the construction method of a wall (see Figure 11.3). A technology to be added consists of:

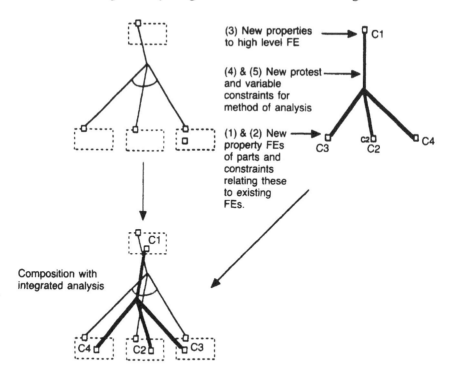

Figure 11.4 The modular extension of a method of analysis to an existing composition. An analysis package may integrate in this manner with multiple compositions

1. The set of FEs that are the part objects of higher level FE, and any other lower level FEs that define necessary detail of the composition elements. (Complex properties, such as shape are also FEs.) These lower level compositions may define lower level technologies within the higher level one. The added FEs will be defined from both new sets of sub-FEs and specializations of others that have already been used, such as geometry.

2. The set of accumulations needed to maintain the integrity relations between different levels of description within the technology. Most of these accumulations will be specific to the technology, such as the structural relations between columns and beams and a structural frame. The accumulations include both the precondition composition rules and also the performance relations between properties.

The integration of a new form of analysis into an existing CAD knowledge base is diagrammed in Figure 11.4. The following linkages are required:

1. New property FEs that are specialized into the contributing FEs, for all the currently used technologies. These FEs carry the properties needed by the analysis that must be available within each contributing FE. Here, contributing FEs are those that contribute to the functionality being evaluated.

2. Constraints that are added into the contributing FEs to maintain the new property FEs' integrity with other properties already defined.

3. New properties of FEs that are added to a higher level composite FE that defines the assumed operating or load conditions under which the analysis is made.

4. An accumulation analysis 'shell' that defines precondition constraints that test whether input conditions are satisfied and that the input data is well formed.

5. Variable constraints that define the integrity conditions satisfied by the output data of the analysis.

This approach has many similarities with the methodology for modular extension of databases developed by Casanova *et al.* (1991). They define a module as consisting of a data schema, operations, constraints and enforcements. The constraints specify the semantics of the operations, i.e. the logical conditions that must be met by the operators. In EDM, the operators are assumed to exist within an external application, with the constraints, as assertions, serving as the operator's representatives. Enforcements are also defined as constraints. In EDM, a modular extension consists of the information model extension and the associated external application, and a data exchange mechanism between them. The application is internal within the extension methodology of Casanova *et al.*.

11.6   SOME CONSIDERATIONS REGARDING SPECIALIZATION

The previous examination identified what specific information was needed to be added to support two common kinds of knowledge extensibility. The examination involved different views involving heterogeneous data and how they may be dynamically added to a building model. It considered the data structures in the large. However, we must also address precisely *how* the information structures are added and how the various constraint relations between them managed. That is, we must address information management in the small.

There has been great interest recently in type extensions and polymorphism (Cardelli and Wegner, 1985), with various forms of extensibility being supported by different languages and database systems. What kinds of extensibility at the object level are needed for these design examples? More specifically, what is the structure of generalization–specialization needed to support design?

We wish to allow views to be added to an object in arbitrary order, while maintaining integrity between the views. We rely on specialization to support such extension. In EDM, like other systems, variables are specialized. However, no distinction is made between classes and instances and any object may be further specialized. Thus variable values are also specialized, resulting in any values assigned to the more general object to propagate to the specialized one(s). The general rule is similar to domain definitions; the value of a specialized variable must be a subset of the values of the more general variable. However, we found it important to treat this integrity condition as a variant constraint, as described earlier. That is, for a proper database, the variable values in any FE must be a subset of values held in the more general FEs from which the variables were inherited. Conflicting assignments may be made, but are managed as an unsatisfied constraint.

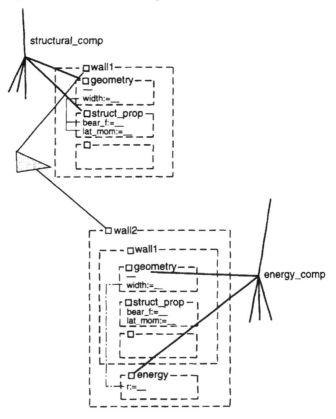

Figure 11.5 One application generates a view of *wall1* and another application generates a view of *wall2*, which is a specialization of *wall1*. Value assignments are assumed to propagate, but may be overwritten

Consider the following example. A concrete wall has been laid out in a commercial building. It is a structural wall, taking up both bearing and lateral forces, and has been engineered with regard to thickness and reinforcing. The wall and the structural accumulation that defines the structural relations are shown in the upper left of Figure 11.5. The loads have been associated with the individual wall, and constraints within the accumulation manage the relationship between a structural FEs loads and the geometric properties that are sufficient to respond to them. The constraints act as surrogates for the analysis program (or the structural engineers calculations) that the loads have been satisfied with the geometric properties defined within the wall. The constraint relation is denoted by the thin solid line.

Initially, this wall had been an interior wall, but later, due to changes, it became an exterior wall and had to respond to thermal requirements. The wall is specialized to accept the thermal properties. That is, all the properties, constraints and values of wall1 are specialized into the new wall2, to which is added the new thermal properties. The new thermal properties are also joined to the thermal accumulation. A desired thermal resistance for the wall is derived by a program for which the thermal accumulation acts as a surrogate. Let us suppose that this program also derives a desired wall thickness, given the selected material, and assigns it to the wall. This new thickness is different from the one that is carried by its generalization. As a result, the specialization of value constraint would flag the inconsistency.

The first point to be noticed is that this mechanism allows applications to be added in arbitrary order. This mechanism allows any number of specializations to be made, each adding a new view of the object. (A cost of this approach is that there is much replication of data. Each view not only adds the new variables it needs, but it replicates all the other views generated before it. We are looking into a less space-expensive way to manage these relations.)

A second point is that a view may simply check values or assign them. Different applications may assign values to the design; the program that assigns values need not be fixed within the building schema. As defined here, the designer can decide whether to use a thicker wall, keep the current thickness but add insulation, or override the thermal program and accept a lower thermal performance for the wall. In general, if different programs (or actors) assign conflicting values to the same design variable, they are all represented, allowing explicit user consideration and trade-off. (Alternatively, an expert system associated with the building model could search for such inconsistencies and apply 'expert' rules to make the trade-offs automatically, such as the program environment developed by Pohls and Meyer (1994).)

This small study shows how specialization, as dealt with by the inheritance schemes of object-oriented systems, can play an important role in the support for dynamic extensibility of a building model. The example here extends the idea that specialization can support version control, proposed by Batory and Kim (1985), to include semantic extensions.

11.7   A SYSTEM PERSPECTIVE OF VIEW TRANSLATION

There have been various efforts directed towards product model integration, in all areas of CAD/CAM. One major class of efforts has focused on the development of backend databases, as described above. In parallel, there have been continuous efforts in developing neutral file translators, such as IGES, as another way to support integration. The goal of neutral file-based translation is to output the data from any CAD application into a neutral file and be able to read in that data to any other application that relies on similar data. While a neutral file relies on batch translation of a data file, a backend database is conceptualized to store in a neutral form all the information from several CAD/CAM applications and to support the applications by providing their data input needs in a more incremental manner. However, both approaches to integration must deal with the common issue of translation between views.

Let us consider the neutral file and database approaches more closely, from the perspective of objects and object types. Both approaches have emphasized a neutral 'canonical' backend representation. That is, each type has a single 'standard' representation with which all applications must communicate. Such a structure is shown in Figure 11.6. Now most applications do not have exactly the same specification for its types as the neutral file. The representation of geometry varies as to the properties they use. An implication of a neutral representation is that all rules regarding how to treat the mismatches of types are incorporated into the code reading from and writing to the neutral file. Even in simple geometry, there are significant differences in representation. Some polylines include arcs and higher degree curves, while others do not. Some curved surfaces are bounded and intersected by exact polynomial curves and others

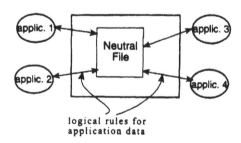

Figure 11.6   Standard neutral file structure

approximated by polylines. In applications, such as energy or acoustic analysis, the geometry of bounded spaces are defined differently and the properties are associated with the geometry in different ways.

In both IGES and STEP, these inconsistencies can only be treated within the code reading to and writing from the neutral model. Thus a translator can be written to serve only a limited purpose, to support a set of application's data needs, as explicitly anticipated. Thus the notion of a neutral file is a chimera; it does not exist.

Efforts in developing backend databases typically follow a better strategy. These databases define a static model, and typically incorporate specified dependencies for view generation, with which various applications communicate. Data derivation via the view dependencies corresponds to the translation process. A graphic depiction of a backend database model is shown in Figure 11.7. Because dependencies are predefined, this scheme provides a specific set of translations, but is poorly suited to handle diverse transaction requirements. To the degree that the rules of translation are embedded into the dependencies (and not the reading and writing operations with the applications), the rules of translation are at least carried explicitly, can be reviewed and are potentially under the control of the user. However, the backend database supports a fixed process of design, based on the built-in dependencies between views. Thus no product databases exist that support design as an iterative and open-ended activity. They all support a limited design process, generally linear.

Translation can be formalized in terms of logical rules defining conversion between types. Parker calls this type of translation transduction (Parker *et al.*, 1992). These rules should allow propagation of values in both directions, if such derivations are known, and to identify inconsistencies, as described in the previous section. Since these rules can be viewed equivalently as (directed) equations or logical constraints, they permit us to put translation on a logical foundation, and can be formulated as the satisfaction of integrity constraints between views.

A portrayal of this transductional approach, as being developed in

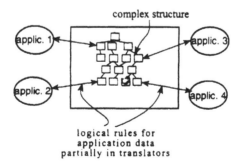

Figure 11.7 Standard database structure

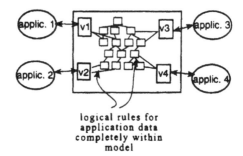

logical rules for
application data
completely within
model

Figure 11.8   Treating translation as a transduction between views

EDM, is shown in Figure 11.8. The object types in one view are related to those in another view or to a central model by constraints, defining the logical mapping between them. All relations are in the backend database and may be accessed explicitly. The rules are used to transduce the data needed in one application from the data generated in other applications. There is no attempt to integrate views or eliminate redundancy. Rather, we wish to express the possibly multiple integrity relations between different views of objects, as declarative representations of translations.

This architecture potentially supports the addition of views in various orders and a design process that is iterative and open to change as design proceeds. It is different from any backend database architecture available today, but one that solves both the issue of dynamic extensibility and the need to maintain integrity across those views, regardless of the sequence in which they are created.

11.8   STATUS

Our earlier work provides both the low level conceptual modelling tool for defining knowledge modules and also addresses the high level interface language for composing new object definitions, structures and relations. It also specifies the needed structures and behaviour for the base system for CAD development. That base system is strongly object-oriented, but with a different flavour of object semantics than exists in any current objectoriented system or database.

We have a prototype implementation of the EDM structure, to explore implementation issues, implemented in Quintus Prolog. We have resolved the naming and unique definition of variables and other internal representation concepts. It supports product model definition, with parsing and semantic checking and storing of structures. It supports translation of data into predefined model structures, for example, from a CAD system or design application. On that data, it allows execution of constraints, checking constraint state and writing formated data from structures to other applications. This direction is consistent with the view of EDM as a back-end information model. Prolog allows quick development and testing of the database concepts. A larger scale implementation is being planned.

We believe that the desirable future will consist of the separate development of a number of stand-alone intelligent design packages, consisting of appropriate objects, operators or methods embedding design rules and evaluation and test applications that assess the standard performances and intentions of the composed objects. We believe that we have developed the conceptual modelling tools for defining such modules.

Second, the modules must be brought together in an environment allowing run-time extension of a product database, especially allowing new object definitions and new specialization of objects into existing ones. Needed to do this effectively is a high level modelling language that will allow privileged users (in the EDM terminology) to compose new object definitions, add new compositions and integrate new accumulations. Of course, the environment must support the required types of extensibility. An initial definition of such a language, called DML has been implemented with our prototype (EDM, 1993).

We believe that it is possible in the future to develop an interactive and graphical means to accomplish both of these tasks, putting these extensions in the hands of users of CAD systems. We envision a graphical information modelling language, similar to EDM's graphical representation, that can be displayed and manipulated directly, allowing equivalent knowledge model manipulations. In such a future environment, a designer wishing to use a new construction technology or new method of analysis would be able to display both the existing information model of the design project and the information model of the technology or analysis to be added. By pointing to the FEs and interactively specializing into them other FEs, the user will be able to integrate properly packaged technologies and analyses in a manner similar in style to the integration of graphical tools using a GUI toolkit.

11.9   SCENARIO FOR CAD USE IN THE FUTURE

EDM is motivated by the following scenario. At some point in the future, intelligent CAD packages for different aspects of architecture and construction will be common. Multiple vendors will have packages for lightweight steel, concrete, wood frame and other types of construction. Other packages will exist for mechanical system design, for exterior cladding, for roofing. These packages will include the parts, material properties, editing and composition operators, with embedded design rules and other kinds of checks. Other packages will be available for different building types: for schools, apartments, laboratories, grocery stores, hotels, etc. They will include workspace design submodules, space layout, circulation and other layout generating and checking routines. Other packages will include analysis packages, those not included in the building technology packages. These will be for lighting, acoustics, energy loads, etc. Such packages will be available in the local construction-oriented software store.

Users will be able to select the kinds and style of knowledge to be used on a project and merge the separate packages into a completely integrated CAD system, using techniques outlined in this chapter. All packages will communicate their interdependencies with other packages. The modules need not be identified and merged all initially, but can be added incrementally as design proceeds. Experts will be able to use the same modelling tools to modify existing knowledge and tailor it to custom use. These capabilities suggest an architecture for CAD systems that can solve the issues of complexity and integration. This scenario of integrated construction information has a technical need for the style of product modelling research described in this chapter.

ACKNOWLEDGEMENTS

The work presented here has been undertaken by the EDM research group at UCLA, including Scott Chase, Hisham Assal, Hing Chan and Nirva Fereshetian. The work is supported by the National Science Foundation, grant no. DDM-8915665.

REFERENCES

Augenbroe, G. (1991) *Integrated Building Performance Evaluation in the Early Design Stages*. Proceedings of 1st International Symposium on Building Systems Automation-Integration, University of Wisconsin.

Assal, H., Cho C. and Eastman, C.M. (1991) An EDM Model of Concrete Structures. *Design and Computation Research Report*. Graduate School of Architecture and Urban Planning, University of California, Los Angeles, CA 90024.

Batory, D.S. and Kim, Won (1985) Modelling concepts for VLSI CAD Objects. *ACM Transactions of Database Systems*. 10(3), 322–46.

Björk, B.-C. (1992) A conceptual model of spaces, space boundaries and enclosing structures. *Automation and Construction*, 1(3), 193–214.

Bond, A.H., Eastman, C.M. and Chase, S.C. (1992) Theoretical Foundations of EMD EDM Product Models. *Design and Computation Research Report*. Graduate School of Architecture and Urban Planning, University of California, Los Angeles, CA 90024.

Brachman, R. and Levesque, H.; (eds) (1985) *Readings in Knowledge Representation*, Morgan Kaufmann, Los Altos.

Cardelli, L. and Wegner, P. (1985) On understanding types, data abstraction and polymorphism. *ACM Comp. Surv.* 17(1), 471–523.

Casanova, M.A., Furtado, A. and Tucherman, L. (1991) A software tool for database design. *ACM Transactions on Database Systems*, 16(2), 209–34.

Eastman, C.M. (1991) Use of data modeling in the conceptual structuring of design problems, in *CAAD Futures '91*, Schmitt, G. (ed.), pp. 207–224.

Eastman, C.M., Bond, A.H. and Chase, S.C. (1991a) A data model for design databases, in *Artificial Intelligence in Design '91*, Gero, J.S. (ed.), Butterworth-Heinemann, Oxford, pp. 339–65.

Eastman, C.M., Bond, A.H. and Chase, S.C. (1991b) Application and evaluation of an engineering data model. *Research in Engineering Design*, 2(4), 185–207.

Eastman, C.M. (1992a) Modeling of buildings, evolution and concepts. *Automation in Construction*, 1(2), 99–110.

Eastman, C.M. (1992b) A data model analysis of modularity and extensibility in building databases. *Environment and Building*, 27(2), 135–48.

Eastman, C.M. and Siabiris, A. (1992) The incorporation of building type information in the conceptual modeling of buildings. *Design and Computation Working Paper*. Graduate School of Architecture and Urban Planning, University of California, Los Angeles, CA 90024.

Eastman, C.M., Chase S.C. and Assal, H. (1993) System architecture for computer integration of design and construction knowledge. *Automation in Construction*, 2(2), 95–108.

EDM Group (1993) DML User Guide (Design Modeling Language). *Design and Computation Research Report*. Graduate School of Architecture and Urban Planning, University of California, Los Angeles, CA 90024.

Fereshetian, N. and Eastman, C.M. (1991) *A Comparison of Information Models for Product Design*. CIB Workshop on The Integrated Building Future, Eindhoven, Netherlands.

Gielingh, W. (1988) General AEC Reference Model (GARM). *ISO TC184/SC4* Document 3.2.2.1 (Draft), October, TNO-IBBC.

Meyer, B. (1988) *Object-oriented Software Construction*, Prentice-Hall, New York.

Parker, D.S., Simon, E. and Valduriez, P. (1992) *SVP – A Model Capturing Sets, Streams and Parallelism*, Proceedings of VLDB Conference, Vancouver, B.C., Canada.

Pohls, J. and Meyer, L. (1994) Distributed Cooperative Model for Architectural Design, this volume.

Smith, J.M. and Smith, D.C. (1977a) Database abstractions: aggregation, *Communications of ACM*, 20(6), 405–13.

Smith, J.M. and Smith, D.C. (1977b) Database abstractions: aggregation and generalization, *ACM Transactions on Database Systems*, 2(2), 105–33.

Stiny, G. (1975) *Pictorial and Formal Aspects of Shape and Shape Grammars*, Birkhauser, Basal, Switzerland.

Yamazaki, Y. (1992) Integrated design and construction planning system for computer integrated construction, *Automation in Construction*, 1(1), 21–6.

CHAPTER 12

Design systems in a computer integrated manufacturing (CIM) context

Godfried Augenbroe, Technical University of Delft, Faculty of Civil Engineering, PO Box 5048, 2600 GA Delft, The Netherlands

12.1 INTRODUCTION

Information technology has been identified as a key enabler of integration of data and processes in industry, leading to efficient increases in project and enterprise management. This is taken to be true for the construction industry as well, although this industry with its highly fragmented nature poses unique demands on the way this integration can be accomplished (Dupagne, 1991). Each one-of-a-kind large scale project starts with the formation of a new 'volatile' project-based consortium structure. The inter-enterprise project management need has to be taken into account along with the intra-enterprise management of (many different) other projects.

Moreover, many different disciplines and skills involved with designing and constructing a building interact either within or across company borders, using disciplinary and traditionally separated computerized tools. Heterogeneity of tools is thus a precondition for the integration approach. This is typical of the CIM approach where different modules (e.g. used by the design, manufacturing and shipping departments) are 'glued' together in an infrastructure which handles the flow of data between tools and lets the different users ('actors') efficiently engage in a collaborative process (Storer, 1992; Gielingh, 1993; Gielingh and Suhm, 1993).

Many ongoing R&D projects take the top–down view that devising and developing such an infrastructure for an existing heterogeneous tool set (CIM modules) is the key priority in CIM. Moreover, to increase efficiency, reuse infrastructures and make CIM modules interchangeable, standardization has received considerable attention. At the present time, the ISO-STEP standard, coming into effect soon, offers (ISO, 1990):

Integrated Construction Information. Edited by Peter Brandon and Martin Betts. Published in 1995 by E & FN Spon, 2–6 Boundary Row, London SE1 8HN.
ISBN: 0 419 20370 2

1. A methodology for establishing integration.

2. Techniques and tools to effectively support data integration (e.g. EXPRESS as a schema language, STEP neutral file format for exchange of instances, SDAI as a schema-based neutral data access specification).

3. Neutral descriptions (data models) of product data (we will refer to them as STEP models). Most models to date concern general resources (e.g. geometry), but branch and product type specific models will be added in the future. Branch-specific models in a certain application domain are called application protocols (AP); they reuse resource models and other APs.

The mainstream of results from these R&D projects and related standardization will be denoted by PDT (product data technology).

Some critical observations can be made:

1. The focal point of PDT is to cover a broad area of industrial activities. This broadness of span has led to a number of choices with respect to granularity, which keep them far removed from use in actual implementations of new systems.

2. The focus is on data integration; no effective support for process integration is delivered as yet. As a result, actual implementations add 'proprietary' process control, in fact negating some of the benefits of open standardized environments.

3. One takes a coarse-grain view in the data modelling, where the main preoccupation seems to be with finding generic concepts in product descriptions, taking the view that CIM modules exchange product states, that should be derivable from common complete product descriptions (product model). Delegating responsibilities for certain states to certain CIM modules with feedback relations is regarded to be a part of process integration. The difficulty in untangling these aspects from the product state descriptions is a point of concern, as it contradicts the usefulness of product models as such.

4. There is some doubt on the AP approach for subdividing the universe into manageable pieces with well-defined interfaces.

5. There is a lack of a methodology for adding semantics, when one wants to refine the product state description to finer levels of detail. This is clearly at stake if we want to move towards increased levels of integration and thus towards increased homogeneity of the system. In that case, the fine grain product description effectively grows into a database design that would supply increased levels of support to all tools (not just static inter-tool communication).

6. There is a lack of semantic richness in the tools of the trade. This applies mainly to NIAM, which is now generally used for coarse-grain modelling in the early conceptual stage. NIAM has gained a lot of popularity because of its ease of use and proven usefulness of product structure descriptions. In essence NIAM supports modelling of invariant relations with a predefined meaning, external to the model, of the 'fact' which is expressed by a relation. NIAM is poor in capturing change-update semantics, inconsistent (design) states and real meaning of relations. There is a clear need for all of these levels of increasing detail, especially where the support of (atomic) design functions within a design tool (Eastman *et al.*, 1993).

Contemplations of a more philosophical nature lead to some additional observations:

1. STEP-standardized PDT models are widely regarded as 'frozen' and static, which might turn them obsolete rather quickly in a rapidly evolving technology ('Frozen into obsolescence'). Obviously, future updates of the standard might guarantee upward compatibility, but there is some doubt in the PDT community on whether this will be possible.

2. Granularity is a key aspect when it comes to the acceptability of frozen technology. Coarse-grain models might find easy consensus and be less susceptible to future change. In the present STEP parts some models show a high level of detail (e.g. in the CONSTRAINT expressions), impinging on the level of detail inside CIM modules (which are not and need not be standardized).

3. Whereas the trend is 'away' from data and towards distributed techniques (i.e. not storing any data nor standardizing it but regenerating the data whenever needed), the PDT's preoccupation with standardizing (persistent) data is suspect. The new trend, strongly motivated by OOP (Object Linking and Object Management) techniques might indeed move us towards a completely different integration paradigm. After all, some of the preoccupation with data stems from the days that computing power was insufficient (Syvertsen, 1992). Of course data models will keep playing a central role in the development (certainly on the fine grain level) but the integration will be function-oriented rather than data-oriented.

4. The tendency to 'integrate everything' is a recurring phenomenon that seems to run through a 10 year cycle, with a new promising solution at the brink of each decade (ISES (1960s), GENESYS (1970s), IGES (1980S), STEP (1990s), etc.) and diminishing interest in the years to follow.

So, what can PDT bring to the development of building design systems? This was, in fact, the question posed at the beginning of the COM-

BINE project. We will introduce the project and our answers to these questions in Section 12.2.

12.2 USE OF PDT IN THE COMBINE PROJECT

The results and deliverables of the first phase of the EC-funded project COMBINE, carried out by 14 partners from eight countries are presented. The first phase of the project, which ended in the Fall of 1992, was a first step towards the development of intelligent integrated building design systems (IIBDS) through which the energy, services, functional and other performance characteristics of a planned building can be analysed (Augenbroe, 1992, 1993). The workplan for the second phase (1993–1995), which is in progress, will be introduced in Section 12.4.

The research in the first stage of the COMBINE project has concentrated on data integration of a number of separate actors. We see 'actors' as anybody or anything (a department) acting as user of the system. In the context of this discussion, we will also refer to actors when we mean CIM modules (or specifically, design tools or just applications, if you like). At this stage the set of actors is still very limited; their selection has been dominated by the need to address energy and HVAC performance aspects in the early design stages. The resulting data exchange system consists of a suite of design tool prototypes (DTP) communicating through an integrated data model (IDM) in a standard format.

The resulting data exchange system serves to test and validate the conceptual approach for interfacing a variety of design tools that would use a common conceptual model for data exchange and an interface through which a design tool can be invoked and its results interpreted. This approach thus follows the coarse-grain product modelling route, where the emphasis is on data exchange capabilities. This will be a basic capability in any future building design system, that would provide adequate intelligent design process support to a set 'design actors', engaged in cooperative design.

The focus of the COMBINE project is on increased awareness of and control over energy-related aspects across all partners (actors) in the design, i.e. enabling multi-criterion design through integration of a range of specific disciplinary energy and HVAC-related tools, and bringing them to the disposal of the design team. In the process, each specialized building performance evaluation (BPE) tool will be embedded in an intelligent design tool, allowing ease of use without requiring specific skills.

By this approach several of the reasons behind the lacking absorption of BPE tools in design practice are targeted to be removed. We are thus integrating the 'CIM modules of the design office' consisting of drawing tools, engineering design modules and evaluation packages. Understandably, our approach is thus very CIM-oriented, although there is some dif-

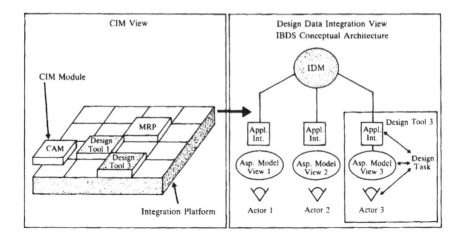

Figure 12.1 COMBINE conceptual architecture

ference in grain size and some additional requirements that relate particularly to design functions. As far as our choice of CIM modules is concerned, COMBINE's first stage has been looking at 'horizontal' integration among actors, within a very limited 'life cycle window'. Figure 12.1 depicts the conceptual architecture of COMBINE's first phase deliverable.

12.2.1 Project definition

The COMBINE approach is based on the following observations:

1. There is a great variety of design support tools since these tools are 'tuned' to a specific design domain or goal. These tools usually perform evaluations (for example, by calling specialized simulation programs) to support design decisions.

2. In no way do we want to imply that future design systems will do 'automatic' design. On the contrary, the designer will retain control over the creative process, with the IIBDS providing the information necessary to make decisions.

3. The notion of a single person, a 'superdesigner', at the controls of the system is by no means implied, nor is it realistic; an IIBDS would normally be used by several team members, each with individual expertise.

4. We must acknowledge the fact that presently available design and simulation tools cannot easily be integrated into IIBDSs.

Translating these observations to a feasible phased approach for arriving at future IIBDS, COMBINE does not target closed-system complete-design

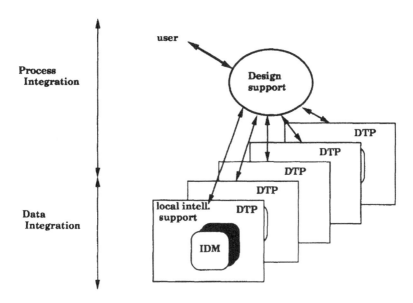

Figure 12.2 Data and process integration levels

solutions. Rather it chooses to emphasize an open loosely coupled set of tools.

In the above approach, the primary emphasis is on the complete description of the design object in order to support all imaginable communication requirements among design actors. No design process model needs to be assumed (at least, not in principle), hence no restricted set of interactions are presupposed. The philosophy behind this approach is obviously less design-oriented in that it targets merely a data exchange tool for actors participating in a design project.

In considering the R&D that is required for IIBDSs, it is useful to distinguish two different areas of integration, reflecting the two approaches described above (Figure 12.2):

1. Data integration: R&D in this area will lead to a standard for describing design objects and methods for making object descriptions available through a neutral format to different design domains, and within each design domain, to different design aspects. This is the main target of the COMBINE approach, based on a variety of actor views, but providing as yet little support for interactions other than data exchange.

2. Process integration: this involves definition of the design context for any aspect-related task, such as performance evaluation. It also involves handling the flow of information and decisions between these tasks, between design domains and between designers.

The first phase of COMBINE was defined such that process integration was outside its scope. One of the deliverables was to show how the data exchange facilities could support future process integration.

This issue of whether we are integrating data or processes is a key concern to integrated construction information generally. It indicates two separate stages that we may have to go through. These two stages of integration also mirror two points on the 'whole' dimension of the research map presented in Chapter 10. Process integration deals primarily with the design meaning, purpose and context of the exchanged data, and the way design functions are called upon and controlled through a dialogue between IIBDS and designer. The design process layer should ultimately help the designer team in conflict-resolving and negotiation, thus steering and supervising the design process as a whole.

12.2.2   Project structure

The project consisted of four main tasks:

1. IDM task: had to define, specify and supply the conceptual building model to the DTP tasks. This task constituted the integrative backbone of the project, its interaction with the other tasks was very intensive.

2. BPE: had to supervise and document a uniform description of the set of BPE tools involved in the DTP tasks.

3. DEMO: had to structure the involvement of design practitioners and subject the demo-prototype to a test.

4. DTP tasks were essentially stand-alone tasks, aimed at the development of the intended design tool prototypes.

12.2.3   COMBINE results

COMBINE has resulted in an off-line data exchange system (DES) for a limited number of design tools. The DES prototype consists of a set of DTPs, logically shared around the database implementation of common conceptual data model (Figure 12.3). The application interface executes the mapping between the IDM and the aspect model of the design tool. The six DTPs that are developed address the following tasks:

1. DTP 1: Construction design of external building elements

2. DTP 2: HVAC design

3. DTP 3: Dimensioning and functional organization of inner spaces

4. DTP 4: Thermal simulation tool in the late design stage

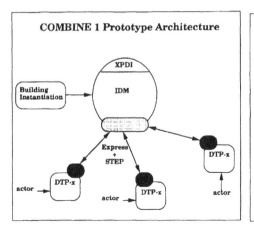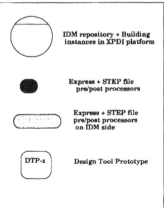

Figure 12.3 COMBINE prototype (first phase)

5. DTP 5: LT method in the early design stage

6. DTP 6: Radiator network design

No attempt is made to include an architectural design actor. This is one of the targets of the second phase.

The data exchange is realized through the following IT tools:

1. A conceptual building model (IDM) that integrates all DTP actor views.

2. Implementation of the conceptual IDM in a powerful software platform.

3. Data exchange facilities through which the DTPs can communicate with the IDM (STEP-based interface kit).

The actual communication is accomplished through ASCII file exchange. A resulting demo-prototype was able to show the operational level of the data exchange. Basically, the results of the project concern general facilities for data integration of a set of tools in early building design.

It should be clear that the present COMBINE deliverables are exclusively on data-integration level, thus inherently very limited in their potential for providing dynamic design support.

There are a number of restrictions when it comes to the design support functions of the present deliverables:

1. No attempt has been made to clearly define a project setting, e.g. participating actors involved in a specific design task. In other words, the resulting tools were not meant to cover a specific design scenario.

2. As a result of this, the resulting suite of tools/actors represents a more or less random selection in a moreover unspecified design process and project environment.

In terms of 'semantic coverage' of the IDM, there are strict limitations as well:

1. Only the external data views of all actors are taken into account and integrated in one common central model ('actors' are the grain size).

2. Extended levels of coverage would have to deal with purpose, intent and other operational issues, associated with the data. Only at these levels would one also be able to provide support and inter-actor control of the dynamic migration and enrichment of data as the design progresses. Support of these levels would obviously, apart from a very rich building model, require data sharing rather than just file exchange.

3. Ultimately, we target full concurrency support where conflict detection and negotiation support are adequately supplied.

12.3   DEVELOPMENT ASPECTS

The results of the first phase of the COMBINE project warrant some reflections on the overall development process. Some lessons learned will be reported.

12.3.1   Integration/granularity levels

Figure 12.4 shows the hierarchy of an IBDS and the components it consists of, down to the level of 'atomic' components (which might be objects in an OOP, subroutines in a traditional programming language or 'events' in an event-oriented programming language or predicates in PROLOG), which represent the basic knowledge components that together will determine the behaviour of the system. PDT is very much concerned with the communication between components (CIM modules) in an IBDS and between different IBDSs, i.e. at the top of the hierarchy. An IBDS itself will be seen as a CIM module in a system on a broader scale, e.g. dealing with design, planning and construction.

Targeting certain levels of the hierarchy as neutral and open external integration levels would necessitate that we, first of all, do an extensive integration task on the input and output data of all modules on that level. This approach will benefit from removing as much as possible of the internal tool semantics from the development of the common generic product description that serves as an integration base on that level. Moving upwards in the hierarchy would thus reduce the extent and semantics in

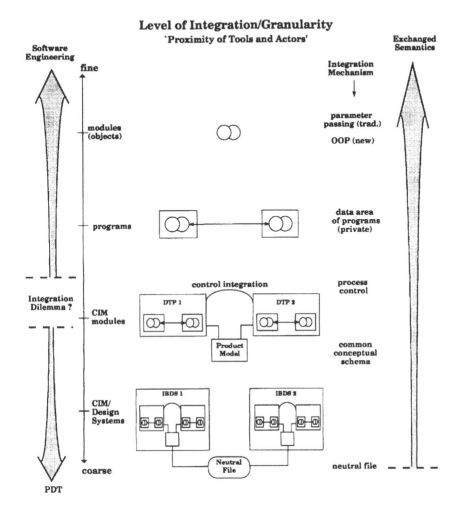

Level of Integration/Granularity
`Proximity of Tools and Actors'

Figure 12.4 Hierarchy of an IBDS

the product model at that level. Naturally, as a consequence, the communication gets more global and less meaningful if we thus move up to coarser grain levels.

12.3.2 Control integration

In doing so, we are not only reducing the level of behaviour we can exchange, but also making implicit assumptions about the product states that are exchanged (i.e. a coarse-grain process model). Figure 12.5 shows an integrated product model of three CIM modules. Imagine that modules '2' and '3' are clients of the data supplied by '1'. We could think of '1' as an

Change Management
of Product Models

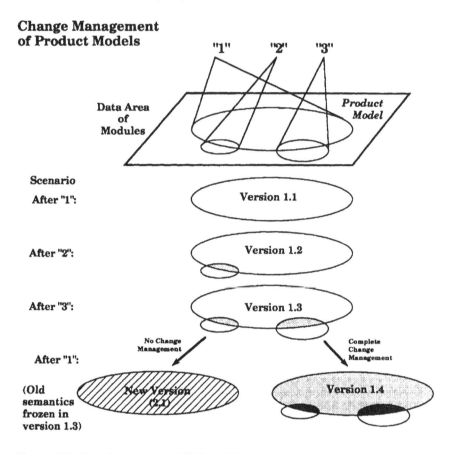

Figure 12.5 Data integration of CIM modules

architectural design tool and '2' and '3' as a duct layout planner and a solar irradiation assessment tool. If we think of the scenario with the tool sequence '1', '2', '3', the updates in steps 1 and 2 are fairly easy to do (we might even assume that the data modelling effort has taken this into account from the very start). Basically, '2' and '3' merely add data to an existing initial state, thus completing remaining parts of the product structure. One can imagine an off-line approach with ASCII STEP file exchange and a one-step update process of the database fulfilling all requirements.

A more realistic scenario with feedback and iterative design refinement would probably show the sequence '1', '2', '3', '1', etc. In that case, the update operations must deal extensively with guarding integrity and consistency of the data during such a design refinement.

Very conservative systems would regard each new design session of '1' as ending in a new state held in a new version in the database. Some version management facilities would let the designer inspect older versions, e.g. to guide him towards design improvements. At the end of the session, a new version will be committed to the database (with loss of all associations with older data supplied by '2' and '3', although possibly still valid). Obviously this would be regarded as a very poor system.

Less conservative control would require that either:

1. All change management is incorporated in '1' giving it access and supervision of the complete model. This is, of course, not attractive, as it would negate all the benefits from the modularization and reuse of modules that is the main reason for the existence of PDT in the first place.

2. All change management is handled in the central data model. We assume that this will be virtually impossible to accomplish in an off-line file exchange mode. This implies that an online incremental update mode of communication between '1' and the implemented data model must be realized. ISO-STEP provides specifications for schema-based online access (ISO, 1992).

In a data-driven context, the latter choice is the obvious one, but severe problems are encountered in realizing this option:

1. The semantics involved have to be assembled through bottom–up integration of the fine-grain semantics of all tools (in our example, case '1', '2', and '3').

2. Present main stream PDT data models (NIAM) lack the expressive power to do so and a gradual refinement of existing models to capture more semantics would require a richer data modelling technique as part of these front-end diagrammatic CASE tools.

3. The bottom–up detailed approach is close to traditional software development and lends itself to CASE tools and rapid prototyping, but as yet a good development strategy and supporting tools are hard to find. Especially, the transition from early conceptual coarse models (front-end CASE) to the refinement of implementation designs (backend CASE tools) is a bottle-neck.

4. It is not altogether clear whether the PDT integration paradigm has additional benefits over emerging OO integration on fine-grain level (object linking and embedding with closely linked data and control integration).

12.3.3 Data and control integration

The above issues deal with giving data more meaning in a design context, which was demonstrated by a scenario where a design is evaluated and consequently refined. Basically, this deals with the intrinsic meaning of data as it is generated and manipulated during the design. It can thus, in principle, be regarded as a refinement of existing PDT models. Control integration can be viewed as an additional (orthogonal) expansion, concerned with the temporal, dynamic semantics involved with the behaviour of actors and their roles in the collaborative group process that design is.

Regarding control integration as orthogonal to data integration suggests that control logic and its implementation can be added to the system in an additional layer, as indeed is suggested by Figure 12.2. Future experiences with conceptual modelling tools with a good gradual interface to CASE tools have yet to corroborate this. The separation (conceptually) would allow us to add control on top of a data-integrated system in an adaptable way. It is expected, however, that ultimate control challenges posed by concurrent design cannot be tackled in this way (Sriram *et al.*, 1992).

The next section will deal with the present (second) phase of the COMBINE project, which targets an actual IBDS prototype in a selected 'project window'. Choices with respect to the above issues will be explained.

12.4   COMBINE: PRESENT PHASE

The second phase (November 1992–June 1995) of COMBINE builds upon the above deliverables by combining them into an operational integrated building design system (IBDS), according to functional specifications resulting from a particular building project–partner setting. This will comprise extension of the suite of interacting design tools to the areas of costing, building regulations and product catalogues. Additionally, an off-the-shelf architectural CAD tool will be incorporated. The central building model IDM will be extended and enriched accordingly.

A robust ODB implementation of the IDM is targeted to be the central DES of the IBDS. Its exchange capabilities will be upgraded from off-line STEP file exchange to shared data access support. Whereas extended intelligent design support features and full coverage of concurrency issues is not within the scope of the project, the resulting IBDS prototype will be primarily configured to support robust multiactor data exchange in specific engineering design office settings. Some added features of future IIBDS will be explored in a demo-prototype, e.g. an intelligent project supervisor will be developed using a blackboard approach. Important feedback from practice will be gathered by doing extensive field testing.

The contribution to R&D and building practice will consist of:

1.   One of the first actual multiactor systems in building design, thereby situating itself in the translation range between product model R&D and practice. This is a necessary step to prepare building practice for future systems. In fact, a recent ESPRIT study identified this as a key priority for the building industry, which is generally slow in absorbing

new IT. The project will emphasize the importance by taking a support office role on board.

2. It will make the first thorough specification of the type of IBDSs, that can be absorbed in practice.

3. Exploration of necessary features of future IIBDSs through field testing.

4. A contribution to the overall management of operational systems, enabled by embedding them in modelling environments allowing full dynamic enterprise integration in the future.

In the JOULE context, the project will provide one of the first precompetitive tools of a new generation of tools for the energy and HVAC consultancy practice.

As 'vertical' extension in the level of IT support is significant with respect to the type of concurrency of interaction in horizontal planes (i.e. between actors), and realizing that the present COMBINE result is 'flat' with respect to concurrent engineering support, it is helpful to distinguish the different levels of concurrency support that could be targeted in this and the next stages. The available level in COMBINE1 can be denoted 'off-line data exchange', realized by file exchanges. Most STEP-linked approaches are on this level.

The data integration is usually built on the premise that the exchanging actors act only on predefined states of the product specification. The semantic coverage in the conceptual model is mostly directed towards (rigid) product description, with little or no design support semantics. Subsequent levels of increased functionality address:

1. 'On-line data exchange': adding some (minor) level of design support through active sharing of data and behaviour.

2. 'Design support': adding extensive design semantics in the common conceptual model, thus providing design support to a range of design operations (on a 'design per view' basis).

3. 'Concurrent design support': adding explicit support for simultaneously operating actors, i.e. explicit interaction support between multiple actor views.

It is important to note that each additional level requires a leap in the semantics of the information model required to support the operational system. The targeted level in COMBINE2 will be on-line data access to the IDM in a ODB data-server environment, paired with standard off-line exchange for those design tools that do not require 'direct' interaction. Adequate data exchange tools will be provided, following the STEP Data Access Interface Specification (DAIS); suitable mapping mechanisms and control mechanisms will be added to support this level of functionality.

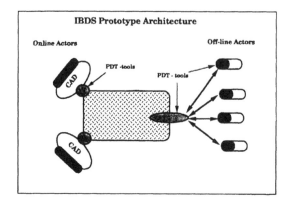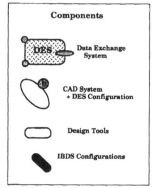

Figure 12.6   COMBINE prototype (second phase)

12.4.1   Project management extensions

No system will find any acceptance in real-life circumstances if smooth integration into a project environment cannot be guaranteed. Special emphasis will be devoted to design tool configuration (e.g. GUI development) towards a desktop engineering and design system adapted to the requirements of the computer-integrated 'Design Office'. Other requirements that have to be dealt with are data interaction management (creation management, security, coherency checking) and support for flexible introduction into project management working procedures.

The global software architecture of DES and IBDS prototype is sketched in Figure 12.6. In view of the available resources, we are targeting at most two CAD-based on-line actors and an approximate number of five off-line actors. The two on-line actors could consist of:

1. A building modeller, based on a CAD system from one of the main CAD vendors and a leading 'geometrical engine' product, to be integrated in the DES. A preselected number of layers are connected on-line to the IDM, other layers are 'free drawing' layers, residing between sessions only in the local native format (no support, no integrity control offered on these layers). The configured CAD system must be equipped with functions to take information from other layers to the assigned IDM layers to start up a project. In fact, this will offer easy integration of the system in existing projects.

2. A HVAC design tool based on another CAD system.

There will be complete integrity control (in the IDM) over the assigned layers in the configured layers of the two on-line actors, but not concurrently. However, between-sessions notifications on affected entities would be posted in configured layers of the other actor.

The following deliverables will result:

1. A robust DES supporting on-line data sharing with an off-the-shelf CAD system and off-line STEP file exchange with a variety of actors.

2. An extended conceptual building model (IDM+).

3. The DES will be built on a ODB implementation of the IDM+.

4. A prototypical IBDS tuned to specific project environment consisting of:
   (a) the DES core;
   (b) configured design tools;
   (c) IBDS configurations such as interaction controls, data management, etc.

5. Experimental facilities on top of the IBDS, exploring future extended design support functionality.

In COMBINE2, there will be a direct input/feedback link to regional application in test beds, supported by industrial end-users.

12.5 EVALUATION

As explained in Section 12.4, the present phase of the COMBINE project is characterized by two complementary views. The first view represents a mainstream PDT approach, very much dominated by the need to integrate the tools currently encountered in the practitioners' design office. External coarse-grain semantics of these (CIM) modules are captured in data models that are consequently integrated into one common conceptual building model (product model). The data view represented in the (NIAM) diagrams is enhanced by IDEF0 activity diagrams and complemented by other (informal) data dictionary forms. This was the leading principle during the development in Phase One. In the present second phase, integration of the off-line actors is treated in the same way. The emphasis is still very much on the conceptual integrated model, which is approached along the lines explained by Nederveen and van Tolman (1992), which is to be regarded as state-of-the-art in product modelling. The model is regarded as an important deliverable and a possible contribution to the international PDT community.

The second view represents software technology where the leading principle is less to find generic data concepts than to find generic objects (data + behaviour). Temporal dynamic aspects are not separated from the data aspect in this type of development. The granularity is much finer from the very start and the whole development tends to be biased by implementation aspects. In view of the challenges of design systems, one sees a strong inclination towards AI techniques. The two more or less opposite views give a good insight into the 'integration dilemma', which reflects that the need for internal cohesion (tight control of behaviour) contradicts the need for decoupling (open interchangeable architecture). Increasing cohesion on one level of granularity would be decreasing decoupling, as seen from the next higher level.

As far as COMBINE's present phase is concerned we make the following observations:

1. PDT product modelling approaches and working styles have claimed a role for themselves on a level which might seem of less relevance to software engineering.

2. Product models are difficult to interface to software engineering; this, in fact, is a general problem in CASE where an important gap between early conceptual modelling towards implementation must be bridged. The present phase of COMBINE struggles in this area. PDT modelling formalisms like NIAM are inadequate to be refined to levels of greater detail due to their lack of semantic richness.

3. Product models are significant R&D outputs capturing part of the semantics on a coarse level. Their re-use in implementation projects should be better supported.

4. Present PDT models are models of certain states of a product. As a communication medium they lack any expression of change management with respect to certain communication scenarios. It is thus left open how and when these process control semantics should be added on the road to real-life implementations.

5. PDT approaches are intrinsically aiming at integration scopes at real-life scale, which qualifies them potentially as relevant practical solutions.

6. Too many implementation projects work on small-scale laboratory problems and moreover 'hide' relevant semantics inside their implementations. PDT's externalization and standardization efforts are valuable alternatives, that could additionally serve as an integration structure across independently developed systems.

The main COMBINE deliverables of the present phase will relate to these issues in the following ways:

1. Definition of 'project-windows' (reduced part of a complete project) in real-life situations with a limited number of modules (mostly existing ones).

2. Data integration:

   (a) coarse-grain product model within the project windows;
   (b) an implementation of the product model in a data infrastructure

supporting off-line exchange in predefined simple scenarios (limited update semantics will have to be added).

3. Control integration: an 'Exchange Executive' module controls the data traffic and performs conservative locking.

4. Additional data and control integration: two 'design-actors' will be linked on-line to the central database. Added functionality to support design functions requested by these actors will be implemented in the OODB. How the link between these fine grain functions and the product model will be maintained is the subject of research.

5. Extensive control integration.

In a separate deliverable, an extended support of system control will be offered in a blackboard architecture, using the approach of Clarke and MacRandal (1991).

12.5.1 Future challenges

The integration dilemma may be resolved by finding the right level where cohesion needs on the one side and decoupling needs on the other can be met satisfactorily. Integration technologies on different levels of granularity tend to be different. It is important that frontend and backend CASE approaches are better interfaced. It will enable software technology and new systems to absorb present PDT models. Product models at present lack the semantic richness required of design data models. New semantic data models (e.g. EDM (Eastman *et al.*, 1993)) will be needed in the future.

Thus far we have not touched upon a third integration dimension: user interface integration. This dimension deals with the broad spectrum of information exchange and management as a group process. In fact, more failures in design occur for human and organizational rather than for technical reasons (Levy *et al.*, 1993). As the earlier chapters by Fischer and Breuer (Chapter 2) and by Powell and Newland (Chapter 5) have argued. An integrated design system should therefore be embedded in a 'co-operative design environment', which must support the integration of many partial designs produced in a group process. N-dim (Levy *et al.*, 1993) is one of the early tools that recognize this organizational context, which is essential if we ever want to put design systems to use on large-scale projects.

REFERENCES

Augenbroe, G. (1993) COMBINE Final Report First Phase. *CEC-DGXII Report* in press.

Augenbroe, G. (1992) Integrated building performance evaluation in the early design stages. *Building and Environment*, Special issue on Integrated Databases and Data Models, **27**(2) 149–61.

Clarke, J.A. and MacRandal, D. (1991) An intelligent front-end for computer aided design. *Artificial Intelligence in Engineering*, **6**, (January).

Dupagne, A. (1991) Computer Integrated Building. *Strategic Final Report*, Esprit II, Exploratory Action No. 5604, CE Commission D.G. XIII, December.

Eastman, C.M., Chase, S.C. and Assal, H. (1993) System architecture for computer integration of design and construction knowledge, in *Automation Based Creative Design: Current issues in Computers & Architecture*, Tzonis, A. and White, I. (eds), Elsevier, in press.

Gielingh, W. (1993) PISA Framework, in *PISA Workshop Report*, Paris, 27–28 January, 1993. ISO (1990) Product Data Representation and Exchange, STEP; *ISO TC 184/SC4*.

Geilingh, W. and Suhm, A.K. (eds) (1993) IMPPAXCT Reference Model, Project 2165, *IMPPACT*, **1**. Springer Verlag.

ISO (1992) ISO-STEP Part 22: SDAI. *ISO-TC184/SC4/WG7*.

Levy, S., Subrahmanian, E., Konda, S. *et al.* (1993) An overview of the n-dim environment. *EDRC-05-65-93*. Carnegie Mellon University.

van Nederveen, G.A. and Tolman, F.P. (1992) Modelling multiple views on buildings. *Automation in Construction* **1**, 215–24.

Sriram, D. *et al.* (1992) DICE: An object-oriented programming environment for cooperative engineering design, in *Artificial Intelligence in Engineering Design*, Tong and Sriram (eds.), Academic Press.

Storer, G. (1992) *Le projet ATLAS*. Les echanges de données informatisées dans la construction. Actes du Seminaire echanges de donnees techniques, October 1992.

Syvertsen, T. (1992) *The Building Industry at a Crossroad. Some Critical Issues at the Threshold of the Information Age*. Symposium on Building Systems Automation-Integration, Dallas, June 10–12, 1992.

CHAPTER 13

Software environments for integrated construction

Patrice Poyet and Anne-Marie Dubois, CSTB, B.P. 209, 06904
Sophia Antipolis Cedex, France

13.1 INTRODUCTION

Objects need to be designed but also constructed or manufactured, tested or commissioned for buildings, delivered to the client, operated by end-users, and finally maintained to extend their life cycle. All along this path and within each phase, a continuous data stream must flow between the various actors (i.e. designers, engineers, manufacturers, etc.) needing the information. Important cost saving can be achieved, not only at the design stage, but during the entire life cycle of the products if the information is modelled in the appropriate manner and moreover survives the particular phases where it has been generated. Modelling in formal ways, representing according to many logical schemas and exchanging information based on open and neutral technologies are key requirements for any software platforms supporting CIME objectives.

For instance, the efficient operation and maintenance of any complex system (e.g. building, aerospace vehicle, etc.) is facilitated or only even possible in some situations if the entire design and construction-related data are available. This optimal situation is not the general case and often large investments are required to derive subsets of the data required in order to operate and maintain the objects (e.g. aerospace industry). In the construction sector, facility management systems are being developed in order to reduce management costs and operating costs. But the real challenge of these environments stems from the fact that these systems should be automatically fed in with the appropriate information in order to avoid re-inputing the data. The observation shows that most of the data useful for management purposes have been created at some time (a significant part during the design phase, but also during construction/manufacturing), but most of them are also lost or not accessible due to the computer systems heterogeneity. Moreover only the relevant part has to be

Integrated Construction Information. Edited by Peter Brandon and Martin Betts. Published in 1995 by E & FN Spon, 2–6 Boundary Row, London SE1 8HN.
ISBN: 0 419 20370 2

extracted from the huge set of data in order to avoid the user to be swamped as Atkin argues in Chapter 18.

As the story starts from design, modifying and improving the design process is a first step on the integration road. Moreover, as design is an extremely complex activity, open software environments supporting the design process by bearing in mind that engineering data need to be modelled, represented and exchanged, also offer a firm ground to address and help tackle the entire scope of the CIME problem. This is a point made by Watson in Chapter 10.

It has been said that design is complex and studying the process shows that it presents at least the following characteristics:

1. Complex and sophisticated. Designing requires the involvement of different disciplines, all having intricate relationships and leading to a wide range of decisions some being non-technical but economical and managerial, for instance. We shall see further evidence of this in Chapter 16 by Platt and Blockley.

2. Co-operative, time-consuming and expensive. Designing is most of the time a co-operative activity that may involve large teams of up to hundreds of individuals over years. Building design is generally for most of the residential or even office buildings of a more limited extent, but large projects can also be impressive with tens of thousands of plans, charts, etc.

3. Design tools consuming. The activity needs to be supported by a vast amount of software tools including technical packages, but also cost control and planning modules, for instance, to monitor the design itself. These tools need to communicate and exchange information, and in the current software context time and money is lost by heterogeneous environments being unable to exchange easily or to exchange at all.

4. Data generative. Designing complex objects generates large amounts of data of various types. These product models should incorporate design information (i.e. components, parts, systems, equipments with their characteristics and representations, etc.), information to validate the design (i.e. test beds, test suites, etc.) and information that documents the design (e.g. engineering documents, approvals, etc.).

5. Dynamic. The process is highly creative even though from place to place the reuse of previous solutions can ease and speed up the design activity. In fact, design is a mixture of stereotyped solutions which have proven to be suitable for particular requirements and innovative answers to unforeseen situations requiring new solutions from

scratch. This leads to an extremely dynamic context where reuse and creation of information is intertwined in a complex manner.

6. Iterative. Design solutions are often met by the exploration of a search path where trials and errors and satisfactory compromises can lead to acceptable designs (that may be quite different from optimal). Iterations are local to specific design phases or even to particular software programs, but often span across disciplines and actors, also contributing to the dynamic aspect mentioned in Point 5. In practice, the costs of the iterative process are such that the process is often shortened. That point provides opportunities for the IT tools to bring significant progress.

7. Concurrent. Often work is done in parallel and teams share access to similar information and should be able to have co-ordinated actions on the project.

8. Co-operative. The consequence of Point 7, leading to the exchange of large quantities of data across teams, and requiring that open product data model interchange facilities be available.

9. Multidisciplinary. The consequence of Points 7 and 8 with the managerial outcomes.

Analysis of the aforementioned characteristics (Points 1–9) shows that complementary approaches are welcome to help solve some of the above difficulties and to help implement software tools bringing real support to such complex activities.

The XPDI approach is to couple two complementary techniques, product modelling for the formal representation and neutral exchange of information and artificial intelligence for the implementation and knowledge representation on computer systems (Poyet, 1993). This leads to defining an AI-based kernel for the basic implementation platform and supplementing it with a universe of systems supporting the product modelling functions. Therefore, the following services are provided:

1. Definition of conceptual data models.

2. Definition of conceptual functional models.

3. Production of implemented logical forms.

4. Management of semantic data models.

5. CAD-based semantic-driven instantiation mechanisms.

6. Numerical tools connections (e.g. building performance evaluation tools).

7. Symbolic tools connections (e.g. decision-making systems).

8. Product databases connections.

9. Access to the Regulations (e.g. building codes).

10. Access to engineering documentation.
    etc.

13.2   OVERVIEW OF THE FUNCTIONS OF A CIC SOFTWARE
ENVIRONMENT

CIME processes are good examples of complex co-operative decision-making activities, involving various skills representing the *aspects* at different periods of time throughout the *life cycle* of the products/projects (van Nederveen and Tolman, 1992). Moreover, the product/project understanding also relies heavily on the actors' cognitive means, themselves strongly influenced and shaped by their intellectual background referred to as *views*.

The actual use of IT in the construction industry relies on the availability of a set of functions able to provide support to the necessary integrative process, giving a company an advantage over its competitors or even improving the competitiveness of the industry as a whole. The real opportunities for IT in the industry increase as the expected benefits and the satisfaction of practice needs increase as Fischer and Breuer demonstrated in Chapter 2.

In that context, an overview of useful CIME functions is listed below.

13.2.1   Modelling

Data modelling techniques are required to express the relationships existing between the entities involved by the aspects, moreover, recording the project alterations throughout the life cycle and serving as a basis for the views (e.g. graphical, topological, logical, numerical, etc.). They should be addressed in two complementary ways: at the conceptual and implementation levels. This follows points made by Eastman in Chapter 11.

The conceptual approach enables the modeller to define, by means of different formalisms, the schema of the data model to be generated and later handled by the engineering tools (Björk, 1992). The implementation is a translation of the conceptual schema into a living version respecting some logical form deemed suitable for manipulation by a computer.

The semantic data models (SMDs) should convey various types of information, including the entities involved by a given project, their content (i.e. the data *stricto sensu*) and their relationships. Thus they represent the work domain modelled by the network of interdependent concepts and carry the semantics, the topological links applying on these entities and addressing the level of spatial dependency, and the geometrical layer storing the set of shapes (2D or 3D) involved in the graphical representation of the concept, if appropriate.

These are the most conspicuous features of the data model, but they need to be supplemented both by the aspect facility, which enables the user with special access modes and browsing facilities of the data network according to their cognitive perception of the project, and by the life-cycle property which labels the project content (i.e. data and structure) according to the relevant project phases. Such advanced view management systems should provide support to the concept of 'perspective' as defined by projects. This is one of the main principles behind the ICON project, described in Chapter 15 by Cooper and Brandon.

SMDs are really the hard core of a truly software-integrated approach and can benefit from other research results. It can provide a means to merge the various isolated conceptual models elaborated within each architectural or technical subdomain, and represents the strong common basis well suited to achieve tools-dependent activities, such as external and graphical displays of the model acting as: (a) one of the side effects of the deep data structure, (b) dedicated computing facilities (i.e. typically what most of the performance evaluator codes do) based on *ad hoc* (re-)interpretation of the entities and of their derived properties and meaning, or (c) technical project management and negotiation functions handling co-operative and inherently conflicting processes.

13.2.2   Visualizing

Graphical displays and visual understanding of construction projects are often an easy common support for human discussion and negotiation. But, for instance, most of the current CAD systems fall short of what is required, as even if they offer powerful drawing capabilities and realistic representations of the future building, they fail to model the semantics and the links between parts of the building or constraints to be fulfilled by the solutions. Moreover, considering the problem-solving capabilities, these tools are frustrating providing only a restricted set of functions and have very little knowledge sharing across the involved disciplines. Therefore, in future, the creation of the data model itself will require advanced semantics-driven CAD systems able to produce the network of objects modelling any project (i.e. the SMD), to generate the topological links applying to the entities and to create the geometrical shapes representing the concepts. Then, as a side on the data structure (SMD), the visualization corresponds to the activation of behaviours associated to the different objects acting as instances of the entities of the model for particular projects.

13.2.3   Reasoning

A common understanding of the data is only a first and limited step on the road to true knowledge integration, and non-procedural models of design activities have also to be devised to handle actors' negotiations operating with common and conflicting resource allocations under ever-changing conditions. In that respect, AI-based techniques provide a sound basis both for data representation purposes (including sophisticated object-based models) and for the reasoning schema offering distributed mechanisms (e.g. blackboard, contract net, etc.), non-monotonic reasoning under evolving and conscious hypotheses, assumption-based labelling of derived properties, etc. Such an environment and these software facilities should enable the actors to gain support throughout all phases of their activities, modelling the work domain thanks to the reachable network of the time-dependent object worlds with their specific views representing as many knowledge indexes as required by the designers' diversity, and providing non-monotonic objective and decision functions for reasoning capabilities. These points are developed further in Chapter 19 by Oxman.

13.2.4   Assessing

The performance assessment relies on numerical tools often called in the construction sector, building performance evaluators (BPEs) able to process the object-oriented and semantically rich data model, helping the user to discriminate between competing solutions according to objective functions (the same as for other industries).

The simulation of systems and of equipment for example, in the construction sector has long been an important activity and can be considered a major task of any design process. For instance, building performance tools are widespread but seldom used as they require a deep understanding both of the techniques and physical models involved and of the parameters needed to operate the underlying numerical codes (Laret and Dubois, 1986). Significant efforts have to be made in order to facilitate access to simulation codes, and user-friendly interfaces play a major role in that respect. Once a design has been devised (given the help of some intelligent computerized assistant) and assessed, thanks to BPEs for example, an incremental process takes place to refine an initial solution until the best compromise can be reached. Finally, the proposal can be transformed in an operational solution thanks to the connection to product databases leading to the recommendation of manufacturers' components.

13.2.5   Browsing documents

Retrieval systems are expected to facilitate the identification of the relevant subset of technical texts for a given project, based on indexing the Regulations, made of unified codes of practice, of calculation rules, of examples of solutions, of technical assessments and standards by fundamental building concepts. Computer-aided conformance checking facilities can also be derived from these legal and technical corpus (de Waard, 1992).

13.3   XPDI, AN EXAMPLE OF CIC SOFTWARE PLATFORM

XPDI appears as a shell generator enabling the developer to select classes of shells before generating instances of these models (Poyet, 1990, 1993; Poyet *et al.*, 1990). For each of the instances of shell produced (these are application generators), the resources required for that particular model are automatically offered to the developer. Some instances of the models have been used to create the subsystems of an integrated environment.

13.3.1   The kernel

The kernel of XPDI is the meta-level where the abstract data types are defined and serve in a reflective way to the definition of the classes of shells to be offered to the user. It also addresses the level of management of the instances of shells to be produced in order to develop later specific applications. It provides a set of utilities ranging from the behaviour of abstract data types to the various interfaces required for handling systems and systems' components.

13.3.2   The models of shells

Six models of shells are available in the current release of the system: the generic model, the design model, the propositional logic model, the node processing model, the documentary model and the simulation model. These models have the following characteristics:

1. Generic is a rather empty core.

2. The design model offers a set of reasoning data structures enabling the developer to create intelligent applications based on the divide-and-conquer paradigm (agenda, strategy, task tree, tasks, rule-bases, first-order logic rules).

3. The propositional model permits one to create shells complying with propositional schemas for their reasoning functions.

4. The node processing model is devoted to the management of hierarchical decisional structures where the nodes are elements of decisional trees.

5. The documentary model of shell proposes a set of data structures suitable for the generation of applications aimed at the management and editing of large sets of documentation.

6. The simulation model is a discrete event schema where the event to be processed is attached to rule-bases according to their model.

13.3.3   The system toolbox

Given the previous models of shells, instances have been generated for the purposes of the realization of an integrated construction and engineering environment.

We can mention COMSET (conceptual modelling software engineering tools) for the development and management of conceptual data models, AMSET (activity modelling software engineering tools) for the development of conceptual functional models, STEPSET (XPDI-oriented STEP Station) for the production of EXPRESS schemas and ISO neutral files, GAMSET and VAMSET (XPDI CAD Station) 2D semantic drafting and 3D semantic display of the SMDs developed with XPDI, DOCSET for the management of the documentation of engineering projects, etc. They represent a system toolbox generated from the XPDI kernel and the associated models of shells. A number of these instances of shells come into play at the system level for the design of a software platform for integrated construction and engineering.

On the other hand, applications have also been developed by the instantiation of models of shells and we can mention QUAKES (design model), ACCESS (node processing model), NOE (propositional model), etc. These represent the application layer finally to be integrated for the future day-to-day working environment of engineering bureaux and architects. They also represent the raw material for our research on integration techniques.

Figure 13.1 illustrates the XPDI Universe of systems. The XPDI kernel (referred to here as the OOL) plus the available models of shells (here called 'reasoning facilities') represent the XPDI core *stricto sensu*. Surrounding this already appear instances of XPDI models of shells used as subsystems for the development of a software platform for integrated engineering and construction. Some of them are dedicated to communication purposes either with the user (CAD station) or with remote actors (STEP station). Others aim at the creation of conceptual data models, at the management of the regulation or of the projects' documentation, at the connection of simulation tools, etc.

13.4   PUTTING XPDI AT WORK ON REAL PROJECTS

13.4.1   COMBINE

The COMBINE (COmputer Models for the Building INdustry in Europe) project was a first step towards the development of intelligent integrated building design systems (IIBDS) through which the energy, services, functional and other performance characteristics of a planned building can be modelled and analysed (Augenbroe, 1992). COMBINE has laid a firm ground leading to a common agreement on a first structure of the build-

The XPDI Station™ and the XPDI Universe

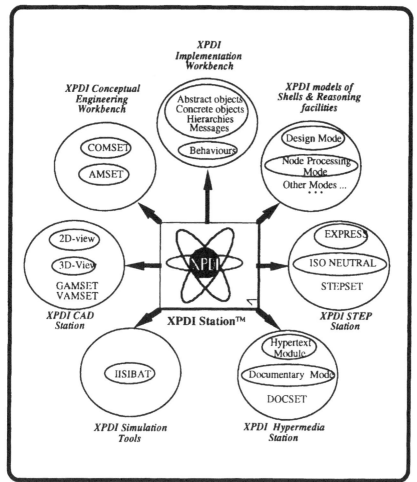

Figure 13.1 The XPDI universe of systems

ing data model, serving as a common medium among different design tool prototypes (DTPs). An overview of the implementation task of the integrated data model of the COMBINE project is given hereafter. This data model is referred to as the integrated data model (IDM), a medium used to support data exchange facilities between the different DTPs involved in the building performance evaluation process of the COMBINE scenario. The IDM can be described as two separate entities:

1. The first one is the so-called conceptual IDM, representing the result of a large modelling exercise (IDM conceptual task), leading to a conceptual description of the entities involved in the project, of their

properties and of their relationships with other entities. This work takes advantage of a modelling language such as NIAM and makes no assumption on the logical form to be used for a later implementation of that model. Nevertheless, management of the coherency of such a large conceptual model showed that without the help of computer-assisted software engineering tools (CASE), it was all but impossible to deliver an error-free IDM even with the careful development methodologies which had been proposed and used for the specification of the IDM (Dubois, 1992).

2. The second one is the implemented IDM, i.e. a living computer representation of the model based on advanced knowledge-based representation techniques (including multiexpertise), object-oriented and frame-based representations, and rule-based and message-based operating functions. The implementation task encompassed two major activities: loading the formal specification delivered as a set of NIAM diagrams (and various complementary information) in order to identify the various inconsistencies that could require modifying the schema, and implementing the specification in a computer environment, i.e. XPDI (Poyet, 1993). Given these objectives, the need for a sophisticated CASE tool, able to handle conceptual representations of the world, able to verify these models against various inconsistencies and to generate various deliverables corresponding to different logical forms of implementations, was obvious (COMSET) (Brisson and Poyet, 1992) (see Figure 13.2).

As far as the implementation task is concerned, given the conceptual specifications and the COMSET software, various logical forms for an operating IDM can be produced. Among the many possibilities offered, we can mention an advanced object-oriented version working in the XPDI environment (see Figure 13.3). This is a relational model which is suitable for creating relational databases (i.e. with SQL create-table order), and the EXPRESS-based model, representing a neutral format for the exchange between the different DTPs of the COMBINE project (including the management of subschemas and the production of the corresponding sub-ISO-neutral files).

However, an integrated design system requires much more than simple data interchange, and real design assistance supposes that complex concepts be represented into the design support system (e.g. current goals to be fulfilled, resources available to reach the objectives, etc.). Therefore, this requires modelling the design status of a building project, the design context of the design tool prototypes and even the process itself. An example of this is shown in Figure 13.4. These objectives should lead to future projects extending the scope of the COMBINE I initiative more towards artificial intelligence (AI) modelling. AI techniques provide support for the representation of sophisticated semantic data models and for the

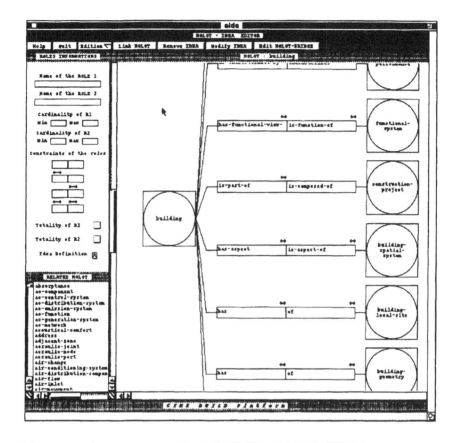

Figure 13.2 A screen open on the COMBINE model (COMSET V1)

modelling of engineering activities (e.g. representation of the actors, of the activities, of competing design solutions, etc.).

13.4.2 ATLES

Large-scale engineering projects for the realization and operation of buildings, process plants and civil works, require the involvement of a variety of specialists. These specialists, representing different areas of expertise, contribute to a project by using and producing heterogeneous sets of information. The complexity of modern construction projects requires increased specialization of experts, but this leads also to a fragmentation of information. Compared with other industrial products, the lead times for design and production in large-scale projects are very short. There is little – if any – time for iteration in the sense of trial and error. Decisions have to be made based on experience and knowledge gained during previous projects. This knowledge is, however, distributed over different

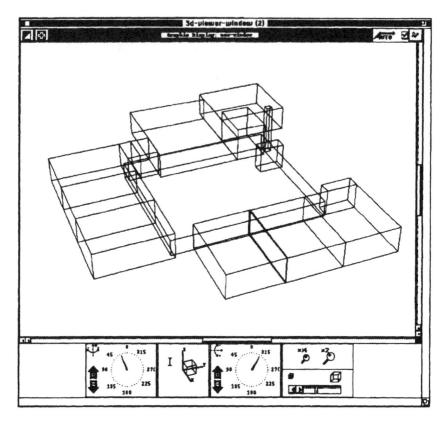

Figure 13.3 The schoolbuilding example of the COMBINE project

specialists with an expertise built up during different previous projects, and is therefore even more fragmented than the information developed during the project itself.

Due to this fragmentation and the diversity of information produced or used by partners collaborating in continuously changing project teams, it is difficult if not impossible to share and exchange information between IT tools. This is a major roadblock for the effective usage of these tools. Another roadblock for effective use of IT is the consequence of advanced integrated systems for the organization of an enterprise. A gradual transition from current paper-based practice to a computer-integrated large-scale engineering (CILSE) environment should be supported.

These problems are tackled by the ATLES project by defining a common data platform, including methods and tools, for the semantic integration of IT tools used by the participants in large-scale engineering projects, in the design, construction and operation stages. This will be supported by prototype implementations and demonstrations, and XPDI is being used at two complementary levels: modelling and implementing the product models by means of AI-oriented techniques.

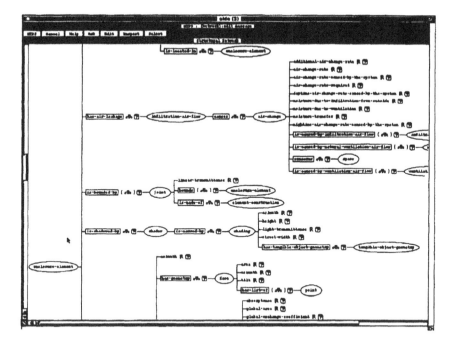

Figure 13.4 Small excerpt of the implemented product model (schoolbuilding)

A CILSE environment should provide construction and other large-scale engineering activities with improved response times to modifications at all stages of the project life cycle, reduction on co-ordination and communication costs, and better documentation handling. It will preserve know-how gained throughout previous projects and facilitate training of inexperienced staff.

In addition to the sort of work done in the COMBINE project, large modelling efforts are also undertaken in the ATLES project to propose a comprehensive functional analysis of the activities taking place during the execution of large-scale engineering projects. The AMSET system offers support for the development of these models and also aims at extending the services provided at the implementation level (see example of Figure 13.5). The functional models are consistent with the IDEF0 modelling methodology, but AMSET targets the incorporation of additional semantics enabling XPDI to deliver AI-based implementation of the functional models (e.g. controls are mapped into daemons, messages or constraints, etc.).

LSE projects presently rely on a large number of dispersed, discrete and non-communicating information channels; some reside on computer systems, but many are substantially paper-based. Examples of paper-based data are bills of material, manual drawings, drawing (document)

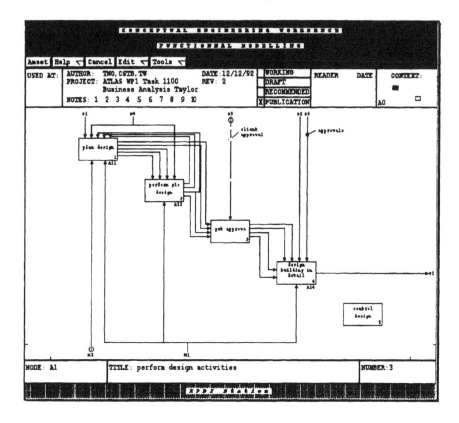

Figure 13.5 An excerpt of the functional model of the ATLES project (AMSET V1)

registries, client brief, maintenance records, bar charts and calculations. Typical computer-based applications are CAD, planning, cost control, payments, structural analysis and member design.

Each of these systems is relatively specific to the needs of one actor and the content highly application-specific. Each has little 'knowledge' about other systems. Although output of data for transfer to other actors and their systems is desirable, the different forms, levels of detail and relationships usually render that impossible or of little practical benefit. Such transfer between systems has even proved difficult in an application apparently as standard as CAD draughting, because of the lack of formal data exchange standards. The current situation is that for a long time, the traditional place of computer applications in the entire construction (building and plants)/engineering sector has been limited to numerical analysis (e.g. structural, thermal or acoustical modelling – engineer work packages), drafting systems (e.g. architectural packages) and managerial environment (e.g. bills of quantity, procurement modules, etc.). In this context, where islands of automation and computerization have emerged, the lack of integrated functions is particularly obvious; it severely hampers the productivity of the sector and the efficiency of the individual actors. Moreover these drawbacks have very negative outcomes on the global quality of the final product, i.e. building and plants especially, producing extra costs for all of the construction/engineering life cycle steps and generating unexpected anomalies.

The exchange of product data models can now readily be supported by STEP techniques. STEP is the standard for the exchange of product model data. STEP is a standardization initiative developed under the auspices of the Technical Committee 184 Subcommittee 4 of the ISO organization (ISO TC184/SC4). STEP reference is ISO 10303 – Industrial Automation Systems – Product Data Representation and Exchange.

STEP provides a framework to describe in a computer-interpretable (and human-readable) format, the conceptual representation of the information associated to a product. This description is made by means of a formal data specification language called EXPRESS, enabling the modeller (or the tools used) to describe the products (the entities) and their properties. It should be made clear that EXPRESS is not a programming language (e.g. no I/Os, no control structures, etc.), but a product modelling language, a language permitting the description of products in a computer-readable format clearly separating the neutral description of the information from the implementation methods used to represent and store the information on computer systems. STEP also provides support for practical file exchange mechanisms, by means of ISO neutral files supporting the exchange of instances of the entities defined by the corresponding EXPRESS files. For further information on the role of STEP-AC see Chapter 10 by Watson.

Various experiments of the STEP technology (even though not mature) made by CSTB demonstrate that it already permits the exchange of engineering data for real projects in an efficient and reliable manner (e.g. the schoolbuilding of COMBINE and ATLES models shown in Figures 13.3–13.5) and the exchange of project data at different stages of the design process (complementary experiences carried out by other R&D centres or industrial bodies should confirm this statement). These experiences have been supported by the STEP station (STEPSET) of the XPDI environment leading to a two-way PDI exchange by means of the mechanisms offered by STEP:

1. Transcription of the product model data structures represented as XPDI abstract objects into EXPRESS (Part 11 Description Methods: The EXPRESS language reference manual), and transcription of the product model data instances represented as XPDI concrete objects into STEP ISO neutral files format (Part 21 Implementation methods: Clear text encoding of the exchange structure).

2. Mapping of EXPRESS constructs into product model data structures managed by the XPDI Station (abstract objects), and mapping of the content of STEP ISO neutral files into product model data instances managed by the XPDI station (concrete objects).

Supporting this two-way exchange, the STEPSET system enables the user of the XPDI station to transfer engineering data in a normalized manner and in full adherence to the STEP framework. The concepts, their semantics and the occurrences of these entities, including their representations (e.g. graphical) can be exchanged across heterogeneous computer systems in a reliable and efficient manner. The Figure 13.6 gives an overview of these exchange mechanisms performed by STEPSET, associating a number to individual functions performed by the system.

13.5   CONCLUSION

The various components of an environment supporting the modelling work to be performed in order to implement the concept of integrated

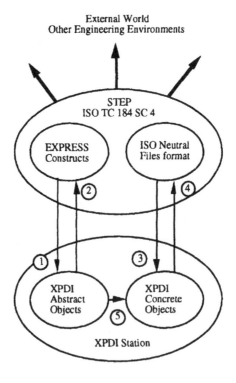

Figure 13.6   Overview of STEP-based exchanges (XPDI STEP station)

construction information have been overviewed. Putting these concept into practice requires much more than the tools presented here. Therefore, CSTB is also involved in the development of a wide range of software systems, all required to demonstrate the effectiveness of the approach. Beyond the implementation of a number of basic software layers and of a comprehensive environment supporting advanced product modelling functions, the Computer Integrated Construction Division at CSTB also develop prototypes of a new generation of CAD systems and of knowledge-based-oriented facilities able to handle and process the product models. Moreover, the management of the engineering documentation and of the regulations is also underway in the wake of national projects.

REFERENCES

Augenbroe, G. (1992), *COMBINE Project: An Overview*. Proceedings of COMBINE Seminar, November, 1992, Fraunhoffer Institute, Stuttgart.

Björk, B-.C. (1992) A conceptual model of spaces, space boundaries and enclosing structures. *Automation in Construction*, 1(3), 193–214.

Brisson, E. and Poyet, P. (1992). Implementation of the Integrated Data Model of the COMBINE Project. Research report by CSTB for the European Commission (DGXII), May 1992, *SIB/DBC/1286*, 79 pp.

de Waard, M. (1992) Computer Aided Conformance Checking. Doctoral Dissertation, Technical University of Delft, Netherlands, 1992.

Dubois, A.M. (1992) *The Construction of an Integrated Data Model for Building Design Information Exchanges*. Proceedings of the COMBINE Seminar, Sponsored by the EC DGXII Directorate, 20 November 1992, Stuttgart, Germany, 12pp.

Laret, L. and Dubois, A.M. (1986) *Knowledge Base and Problem Solving – Model Library Contribution*. Second International Conference on System Simulation in Buildings, Liege, December 1986.

Poyet, P. (1990). Integrated access to information systems. *Applied Artificial Intelligence*, 4(3), 179–238.

Poyet, P. (1991) Artificial intelligence and construction: Research in Europe. *Microcomputers in Civil Engineering*, 6(4), 263–5, Introduction to the Special Issue.

Poyet, P. (1993) XPDI: The eXpert Product Data Interchange Station Documentation. *CSTB ILC/93/1314/PP*, 456 pp.

Poyet, P. and Brisson, E. (1992a) *Des Outils Logiciels pour la Conception Intégrée – Un Sistema Inteligente Integrado del Proceso de Conception en el Sector de la Construccion – CASE Tools for Integrated Design*. Proceedings of Jornadas COTEC Sobre las Nuevas Technologias Y el Sector de la Construccion, Madrid, Spain, 24 March, 1992, 69–99.

Poyet, P. and Brisson, E. (1992b) *COMBINE Project, Implementation of the Integrated Data Model*. Proceedings of the COMBINE Seminar, Sponsored by the EC DGXII Directorate, 18–20 November, Stuttgart, Germany, 15 pp.

Poyet, P., Dubois, A.M. and Delcambre, B. (1990) Artificial intelligence software engineering in building engineering. *Microcomputers in Civil Engineering*, 5(3), 167–205.

van Nederveen, G.A. and Tolman, F.P. (1992) Modelling multiple views on buildings. *Automation in Construction*, 1, 215–24.

Process and information modelling

Part Four of the book is the second to consider some of the latest technologies that are emerging which might potentially result in solutions to some of the integrated construction information problems we posed. This part moves on from the earlier concern with product modelling to the related issues of information and process modelling.

Chapter 14 by Lockley, has some links with product modelling technology but develops into a broader discussion of information modelling issues. It proposes that a key contribution to the integrated construction information problem is the need for an industry server to reconcile differences in syntax and access to shared information.

The second chapter by Cooper and Brandon (Chapter 15), describes the most substantial UK information modelling work under way at present. This is ICON, or integrated information for the construction industry, work based at Salford University. This is funded by the UK government but is jointly directed by the RICS, the RIBA and the CIOB. As an organization structure, this illustrates the interprofessional collaboration and professional deregulation that is likely to occur as a process change to mirror our attainment of integrated construction information. The ICON philosophy is based on the concept of multiple perspectives being possible to suit alternative information needs and application interfaces to be found in construction. The object-oriented information models built within the project can be argued to be a substantial contribution to the integrated construction information problem.

Chapter 16 moves us from information modelling to process modelling. The work by Platt and Blockley describes in some detail how process modelling and the concepts of responsibilities, objectives and roles can be used to model the way information is currently communicated. Their computer-based process modelling applications would contribute further by managing how information would be transferred and maintained in an environment that fully supported integrated construction information.

The final chapter is by Sanvido from Pennsylvania State University in the United States, one of the leading process modellers. Chapter 17 describes his integrated building process model and then uses that description to compare a series of related research initiatives at Pennsylvania State University that fall broadly within the integrated construction information field. Many of these other initiatives in this chapter are taken to the stage of potential practical applications of integration, forming a neat link with the final part of this book.

A building server for a construction industry client

Stephen Lockley University of Newcastle, Newcastle, UK.

14.1 INTRODUCTION

A consequence of the arrival of low cost sophisticated computer-aided design systems as everyday tools in the construction industry, is the increased perception that all the information that needs to be collected about a building can be integrated through the CAD package. This expectation has led to a miscellany of software tools with extensions to allow them to extract information from the CAD tool, perform some evaluation and return the results. However, it is now becoming clear that the results of this activity are far from satisfactory when examined from the perspective of integration of construction information. Each of the software tools uses different conceptual information models resulting in a very low level of overall integration. This has not gone unnoticed and perhaps the most positive outcome is the increased awareness of the need for standards in the area of integrating building information systems.

There is clear need for a mechanism which can not only represent construction information in a standard format, but also integrate both software tools and industry working practice with that information. This mechanism is conceived as a building server for use by a construction industry client. The analogy with the client/server architecture, now an established part of the information technology industry, is not incidental. In the field of computer science, a server is a software tool which provides services to a range of other software clients. A typical example is a generic database engine serving a set of different database applications or clients. Others include graphical user interface servers, such as the XWindows environment, the object linking and embedding facilities of the IBM PC, and the publish and subscribe protocols of the Apple Macintosh. The approach and philosophy of the building server have similar goals to these IT solutions.

Integrated Construction Information. Edited by Peter Brandon and Martin Betts. Published in 1995 by E & FN Spon, 2–6 Boundary Row, London SE1 8HN.
ISBN: 0 419 20370 2

The overall objective of the server is to provide support for the more common information-related activities in the construction industry. It is conceived as an extendible object-oriented model built upon an object-oriented database, providing an application program interface (API) to support a range of client applications.

Adopting the client/server approach has several advantages:

1. It encourages the development of a common conceptual framework for representing construction industry information leading to greater standardization.

2. It leads to reuse of resources. Many clients can share the server functionality.

3. It facilitates quicker and easier development of new client applications.

4. It supports data integration and sharing between different client applications.

The server reflects and incorporates the work of four separate research programmes, the scope of these programmes span the building life cycle and include:

1. Early but post-conceptual building design with an energy focus (DTP 1, The COMBINE project 'Combined building models in Europe', a CEC-funded research programme completed 1992)

2. Quality assurance of the building design ('ITE: A quality assurance infra-structure for the construction industry', a SERC-funded research programme to complete in 1995)

3. Production of contract documentation, specifications, drawings, etc. (collaborative research and development programme with National Building Specification and the Royal Institute of British Architects Services Ltd)

4. Building stock condition and maintenance ('Housing Property Information System', a joint research programme with the City of Newcastle-upon-Tyne, begun in 1990)

Background to the Building Server

The need for a building server was defined during a research programme to investigate the application of expert systems to design evaluation in 1988 (the SERC-funded research programme 'Design and evaluation of cavity walls in framed buildings'). It became clear in this programme that the existing expert system technologies were not scalable, that is, expert systems could be built and validated for small, well understood and bounded knowledge domains. However, the validity of these systems could not be assured as the domain was expanded in scope to meet the real demands of design analysis. A proposed solution was to integrate the information needs of many small expert system tools through a central information manager, a strategy that has since been observed in the 'blackboard architectures' of projects such as the Intelligent Front End proposed by Clarke and MacRandal (1991)

Work initially began on a prototype of the building server, funded under a UK Science Engineering and Research Council programme, 'Interactive Information Systems for the Construction Industry' (completed in 1991). This investigated the information requirements of designers in the industry and how that information could be provided interactively as part of the design process. The conceptual framework was a central building information manager (BIM) holding an object-oriented model, serving a range of data requests from different applications (Figure 14.1). A part of the BIM is now a commercial product dealing with project documentation. This is 'Specification Manager', a software tool to manage contract specifications. It is a joint product from Building Design Software Ltd, National Building Specification Ltd and the Royal Institute of British Architects Services Ltd, released in 1991.

The work so far on the BIM has demonstrated the validity of the object-oriented approach in this area and highlighted some of the problems of constructing complex information structures. These problems include the need for and lack of formal modelling and analysis techniques. Without implementing the information structures in a specific object-oriented

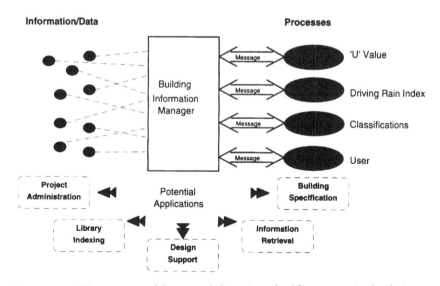

Figure 14.1 The conceptual framework for using a building server in the design process

language, it is difficult to define or explore the information model. In recent years, possible solutions to this problem have emerged; these include modified forms of NIAM (Nijssen and Halpen, 1989), Yourdon Coad, EXPRESSG and Booch. These have been found to greatly improve the ability of the modeller to communicate information structures to the client or end-user, prior to implementation.

The use of formal modelling methods was proposed by Augenbroe in 1992 from the COMBINE (Combined building models in Europe) project funded by the CEC. The NIAM formalism was used to describe the static data requirements of each of the design tools that were to be integrated with the integrated data model, IDEFO (1981) used to describe the dynamic, or process-related activities associated with this data. In terms of communication, these diagrams were generally successful in allowing the different partners to express their data models. However, they were only paper models and no electronically readable form was produced. Therefore it was not possible to automatically check the models' logical consistency and the task of integrating them into one model had to be performed manually. Currently new tools, such as the Configurable Graphical Editor (CGE – a software tool from TNO, Delft), have become available which not only provide a computer-aided drawing tool for these graphical formalisms, but also translate these graphics into EXPRESS schemas (ISO, 1991). This facilitates better model validation and integration. This same strategy is being used in Phase II of the COMBINE project.

The first phase of COMBINE provided the opportunity to produce a demonstrable prototype of the Building Server, called MultiCAD. This was developed collaboratively with the UK Building Research Establishment. MultiCAD is a design support package which includes a set of design tools for evaluating the energy-related aspects of a building during the early design stages (Figure 14.2). Each of the tools obtains the information about the building from the server which has been implemented in an object-oriented language and database. As the designer makes changes to the building in one tool, these changes and their consequences are automatically reflected in the other tools. This allows multiple design evaluations to be carried out concurrently. A full description of MultiCAD is given by Lockley *et al.* (1993). Figure 14.2 gives the overview of the tools in MultiCAD sitting over an aspect model comprising a generic kernel for geometry and topology, and a specialized layer for Building Performance Evaluation in the energy domain.

The building server element of MultiCAD has an EXPRESS/STEP interface which allows the current state of the building model to be either instantiated from, or exported to, an external data model such as the COMBINE Integrated Data Model. This interface was implemented using the STEP/EXPRESS tool kit developed by TNO (1992). Two aspects of particular interest emerged from this work relating to the 'on-line' and 'off-line' exchange of complex information models.

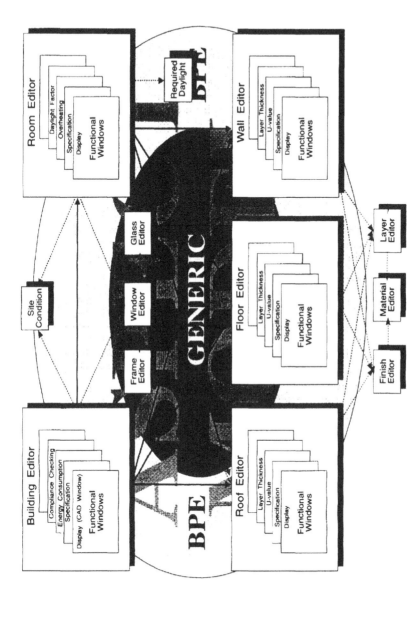

Figure 14.2 MultiCAD package for evaluating energy-related aspects of a building in early design stages

14.1.1 Off-line data exchange

This is the exchange of data between two different applications through a neutral static representation, such as the STEP physical file. In this scenario neither tool has any knowledge of the other, all information for the exchange is contained in the EXPRESS schema and the data file. In order to achieve this each tool must map their internal representation to, and from, an EXPRESS schema. In the case of the building server, this required the complex object-oriented representation in MultiCAD to be translated into an EXPRESS representation and then for this representation to be mapped onto the equally complex, but contextually different, EXPRESS model in the IDM.

In general, it was found that the object-oriented representation could be translated into the EXPRESS language without great loss of information. However, as EXPRESS does not fully support behaviour at present, a great deal of the information had to be translated from behaviour to data. For example, the area of a wall may be held in the OO representation as a formula which can be evaluated during the translation to a number. Clearly though, the reverse process is not possible, a formula cannot be derived from a data value. There is a need to better support behavioural representation in the EXPRESS language and work is currently underway in this area.

The EXPRESS schema resulting from this translation was then mapped to the IDM EXPRESS schema. This is a non-trivial problem and highlights the need for tools to assist in the automated mapping of schemas. Specifically, these tools are needed to maintain the context and coherency of the data model between several mapping transactions. This is exemplified in the problems of garbage creation and collection. Garbage is data that is passed from the IDM to MultiCAD, is deleted in MultiCAD and subsequently not passed back to the IDM. However, the original version still resides in the IDM but is no longer referenced as part of the model. Mapping tools are needed to detect this loss of information and tidy up the model after the translation.

14.1.2 On-line information exchange

On-line data exchange is the ability for several applications to share complex data objects and to alter the state of these objects during their life cycle. Microsoft Windows Object Linking and Embedding and Macintosh Publish and Subscribe are commercial examples of this facility. In essence, complex objects such as a graph or a spreadsheet can be associated with a server which provides their behaviour such as 'draw' or 'edit'. Client applications can embed within themselves these objects and use the facilities they provide. This allows authors of software tools to quickly and easily incorporate sophisticated objects into their programs. MultiCAD is an example of this approach applied to the construction domain. MultiCAD is a Building Server providing support for complex objects such as 'Wall', 'Space', 'Window' together with their behaviour, 'draw', 'volume', 'adjacent to', etc. A range of tools were then produced which utilized these objects to evaluate average daylight factors, annual energy consumption and other building energy-related tasks.

This strategy allowed a designer to define a building using any one of a number of tools and to simultaneously evaluate the performance of that building as the design evolved. It highlighted several problems specifically relating to management of data transactions in an 'on-line' environment. An example of one such problem is the prioritization of changes, e.g. when two tools concurrently transform the data model there is a need to manage these interactions to ensure that the result is still a coherent model. Another example is the appropriate use of 'shallow' and 'deep' copy mechanisms during an editing session. These and similar problems are being explored further under Phase II of COMBINE. Figure 14.3 gives an outline of the data exchange system being produced for COMBINE II.

14.1.3 Summary

The concept of a Building Server for the construction industry is not far from realization. During its development, several issues have been identified that are believed to be generic to the development of information systems in the construction domain. These can be grouped into two areas relating to information representation and information access.

14.2 INFORMATION REPRESENTATION

The process of constructing an information system can be thought of as being analogous to defining a natural language. It requires a vocabulary, a grammar and the protocols necessary for a conversation. The vocabulary is often held in a dictionary and the grammar represented as rules. In our work to date, we have investigated the vocabulary or terminology of information systems and the grammar or relationships which bind this terminology. Future work will examine the protocols of conversation; under COMBINE II the STEP Standard Data Access Interface Specification (SDAIS) is being considered.

14.2.1 Terminology and relationships

Whether we are dealing with object-oriented models, entity relation models or prolog-type tuples, it is necessary to give each information item a meaningful name and then to describe its relationships within a larger system. If we consider first the application of terminology or information

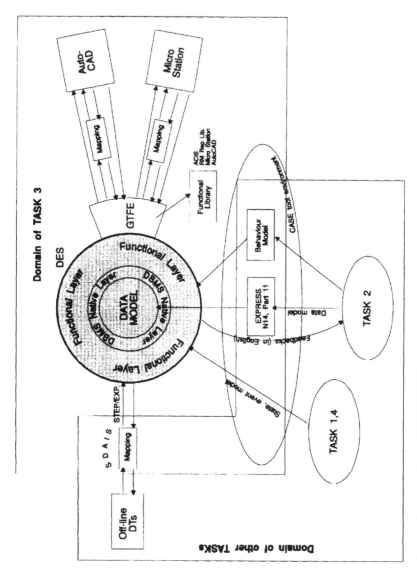

Figure 14.3 The data exchange system of COMBINE II

naming, in the research community there are a range of approaches to this apparently simple process. The approach taken at EDCAAD in the MOLE implementation is to devolve the responsibility of information naming to the user, who then uses the terminology which is most meaningful to themselves or their organization. As the model of the building evolves, so the information items are defined and named. In this way, the terminology and the relationships between that terminology are compiled at the work place. At the other end of the spectrum is the COMBINE approach, where a data dictionary is compiled and the terminology agreed by those who will then make use of it.

This definition of names is not a new process nor a new problem, it was first identified in 1952 by the International Building Classification Committee, who then went on to develop the SfB classification which is a terminology accepted and used throughout the European construction industry. The danger is that, with the excitement of new information modelling techniques, researchers will ignore the existing substantial body of work in this area and either re-invent or redefine the terminology and in so doing isolate themselves from industrial acceptance and at the same time wasting considerable effort.

At Newcastle University we have been investigating this area. This work initially began with the development of an expert system to assist in the design of the external building envelope; this highlighted the need for a common representation for building information. The foundation of this common representation was proposed as a 'basic syntax' following from the work of the Department of the Environment on information systems (1971). This acknowledged that the various 'actors' in the construction process had developed over time 'special purpose syntaxes' to handle information representation for specific roles or tasks. The basic syntax is not an information structure as such, but a resource to use when defining new information structures. In essence it holds the industry's accepted data dictionary, rules of grammar, thesaurus and synonyms. It allows information modellers to identify the correct names for their information attributes which will receive industry acceptance. It is a building block for special purpose syntaxes that are required for specific aspect models. The problem with these special purpose syntaxes is that they imply a certain context or aspect through which the information they contain must be interpreted. Consider a special purpose syntax such as the COMBINE data model for energy in building. The term 'conductivity' is understood by all as 'thermal conductivity'; pass this information from the heating and ventilating engineer to the electrical engineer and 'conductivity' becomes electrical conductivity. This problem has long been understood by the industry as Part Two of this book, dealing with historical classification work, shows. We have only to look to the CIB Master Lists (1964, 1972, 1983) for the proper definitions for these items. Reviews have been carried out of the various formal methods currently used by the construction industry to represent information (Lockley *et al.*, 1989a,b). Based upon these reviews, an approach to develop a basic syntax for representing construction industry information was proposed.

14.2.2   Constructing a basic syntax

The two key elements of the basic syntax are the names of entities (vocabulary) and the rules for combining these names (grammar). It is proposed that the names are identified in accordance with ISO 2788 (1987). This divides names into three categories:

1. **Concrete**: e.g. brick, window, door. Concrete names can initially be derived from the SfB (RIBA, 1961), CI/SfB (RIBA, 1974), Common Arrangement (1987) and EPIC (1991) classification systems, which provide a good coverage of building materials elements and components.

2. **Abstract**: e.g. thermal conductivity, rain penetration. Abstract names can be derived from the CIB Master Lists and ISO 6240 (1980), 6241 (1984), 6242, 7162, which provide good coverage of performance considerations and the common arrangement which covers work and workmanship.

3. **Individual**: e.g. Pilkington, Red, Dense. Although individual names are not really part of the basic terminology, there is a need to identify them to form databases or libraries of component data, for example, libraries of plant components. The foundation for these libraries can be found in such documents as the RIBA Product Selector and its European/international equivalents.

The components that relate these names together are the rules or grammar of the basic syntax. Seven types of relationship are proposed (see Figure 14.4):

1. Broader

2. Narrower

3. Equivalent

4. Attribute

5. Value

6. Paradigmatic

7. Syntagmatic

Broader and narrower relationships are analogous to the super-type and sub-type roles in EXPRESS, or the ancestor and descendent roles of object-oriented languages. However, the difference is they suppose no

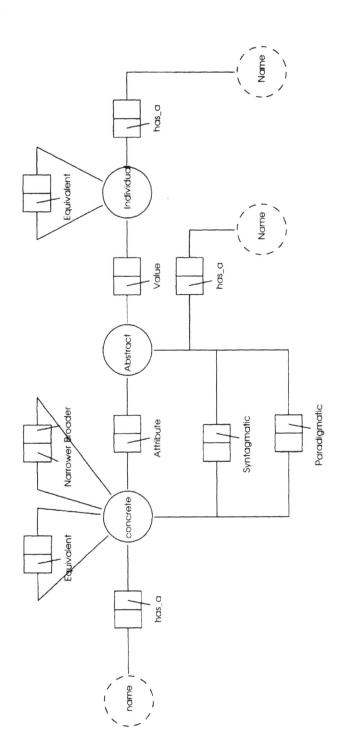

Figure 14.4 Constructing a basic syntax

implementation, therefore they do not fit into a single coherent entity or class hierarchy as would be the case if they were part of an EXPRESS or C++ schema. They describe relationships between 'concrete' names, e.g. 'window' is a narrower definition of 'glazed opening'.

Equivalent relationships allow transposition between two different conceptual frameworks or even languages. For example, a partition wall in a designer's view of the world may be named as 'a timber frame partition', but in a carpenter's view it would be 'a stud wall'. They are identical definitions but are named differently in the two frameworks. This type of equivalence is the same as a synonym, but equivalence relationships can also be used to transcend language barriers. Equivalence can be applied to 'concrete', 'abstract' and 'individual' names but not between 'concrete' and 'abstract' or 'individual' names.

Attribute relationships link 'abstract' and 'concrete' names. Each 'concrete' name can have many attribute relationships with an 'abstract' name. For example, a 'brick' has attribute relationships with 'colour', 'density' and 'size'. Again these are not to be confused with the attribute definitions in entity–attribute models. These attributes are only linking two 'names', they are not linking two entities. The basic syntax is a resource from which entity schemas can be created; therefore to compile a *schema* for an *entity* 'brick' the basic syntax would offer possible attribute 'names' to build the data model of the brick.

Value relationships link 'individual' names with 'abstract' names. A value relationship exists between the abstract name 'colour' and the individual name 'red'. They are analogous to the 'role' element in the NIAM paradigm, but only when the role exists between a lexical type and a non-lexical type. Value relationships do not exist between two 'concrete' names.

Paradigmatic and Syntagmatic relationships defined by Lyons (1977) allow the construction of sentences or phrases as well as the association of names within a class. Syntagmatic relationships exist between two 'concrete' names and one 'abstract' name. For example, a 'wall' 'is made of' 'brick'. Wall and brick are 'concrete' names, 'is made of' is an 'abstract' name. Paradigmatic relationships are between 'concrete' names in the same or similar family. For example, 'Brick wall', 'Stone Wall', 'Glass Wall' are all paradigmatic, being part of the family of 'External Envelope'. Paradigmatic relationships bind together names in a domain and describe relationships which cross the hierarchical structures of the 'broader' and 'narrower' definitions.

14.3   INFORMATION ACCESS

A common misconception is that information access follows wholly from the development of appropriate information structures. In reality, the way information is accessed rarely flows totally from the way the information is structured. Consider a library of books, each book in the library has one of several information structures depending on the class of publication it falls into, e.g. a novel may comprise

1. a title;

2. an author;

3. a publication date;

4. a filing code.

All of this information is held on or within the book. The librarian organizes all books in the library by their unique filing code. Accessing the information in the book is simple, just know the filing code and retrieve it from the library. This clearly is impractical as we are hardly ever likely to know the exact number of the book we want to read.

An alternative would be to ask the librarian to file all books in order of their title or their author. However, this provides no solution when we wish to access books on a particular topic, e.g. all books dealing with computers. The obvious solution then is to attach to each book a subject classification. Unfortunately though, some books cover many subject areas. The classic solution to this problem is to create a card index of all the subject classifications and to list on these cards the filing code of each book which has this classification. The reader may now identify where all books dealing with computers are filed in the library simply by looking them up in the card index.

An access path has been created to the books, but with this access path has come a new information structure, a subject index filing system. This has nothing to do with the way we produce information models for books. It exists only in the context of a library. Furthermore, it is not the only extension that must be provided, there is a clear need to access the books by an author index, a title index or a free text index.

In the context of the Building Server, these are generic information access mechanisms. A range of generic access mechanisms have been identified and several have been implemented. In doing so, it has become clear that access mechanisms have a special kind of relationship with the entities to which they provide access. This affects the way in which they are modelled.

14.3.1 Modelling access entities

The library book example, simple though it is, can lead to fundamental problems if it is incorrectly used in an information model. The card index system we have now identified could be represented as Figure 14.5. In this scenario, a subject card is identified and the set of books indicated by that

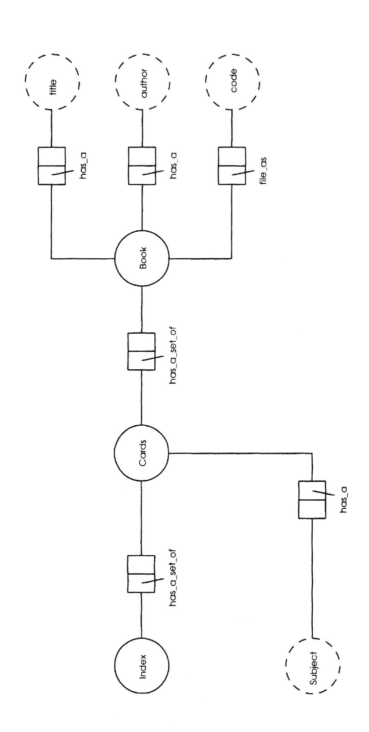

Figure 14.5 Forming a card index system

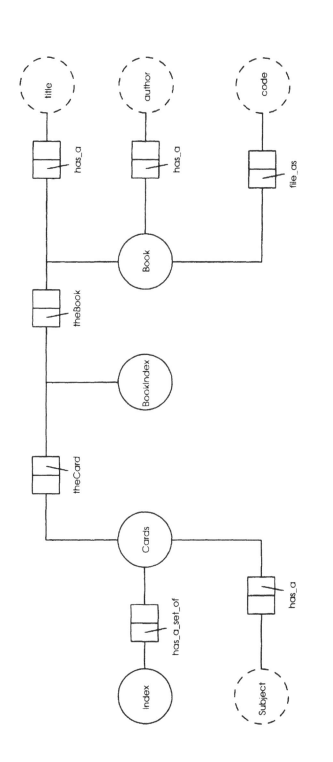

Figure 14.6 Associating one class of entity with another

card can be identified. Although this will work, it does not lead to the kind of generic information tool that we need in the server, because we have, in effect, specialized the card index to be a book card index. It cannot be used to index journals or videos, etc., unless we extend the model to allow these additional types. In doing so, we would predefine the possible uses of the card index and make future extension more difficult.

Alternatively, we could add to the book information structure a reference to the card index; this would specialize the book to become a library book organized by a card index – an unnecessary overhead if the book is not part of a library. The preferred solution is to create 'access entities' that associate one class of entity with another, Figure 14.6. Here the book index holds the relationship a book and a card, neither book nor card have any additional data members.

In this way the card index becomes a generic resource in the server that can be used to access a range of media, and the book does not carry any unnecessary data overheads. It also becomes much simpler to extract instances of any of these entities and exchange them with other information systems. Figure 14.7 gives an example of this in a construction context. A building in the COMBINE I context can comprise a set of spaces and/or a set of zones. In reality, zone and space are concepts to access subsets of the building and their meaning depends upon the context in which they are used. Therefore they should not be implemented as attributes of a building because in doing so we specialize the building entity into a 'zonal building' or a 'spatial building'. By using the discrete 'ZoneLink' and 'SpaceLink' entities, an application which requires either type of specialization of building can achieve this and still use the generic building

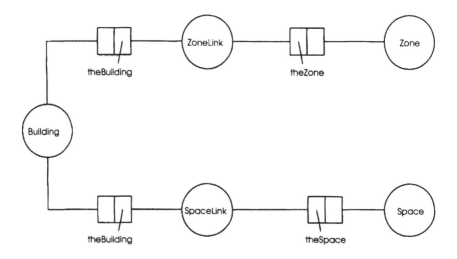

Figure 14.7 'Access entities' type card index as applied to construction

resource. This would allow greater flexibility in implementation, allowing a range of different space/zoning regimes to be used without the data model of the building exploding with the overhead.

14.4 CONCLUSION

Data model schemas, by definition, are defined for a specific purpose and therefore contain relationships which reflect a particular context. It is necessary to have a basic syntax from which all model schemas can be composed, but assumes no implementation. Such a basic syntax would help to standardize the development of new domain models and improve future integration of such models. Furthermore, there exists a large body of work carried out by the construction industry related to the terminology of describing buildings. This work is relevant to information modellers; its use would make it easier to develop integrated data models that gain industry acceptance. Its avoidance can only lead to duplication of effort and misinterpretation by industrial users. Much of the essence of this work is reported in Part Two of this book, concerned with historical classification efforts.

It is argued that accessing data in complex information models requires access mechanisms more sophisticated than those available solely from the relationships in the data model and through indexes of specific attributes in the data model. These access mechanisms need to be part of the data model and only weak links should be provided between them and the data they access. In this way, they can be a generic resource allowing high levels of reuse without imposing an unnecessary overhead or specialization of the data to which they provide access.

Experience of applying the object-oriented paradigm to building modelling indicates that this could create a way for a new generation of building software tools. Engineering has benefited from computer-aided design and manufacturing because it requires precision; it is suited to mass production and, in general, the final product is made of a relatively small number of engineered products from controlled sources. Construction, on the other hand, is not a precision industry; most of the products are unique, and the resulting building is a complex aggregation of many parts from different sources that are often not engineered. Construction is therefore well suited to take advantage of the developments in data modelling and the new generation of object-oriented languages and databases. These support data-rich domains where there are many complex interactions between data entities, and the information models evolve with time. Hopefully these techniques will help to deliver integrated information systems to the construction industry.

REFERENCES

CIB (1964) CIB Master Lists of Properties for Building Materials and Products. *CIB Report No. 3*.

CIB (1972) CIB Master Lists for Structuring Documents Relating to Building, Building elements, Components, Materials and Services. *CIB Report No. 18*, Rotterdam.

CIB (1983) CIB Master List of Headings for the Arrangement and Presentation of Technical Documents for Design and Construction.

Clarke, J.A. and MacRandal, D. (1991) An intelligent front-end for computer aided design. *Artificial Intelligence in Engineering*, **6** (January).

Common Arrangement of Work Sections for Building Works (1987) Co-ordinating Committee for Project Information.

Department of the Environment (1971) *An Information System for the Construction Industry*, Directorate General of Development (Housing and Construction), London.

European Product Information Committee (1991) *Construct Product Grouping – Progress Report*.

IDEF0 (1981) ICAM Architecture, Part II, Volume IV – Functional Modelling Manual (IDEF0), *Report number AFAWL-TR-81-4023*, June.

International Standards Organization (1980) *ISO 6240*, Performance Standards in Buildings – Contents and Presentation.

International Standards Organization (1984) *ISO 6241*, Performance Standards in Buildings – Principles for their Preparation and Factors to be Considered.

International Standards Organization *ISO 6242* (1992) Building Performance – Expression of Function of Requirements of User.

International Standards Organization (1987) *ISO 2788*, Guidelines for the Preparation of Monolingual Thesauri.

International Standards Organization *ISO 7162* (1992) Performance Standards in Buildings – Content and Format of Standards for the Evaluation of Performance.

ISO (1991) Project 3 of Working Group Five of Subcommittee Four of ISO Committee 184, 'EXPRESS Language'.

Lockley, S.R., Hardy, A.C., Wiltshire, T.J. and Dudek, S.J.M. (1989a) *Science Engineering Research Council Report, Design and Evaluation of Cavity Walls in Framed Buildings*, Appendix 2, Classification Systems, March.

Lockley, S.R., Hardy, A.C., Wiltshire, T.J. and Dudek, S.J.M. (1989b) *Science Engineering Research Council Report, Design and Evaluation of Cavity Walls in Framed Buildings*, Appendix 4, Information Activities, March.

Lockley, S.R., Ming, S., Parand, F. and Roche, L. (1993) The Integration of Multiple Design Tools with an Object-Oriented Building Model. Paper presented at the CIBSE National Conference 1993.

Lyons, J. (1977) *Semantics*, Cambridge University Press.

Nijssen and Halpen (1989) NIAM, 'Nijssen's Information Analysis Method', in *Conceptual Schema and Relational Database Design: A Fact Oriented Approach*, Prentice Hall.

RIBA (1961) *SfB/UDC Filing Manual, Recommendations for Standard Practice in Pre-Classification and Filing*, RIBA, London.

RIBA (1974) *CI/SfB Construction Indexing Manual*, RIBA, London.

TNO (1992) *The EXPRESS/STEP Interface Kit*, TNO/TUD, Delft, May 1992.

CHAPTER 15

Integrating points of view through information modelling

Grahame S. Cooper, Information Technology Institute and Peter Brandon, Department of Surveying, University of Salford, Salford, M5 4WT, UK.

15.1 INTRODUCTION

Traditional methods of modelling information often suffer from a need to make compromises in order to reconcile the differing information requirements of various users of the information. In the construction industry itself, a number of very carefully thought out information models (e.g. the Common Arrangement, SfB, and others described in earlier chapters of this book) have nevertheless suffered in the attempt to make them all things to all persons. The ICON project is concerned with the creation of a generic information model for the construction industry which, aided by object-oriented modelling techniques, will allow the individual perspectives of different disciplines to be accommodated within a single information model. It is also important to recognize that the models produced do not simply describe *computer-based* information and processes, since a total information system generally contains some computer-based parts and some non-computer-based parts. Indeed, the models should be completely independent of any particular implementation method. This chapter describes the philosophy behind the modelling methods being applied in ICON.

Integrated Construction Information. Edited by Peter Brandon and Martin Betts. Published in 1995 by E & FN Spon, 2–6 Boundary Row, London SE1 8HN. ISBN: 0 419 20370 2

15.2 ABSTRACTION AND PERSPECTIVES

15.2.1 Abstraction

Abstraction (Howard, 1988; Sanvido, 1992) is very important in information modelling, both in general and within the construction industry in particular. Abstraction is an important mechanism for managing complexity because it allows irrelevant detail to be hidden when considering information within a particular context.

The two most important mechanisms for abstraction are composition and generalization. Composition is a mechanism for grouping together a number of concepts and referring to them as a single item. For example, the word *'house'* encapsulates the idea of a structure with rooms, passages, stairs, walls, ceilings and floors, and also ideas such as ownership, occupancy, usage and functional requirements, legal rights and obligations, etc. Generalization allows information to be represented regarding a whole range of objects that are different in many respects, but are similar or identical in relation to the requirements of the particular context in which an abstraction is used. Referring again to the word *'house'*, this word may be used, when appropriate, to refer to houses in general in those contexts where the number of floors, the particular construction method used, the type of ownership, etc., are not important. Object-oriented approaches support the use of abstraction by providing mechanisms for representing composition and generalization directly.

It is important to recognize that the use of abstraction is not confined to information modelling. People use abstraction in all their communication and reasoning activities. If I were to give directions to allow a visitor to find my house, I might say that 'it is the third house on the left after the crossroads'. This would normally allow a person to find the house, even though they might not know what kind of house it is until they actually see it. In giving the directions, I also assume that the visitor will ignore the small building that contains the electricity substation, situated between the first and second houses. The successful interpretation of the directions requires that the visitor has prior knowledge of what constitutes a house, and that the visitor's understanding coincides reasonably well with mine, at least within the context of the problem.

15.2.2 Perspectives

The prior knowledge referred to above comprises the rules of interpretation that allow one party to understand the meaning of abstract information passed by another. These rules of interpretation are shaped by the social, cultural and technical background of the interpreter. This has been referred to as the *'Weltanschauung'* (Checkland, 1981) or 'World View' of the particular person whose information requirements are captured in an information model. It is important, therefore, to recognize the existence of the *Weltanschauung* when attempting to model the information requirements of an individual or group of individuals and, in particular, to recognize that these may be different from one individual to another. This issue is at the heart of the problem of integrated construction information.

Because of this, the extensive use of abstraction in information modelling presents a problem where the information requirements of a range of different groups of users are to be captured in an integrated manner. Different users of information will often wish to aggregate and classify concepts in different ways, and will interpret information in different ways according to their own terminology, culture and background. Even as simple a concept as the width of a room may vary from one participant to another. These are certainly very different for the carpet fitter as opposed to the architect.

In order to represent accurately the requirements of the different groups of participants in a construction project, the method being used in the ICON project involves the construction of a number of separate information models, each describing a different perspective on the information. It is from these domain perspectives that the final object model is synthesized. The use of such perspectives allows complex information models to be built up in a manageable and understandable form, whereby each individual perspective may be understood in its own context.

15.3   TYPES OF PERSPECTIVE USED IN ICON

A number of different types of perspective have been identified to allow modelling, implementation and use of a shared information system. These allow an ICON system design architecture to be defined in three main parts (see Figure 15.1): domain perspectives, the implementation perspectives and the application perspectives. Domain perspectives are concerned with modelling the information as viewed from the application domain. Implementation perspectives model the information from the point of view of the system implementor. Application perspectives provide a convenient way to map the implemented information system onto the software and standards used in real life construction projects.

15.3.1   Domain perspectives

The primary focus of ICON to date has been concerned with the modelling of domain information as viewed by construction project participants, and the integration of those models to form a single object-oriented information model. There are two types of domain perspective: user perspectives and integration perspectives.

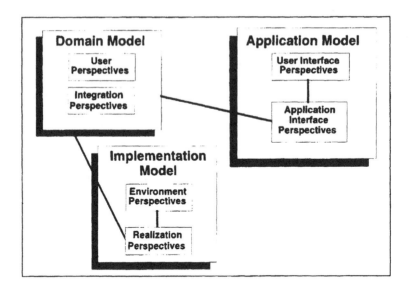

Figure 15.1   The three-part architecture of the ICON models and system

User perspectives

A user perspective is a model of information as seen from the perspective of a particular user, or a well-defined type of user, performing a certain type of task or activity. It is important that the different user perspectives are analysed as independently as possible, so that the models capture a true representation of the information as understood by the particular users whose information requirements are being captured. Ideally, a user perspective will cover the information needs of a well-defined activity, so that it may be easily understood in relation to that activity. It would be expected to contain in the region of 10 to 20 distinct entities, although this should not be taken to be a hard and fast rule. Figures 15.2 and 15.3 show the ICON object models for construction planning and physical design respectively.

Integration perspectives

Once the user perspectives have been modelled for the areas of interest, the integration of the models is addressed. Integration perspectives are used to show where the overlaps occur in the user models. A very much simplified example will serve to illustrate this. Figure 15.4 shows part of an integration perspective for construction planning and physical design. Here, a new abstract concept 'building element' has been created in recognition of the fact that the construction planning concept 'building element' and the physical design concept 'physical space separator' have some features in common. Note that the concept 'physical space separator' is already a

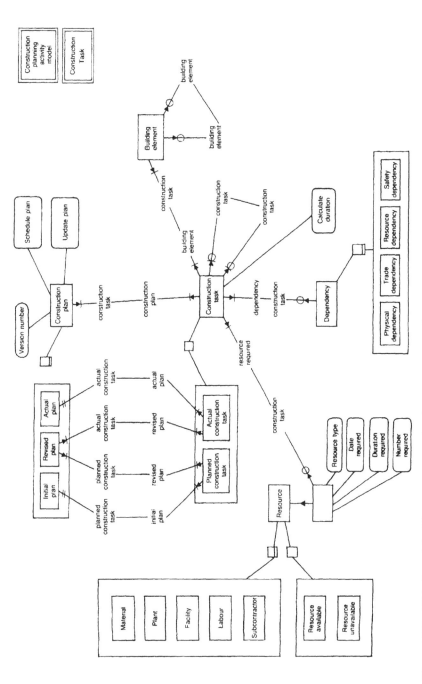

Figure 15.2 Object model of the user perspective for construction planning

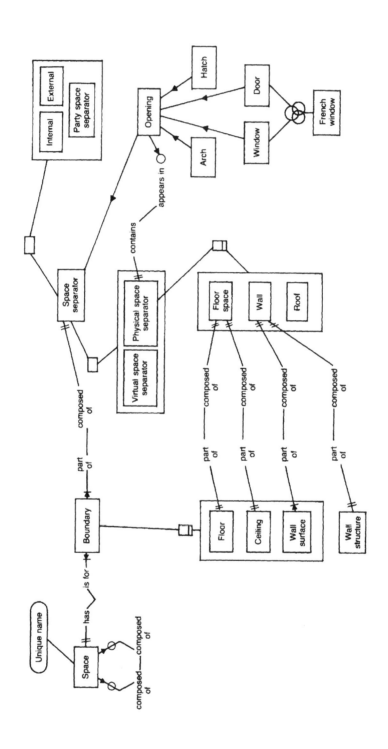

Figure 15.3   Object model of the user perspective for physical design

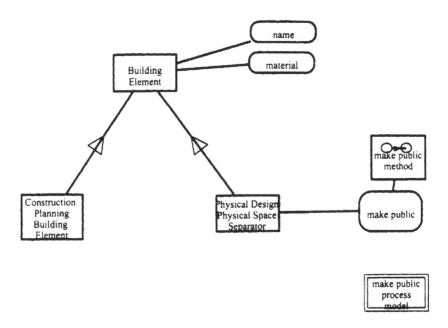

Figure 15.4 Part of an integration perspective showing the overlap between construction planning and physical design

generalization of walls, floor/ceilings, doors, etc. Suppose a wall is created by the architect or engineer. It may be classified both as a physical space separator within the physical design model and as a building element within the construction planning model. In this case, the new wall will inherit the properties of name, material and size from the generic building element class. These properties will be accessible both from the construction planning perspective, through its building element class, and from the physical design perspective, through its physical space separator class.

Integration perspectives may relate object classes from different user perspectives in a number of ways: a class in one perspective may actually be the same as a class in another; a class in one perspective may be a supertype of a class in another area; or a class in one perspective may share a supertype with a class in another. It is important to recognize that the different names might be used to refer to the same concept in different perspectives, and the same name might be used to refer to different concepts in different perspectives. The latter case is illustrated by the example given above, in which the generic 'building element' class has the same name as the 'building element' class from the construction planning perspective, even though they represent different concepts. Because of this, it is important that the techniques and tools used for modelling provide some form of name-scoping mechanism.

Identifying user perspectives

An appropriate set of user perspectives is identified primarily by means of a top–down decomposition of the function and process types carried out within a construction project, indicated by the hierarchy at the top of Figure 15.5. This breakdown is carried out to the point where a relatively simple object model may be constructed for each of the basic processes or tasks identified, in which sufficient detail is defined to support any required applications. Each of these object models represents a perspective on the overall information model.

In some areas, such as the management of construction, it is relatively easy to describe a process model as a set of subprocesses that must occur in some order, and in which well-defined information flows exist between the subprocess, although the order may change depending on the particular circumstances (e.g. a particular company or procurement path). However, other areas, such as design, are not so easily described in terms of a process flow. In such areas, a more direct approach may be appropriate, in which perspectives are chosen according to some other logical or conventional breakdown. In the design area, perspectives such as *spatial design, structural design, technical design,* etc., are used in ICON. This is one reason why we feel that information modelling, as opposed to either process or product modelling, are appropriate to the integrated construction information problem.

Note that object models need not only be produced for those nodes at the leaves of the hierarchy. Indeed, object models should be developed for the perspectives high up the hierarchy to provide a high level overview of the objects involved in processes. The very highest level perspective in ICON is based on IRMA (Information Reference Model for AEC) (Bjork *et al.*, 1993; Luiten *et al.*, 1993), which was developed at a meeting in Helsinki in October 1992. Work is currently in progress to integrate IRMA into the ICON model.

15.3.2   Implementation perspectives

Implementation perspectives model the information from the point of view of the system implementor, and are used to map the domain perspectives into some implementation language. In the ICON project, the prototype system is an object-oriented database implemented in ONTOS, an object-oriented database management system. There are two types of implementation perspective defined in the ICON method: environment perspectives and realization perspectives.

Environment perspectives describe the abstractions that are provided by the language or environment in which the system is to be built, along with any class libraries that have been added. For example, in the ICON project, the language used is C++, used in conjunction with the ONTOS

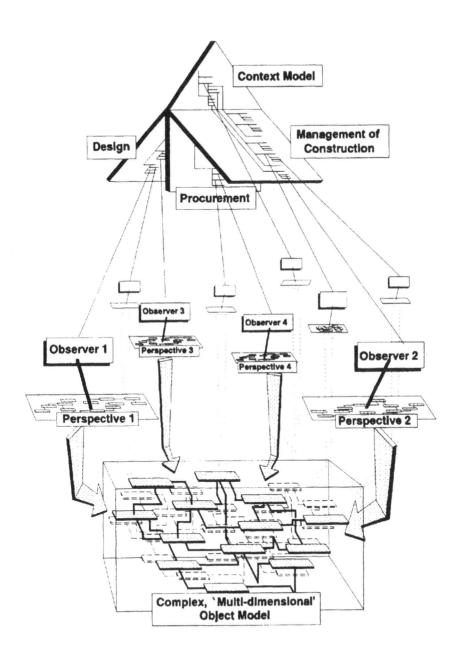

Figure 15.5 Identifying user perspectives for the ICON domain model

database system. Together, these provide abstractions ranging from integers, floating point numbers and character strings, to list management objects such as 'dictionary' and 'set'. In general, the environment perspective is not generated by the analyst or designer, but is provided by the definition of implementation language (C++ and ONTOS) and any additional class libraries that have been purchased or created.

Realization perspectives show how the abstract concepts represented in the domain perspectives are actually realized in terms of the object classes defined in the environment perspective. For example, the domain perspectives may contain a class 'bill of quantities', which might be implemented by means of the ONTOS 'dictionary' class mentioned above. In this case, the realization perspective would show that the class 'bill of quantities' has a subclass: 'bill of quantities implemented as an ONTOS dictionary'.

The aim of this approach to modelling and implementation is to provide a mechanism for implementation of a working system, while keeping the domain models independent of any particular method of implementation. This should permit standardization to occur at the domain model level while still allowing freedom over the choice of implementation language.

15.3.3 Application perspectives

Application perspectives provide a convenient way to map the implemented information system onto the software and standards used in real life construction projects. There are two types of application perspective: user interface perspectives and application interface perspectives. User interface perspectives provide a model of the concepts that are supported by applications that are linked into the ICON database. In the case of AutoCAD AEC, these would include concepts such as spaces, rooms, windows, doors, walls, etc. In the case of a project management tool, these would include: tasks, dependencies, resources, etc. Like the environment perspectives described in the last paragraph, these are not primarily created by the modeller, but are provided by software and standards that exist already.

The user interface perspectives are mapped into the domain perspectives by the application interface perspectives. Each of these is expected to consist primarily of an implementation of the classes defined in the user interface perspective, defined in terms of the classes provided in the domain model. It will be the responsibility of the classes defined in the application interface perspectives to carry out the necessary operations on the database objects, to reflect changes that occur in the application software that is operated by the users. This aspect will be addressed in more detail in a forthcoming publication. Note that the application interface perspectives are defined in terms of the domain perspectives rather than the implementation perspectives. This is important if a logical independence is to be maintained between database implementations and application software. Such logical independence is made possible in the ICON model because the domain models may be thought of as providing a standard interface to the implemented database, which will not change from one database implementation to another. It is intended that this approach will be compatible with the 'Object Request Broker' concept being developed by the Object Management Group (Hollowell, 1993) as a standard for interworking between object-oriented applications and databases.

ACKNOWLEDGEMENTS

The ICON project is funded by the UK Science and Engineering Research Council under its 'Information Technology in Engineering' initiative. Thanks are due to the other investigators involved in the project and also external members of the ICON project steering group for their support and assistance with the creation and verification of information models. The other investigators include: Terry Child, Serena Lord, Ghassan Aouad, John Kirkham, Frank Brown and Rivka Oxman. Rivka Oxman's own work on Integrated Construction Information appears as Chapter 19 in this book.

REFERENCES

Björk, B.-C., Cooper, G.S., Froese, T. *et al.* (1993) *IRMA: An Information Reference Model for Architecture, Engineering, and Construction*, in Proceedings of First International Conference on the Management of Technology in Construction, Singapore, August 1993.

Checkland, P. (1981) *Systems Thinking, Systems Practice*, John Wiley.

Hollowell, G. (1993) *Handbook of Object-Oriented Standards*, Addison-Wesley.

Howard, R. (1988) An object lesson in modelling. *Building* (November).

Luiten, B., Froese, T., Bjork, B. *et al.* (1993) *An Information Reference Model for Architecture, Engineering and Construction*, in Proceedings of the International Workshop on Models for Computer Integrated Construction, Espoo, Finland, October 1992, VTT, Finland (in press).

Sanvido, V.E. (1992) Linking levels of abstraction of a building design. *Building and Environment*, **27**, 195–208.

CHAPTER 16

A process support environment for design management

David G. Platt and David I. Blockley, Department of Civil Engineering, University of Bristol, Bristol, UK

16.1   THE PROBLEM

Construction projects require the handling of large amounts of information. Managers of the design and construction process have to take on a wide range of roles to manage this information. The aim of ongoing work described here is to provide an integrated computer-based environment which will support managers in fulfilling these roles and provide real business benefits. The intention is to help co-ordinate the activities of people and the ways in which information is transferred in and between companies. In particular, the feasibility of using process support (PS) systems to aid quality management (QM) in civil engineering design is being examined.

In order to achieve these objectives, a model of processes within a construction organization has to be produced. The purpose of this paper is to introduce a preliminary model based on the reflective practice loop, which is being implemented in a generic self-replicating software object.

A process is simply a task, albeit a long and complicated one and separable into many activities, carried out using resources and resulting in deliverables. Process modelling is a way of representing on a computer certain kinds of activities in the actual world. These activities occur within an organization in which there are people who have obligations to fulfil, tasks to carry out and tools to help them. The idea is to capture the organization of the processes of meeting responsibilities and carrying out activities using a modelling language. Execution of the model prompts users to perform activities in accordance with their roles in the organization and provides them with accurate and up-to-date information. The process model handles the sequencing of activities, and is continually changing as

Integrated Construction Information. Edited by Peter Brandon and Martin Betts. Published in 1995 by E & FN Spon, 2–6 Boundary Row, London SE1 8HN.
ISBN: 0 419 20370 2

people define and redefine activities in response to changes. For example, a new project will spark off a new set of activities. A process model must be dynamic, its existence and evolution are dependent on interactions with the people whose responsibilities it models.

In previous implementations of support systems, the processes have generally been analysed in terms of activities. This has been possible because the scope of the support systems has been fairly limited. These previous processes have tended to be:

1. predefined and contain relatively few decisions which depend on context;

2. fairly linear and sequential;

3. not concurrent or with very few concurrent activities;

4. monitored and managed to ensure objectives are achieved;

5. subject to little change over fairly long periods, and when change is made it is well controlled;

6. time critical;

7. typically high volume;

8. fairly easily represented in terms of some form of work chart, such as a role activity diagram (RAD), data flow diagram (DFD) and IDEF0.

Conventionally the differences in processes have been shown on a single dimension of repetitivity of product type, from one-off to high volume mass production. There are, however, many possible ways in which processes could be characterized. Figure 16.1 shows a spectrum of processes in the three dimensions of context dependency, volume and rate of change which is useful for the purposes of this chapter. It is clear that many of the processes previously studied are distinctly different from the construction industry. For example, the customer services and the Inland Revenue models implemented by ICL are shown. A metaphor for the customer services model is that of a work stack or work pool. The critical business benefit to the organization is the achievement of a co-ordinated and controlled flow of work from the pool with an account of its progress. A metaphor for the model developed for the Inland Revenue could be that of a branched chain. Each person is involved in a progressive chain of events, receiving inputs, processing them and outputting the results to form the input to the next person. The routeing of the work is usually controlled by simple rules or preferences.

In this project, studies of quality assurance in two civil engineering design offices show a quite different situation. For example, design and build contracts and fast-track construction have been studied. The processes required to service these contracts are concurrent. At a high

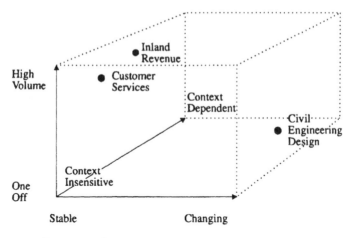

Figure 16.1 Spectrum of process

level of decision-making the responsibilities are relatively stable, contracts are negotiated and agreed, work is performed, products made and delivered to the client in return for payments. As the granularity of the processes becomes finer then the degree of change becomes greater. Each new contract often requires a new set of modifications and as a result the processes become highly dependent on context. It becomes impossible to capture the possible routes by which the objectives may be achieved.

16.2   THE IPSE 2.5 PROJECT AND PROCESSWISE

The aim of the Integrated Process Support Environment (IPSE 2.5) Alvey project (Snowdon, 1990), out of which the current work springs, was to develop a technology which can be applicable across a variety of processes and to explore the use of the technology in support of known processes. The project consortium comprised several UK companies and universities, and the programme cost in the order of £9 million. The target was the problem of how computer systems can be used in the development of software-based information systems. The idea has been to provide a coherent set of facilities to support people who develop computer systems. IPSE is a process support system which is intended to support:

1. the processes of the company;

2. the new techniques aimed at improving the products themselves and;

3. the management of change and the reengineering of the business.

PROCESSWISE[1] is process support technology which derives from the IPSE 2.5 project and is being used in the work described in this chapter.

[1]PROCESSWISE is a range of ICL software and service products that enable the modelling of business processes and their redesign for optimum performance followed by the generation of the necessary process support systems and their integration into the enterprise information systems infrastructure.

16.3 ROLES AND RESPONSIBILITIES

The conventional methodology is to focus on the activities in a process. Thus, for example, roles have been defined in the IPSE 2.5 project as sets of tasks. For processes characterized as being towards the top left-hand corner of Figure 16.1 this is probably satisfactory as previous work has shown. However, the resulting model for a construction process is difficult to capture, is highly dynamic, context sensitive and probably unworkable using current technology.

In order to tackle this problem engineers, directors and managers in the construction companies were interviewed and the data was analysed using Grounded Theory (Pidgeon *et al.*, 1991). The analysis showed that a relatively stable view of the processes within the company could be obtained by focusing on the responsibilities and objectives of the roles contained in those processes. People frequently change the procedures by which they achieve their objectives, due to changes in context, for example. Change can be accommodated provided the frequency and amplitude is manageable by those implementing and using the system. However, the central point is that their responsibilities do not change nearly so much within a role; they tend to be the same from one project to the next.

A major premise, therefore, in process modelling is that roles and interactions are an effective way of describing what people do. A process model consists primarily of a set of roles which interact in a meaningful manner. Such a model of a world of roles is intended to exploit the notion of concurrently executing agents all co-ordinating to achieve a common goal. The process model itself is subject to change, for example, as in-house procedures for handling orders change. The modification of the process model is a management role and this is part of the process model itself.

A role is defined in this work as a set of responsibilities rather than as a set of tasks. In order to design a technology to support company processes we will now define some basic terms. The central concepts to the approach described in this paper are those of responsibility, role and objectives. They are defined as follows.

1. *Responsibility* is the state of being accountable for something or someone that is or who is in your control. In particular, it is being accountable for the achievement of a particular (set of) objectives.

2. An *objective* is a state of the world which someone or something (a computer) desires to achieve. It must be defined so that it is clear when it is attained, for example, by it being measurable.

3. A *role* is a collection of responsibilities with consistent objectives.

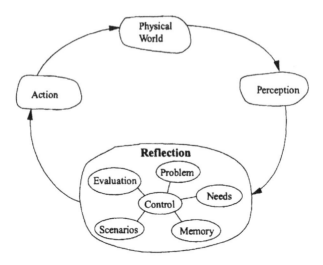

Figure 16.2   Reflective practice loop

16.4   THE SELF-REPLICATING REFLECTIVE PRACTICE LOOP

The reflective practice loop (RPL) was introduced by Blockley in 1992 to model the basic processes in problem-solving and decision-making in science and engineering. The RPL is being applied to the modelling of roles within an organization. Consider the model in Figure 16.2 as the process for each role. The basic and simple idea is that roles receive messages from other roles, the message is filtered through a perception module, passed into a reflection module which operates on it and outputs a message through an action module to the outside world or to another role.

The RPL is currently being made self-replicating by enabling it to create subroles, identical in structure but void of content until occupied by man or machine and to which it can then delegate some of its responsibilities.

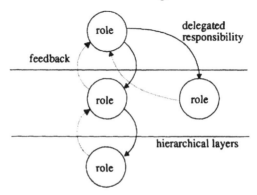

Figure 16.3   Outline of basic structure for the delegation of responsibilities and their subsequent monitoring

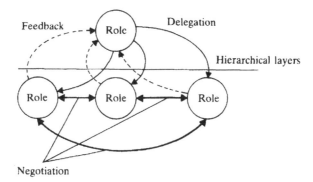

Figure 16.4 Model of the interconnected relationships between roles

This initiates a further process in which the two roles will negotiate new objectives. Progress towards these objectives can be monitored by both roles. It follows that in their turn subroles can create further subroles generating a hierarchy of responsibilities. The result is a series of delegation and feedback RPLs as shown in Figure 16.3. The RPL is being programmed into a simple role object.

16.5   THE MODEL OF A DESIGN ORGANIZATION

The model of the structure of the organization is developed by 'growing' the self-replicating unit through all of the roles within that organization. It is important to note straightaway that the resulting hierarchy is not a hierarchy of authority (i.e. the structure through which legitimate power is exercised), except for that type of authority which is associated with higher levels of responsibility (Handy, 1985). The hierarchy should not be seen as the setting up of an authoritarian power structure. The culture of the organization is a separate issue largely set by senior management. At one extreme it may be hierarchical (as in the military) or loose and flat in the sense that all workers are colleagues (as in the universities). The hierarchy suggested here is one of delegated responsibilities which can be, but is often not in practice, set up independently of the organizational culture.

Communication between roles for purposes other than delegation and feedback is also required. Groups are composed of members who must negotiate among themselves the objectives that they collectively must achieve. They must also negotiate how such objectives will be monitored. The basic role object must be able to support such a process of negotiation and monitoring. Therefore, at any level in an organization the roles may collaborate as shown in Figure 16.4 but will not have the authority to delegate responsibilities to one another. However, they may have the authority to negotiate the exchange of responsibilities. This would involve the

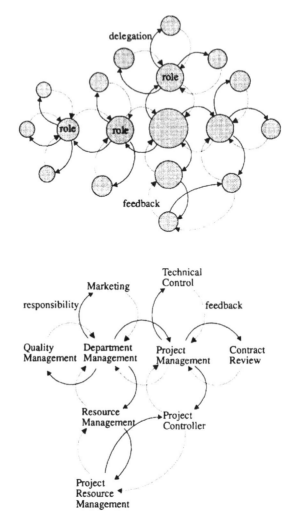

Figure 16.5   Emergence of an organizational model

obtaining of consent from the role from which the responsibilities were delegated. The way in which this is done will depend upon the culture of the organization. Figure 16.5 shows a model of the organization that emerges as different roles are identified and populated.

If certain responsibilities are delegated by a role to a subrole, then those responsibilities are not lost, rather they are met at a distance – through the activities of the other role. It is therefore important that the delegation is seen as a two-way process. The delegating role may propose certain objectives but the subrole is accountable for setting and achieving its objectives; the process is one of negotiation and monitoring of subsequent performance.

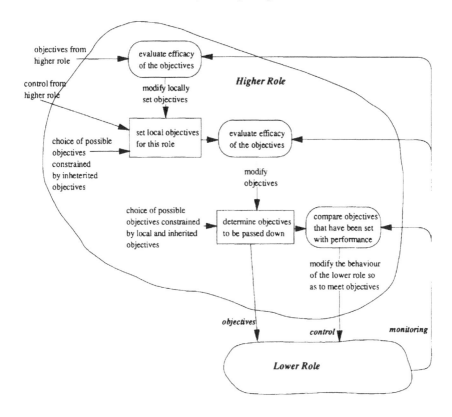

Figure 16.6 Model of double loop learning between roles

Of course, the objectives set under any given role may not be met for various reasons. For example, it is possible that the objectives were inappropriate. In such a case the delegating role needs to initiate a renegotiation of objectives in order to improve performance. The system is therefore one of a double loop learning where the delegating role not only monitors the performance of the subrole but also evaluates its effectiveness and helps to diagnose remedial actions. This view of management and leadership complies with that promoted by Senge (1990) who argues that the new view of leadership is that leaders are designers, stewards and teachers. They are responsible for building organizations where people continually expand their capabilities, clarify vision and improve team learning, i.e. they are responsible for organizational learning. An outline of how this double loop learning may operate is shown in Figure 16.6.

Each role, therefore, has the potential to be responsible for many subroles, a one-to-many mapping in a tree of delegation. However, although this would be natural in a hierarchical organizational culture, in many organizations the situation may not be as simple; there may be many-to-

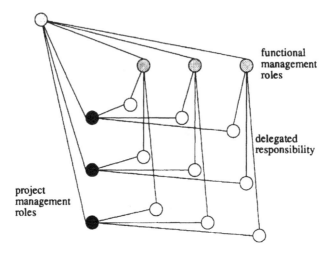

Figure 16.7 Matrix management structure

many mappings of responsibilities in a flatter organizational culture. In this case, a role may be accountable to more than one superior role. For example, a role may be accountable to both departmental management and project management as shown in Figure 16.7. In this situation there exists a potential for conflict. The boundaries of each of the delegating roles will need to be defined rather carefully and if there is an overlap a system needs to be established whereby conflict can be negotiated away.

16.6   THE ROLE OBJECT

The minimum requirement of the simple role object is that it should be capable of:

1. interpreting messages;

2. reflecting upon that information so that it can decide on action;

3. sending messages to neighbours;

4. replicating itself;

5. accessing and using tools such as spreadsheets and databases.

The internal structure of the role object is shown in Figure 16.8 and directly corresponds to the RPL. However, the role object should be able to create and delegate responsibilities to subroles, monitor the progress of subroles and send monitoring information to the role to which it is accountable.

The majority of these requirements are met reasonably easily in a computer-based PS such as PROCESSWISE. The exception is ability to reflect.

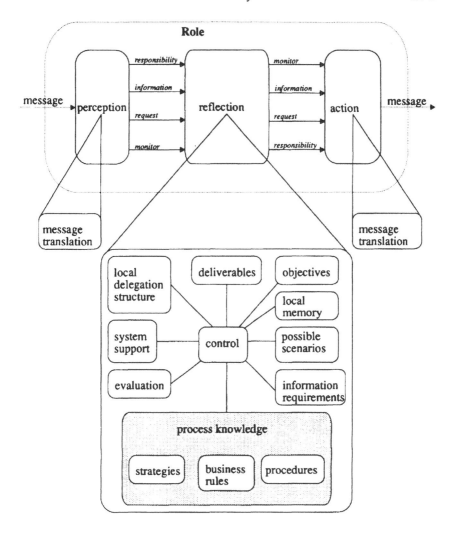

Figure 16.8 Knowledge requirement of a role object

The objective of reflection is to determine the most appropriate response to a message. This is a complex requirement which will involve considerable interaction between the person exercising the role and the computer supporting it. In the initial stages of this work, much of the detailed knowledge specific to the domain of the role is assumed to reside in the person responsible and not the computer. As the system develops, certain roles will be delegated to the computer, but of course the computer must be programmed to respond to the inputs it receives. The programs are effectively mappings from input-to-output messages which can be mathematical functions, rules or other transfer functions in a local knowledge base. In passing, it can be noted that this provides a way of distinguishing safety critical systems from other safety engineering systems since in a safety critical system a major safety role, such as the control of a reactor or a 'fly by wire' aircraft, will be delegated to a computer.

Support for a process involves anticipating the next stage of the process together with the information and the tools required. For example, a resource allocation role may receive a message that Project X requires more structural engineers. The system will not normally make the decision as to whether structural engineers are available and if so who they should be: that is the decision of the person exercising the role. The system can support the responsible person by enabling access to staff databases, programs and progress charts of the various projects on which structural engineers are employed. The person will need to negotiate with the managers and engineers of projects from which individual engineers will be transferred. This may be done formally through the computer process system or informally quite outside of the system, say by telephone or personal contact. Once a decision between the various parties has been agreed, the role of Resource Allocation will transfer the structural engineers from their present projects to Project X. Roles within the system may then perform a number of tasks automatically. For example, the structural engineers may each be sent a new brief; the role of Project Controller may be informed of the transfer and be prompted to delegate the work of the engineer; details concerning the engineers may be added to the project resource database. Finally, the role of Project Management may be prompted to update the program. Thus the transfer of an engineer triggers a series of concurrent activities which are handled by the computer role objects.

Knowledge Category	Location
message translation	perception and action
delegation structure	reflection
objectives;	reflection
information requirements;	reflection
deliverables;	reflection and action
possible scenarios	reflection
message and problem evaluation	reflection
system support	reflection

process knowledge
   – strategies
   – procedures
   – business rules
	reflection

Figure 16.9 Categorization and location of knowledge in a role

16.7 THE STRUCTURE OF KNOWLEDGE IN THE ROLE OBJECT

The objective is to provide a template in PROCESSWISE for structuring the different types of knowledge the role object will own and to which it will have access.

Knowledge is initially structured in one of a number of categories, an outline of which is shown in Figure 16.9. The authority for a role is for the whole life cycle from creation through its working life to its final termination. The delegation structure shows which roles have the authority to delegate responsibilities to other roles. It represents the hierarchical tree of responsibilities in the organization as discussed earlier. The resulting structure may be very flat and highly interconnected as in an organization using a matrix style management structure, or it may be highly layered in a vertical form as in the military.

The objectives of a role may be at four levels. First, there will be the statement of the mission, i.e. the central purpose of the role expressed in one sentence or short paragraph. Second, there will be a set of aims which are more detailed than the mission but are still rather vague general expressions of intent. Third, there will be detailed objectives which will be, as defined earlier, the precise measurable states that must be achieved in order to attain the aims and fulfil the mission. Finally there will be some statement about the broad way in which the role will go about achieving the objectives; this is in some sense a preliminary programme of work.

The information requirements are a list, as complete as possible, of information necessary or desirable for a role to meet its objectives. The deliverables represent the output of a role. The completion of a deliverable provides the delegating role with a measure of the success of the role in achieving its objectives.

The philosophy of the design of the roles is that of minimum tasking. Thus the responsibilities are kept to a minimum while preserving completeness, coherence and consistency. Each role fulfils only the objectives strictly necessary to meet its responsibilities. Deliverables are recorded only once in the appropriate database. Thus it becomes the responsibility of roles to acquire the information they need and to ensure that it is the latest release. Therefore, it is not the concern of a role to know what other roles may want to use the information generated by that role. This is because such a system would be very difficult to maintain in an environment such as construction where the structure of teams is often changing. However, as pointed out earlier roles can be defined in the computer to carry out certain transfers of information automatically so as to provide PS support. This support will include some form of generic access to databases and a tool interface which will allow interrogation by both the computer part of a role and by the human user responsible for that role.

Process knowledge available to a role may be local or global. Local knowledge is contained within the role and is known only by it. Global knowledge is available to many roles. It is stored outside the role and it is the responsibility of some role to maintain it, for correctness and availability. For example, the procedures for handling design change requests are contained in the Quality Handbook. They are available globally to the whole design organization and their maintenance is the responsibility of the Quality Manager.

Process knowledge has been classified into three types: strategic, procedural and business. Strategic knowledge is important for context dependent processes. Procedural knowledge is concerned with the order in which information is processed within the reflective phase of the role. Business knowledge contains the rules by which the role operates.

Role objects communicate by passing messages. The perception module interprets the messages for reflection. It is worth noting that entities such as drawings, time sheets, invoices, etc., are not messages – they are role objects. In other words, they are files of information which have 'responsibilities' to function in a certain way. For example, a time sheet could 'know' when it is expecting to receive information from another role and request that information at that time.

Messages are strings of code with various defined fields which are used to request, inform, monitor or delegate. They consist of the form

(Identifier) (Status) (Type) (Message)

1. Identifier: subject
                    source
                    destination

2. Status: urgency
                    importance

3. Type

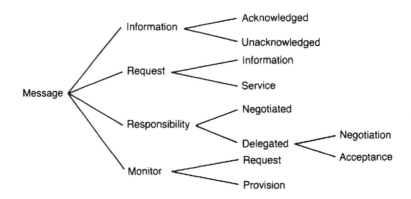

4. Message: The details contained within the message will be
                 unique. There will be templates that will determine
                 the structure of the information and the attributes.

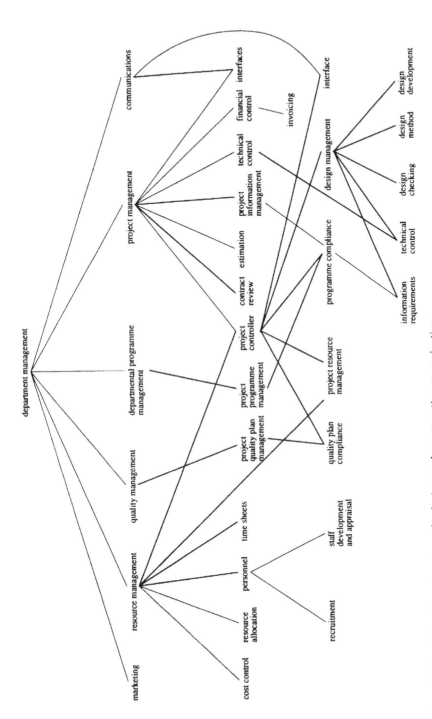

Figure 16.10 Delegation structure of a design and construction organization

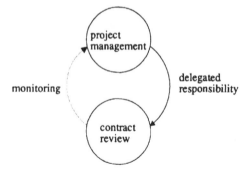

Figure 16.11   Delegation structure for contract review

16.7.1   Examples

Figure 16.10 shows the structure of delegations for a typical departmental manager and a project controller. Some examples concerning the role of contract review will now be given.

The project management role has the responsibility of initially setting up the role of contract review for a specified project. In most cases it will be the same person who is responsible for both roles. The delegation structure is shown in Figure 16.11. The mission and aims of the contract review role are 'To review the final contract as agreed with the client with a view to producing a strategy as to how the contract will be tackled and assessing the capabilities of the department to perform the task.' The immediate objectives are:

1. To become familiar with the terms of the contract paying particular attention to the client's letters of conditions and changes to the contract.

2. To nominate a Project Controller.

3. To identify milestones and key deliverables.

4. To identify major information interfaces for both the receipt and provision of information.

5. To assess the level of resources required to complete the project.

6. To identify any special services that will be required.

7. To provide a high level programme of work.

8. To assess the capabilities of meeting milestones and key deliverables.

9. To identify major regulatory constraints.

Some information requirements and deliverables are shown in Figure 16.12.

Information Requirements

Description	Source
all contract documentation	• project data base
letters issued and received whilst negotiating the contract	• project data base
changes made to the contract	• project data base
changes made to the pre-contract design	• project data base
agreed costs	• contract • clients brief
clients programme	• contract • Pre-Tender Management role • clients brief
regulatory requirements	• external sources • local authority
resource availability	• Department Resource Management
other programmes	• other disciplines

Deliverables

Description	Information Content	Sent To
report on the departments capabilities in respect to the agreed contract	• nominated Project Controller • high level programme • regulatory requirements • identify need for assistance or special services • high level resource requirements	• Project Director • Project management • Department Resource Management • project data base

Figure 16.12 Example of typical information requirements and deliverables for the role of contract review

16.8 CONCLUSIONS

1. Processes within civil engineering design are quite unlike previous applications of process modelling in that they are context dependent, concurrent, subject to major changes, one-off and difficult to represent using work charts.

2. Process modelling is based on the premise that roles and interactions are an effective way of describing what people do.

3. A role is defined as a set of responsibilities rather than a set of tasks as in previous work.

4. A self-replicating role software object has been designed and is currently being implemented in PROCESSWISE.

5. The structure of knowledge in the role object has been outlined.

6. Some of the knowledge in a contract review role object has been outlined briefly.

REFERENCES

Blockley, D.I. (1992) Engineering from reflective practice, *Research in Engineering Design*, **4**, 13–22.

Handy, C.B. (1985) *Understanding Organisations*, 3rd edn, Penguin Books, London.

Pidgeon, N.F., Turner, B.A. and Blockley, D.I. (1991) The use of grounded theory for conceptual analysis in knowledge elicitation, *International Journal of Man–Machine Studies*, **35**, 151–73.

Senge, P. (1990) *The Fifth Discipline: The Art and Practice of the Learning Organisation*, Century Business Books.

Snowdon, R.A. (1990) An Introduction to the IPSE 2.5 project, in *Software Engineering Environments*, Long, F. (ed.), Springer-Verlag, Lecture Notes in Computer Science 467.

CHAPTER 17

From modelling to applications

Victor Sanvido, College of Engineering, Department of Architectural Engineering, The Pennsylvania State University, 104 Engineering Unit 'A', University Park, PA 16802, USA

17.1 INTRODUCTION

Computer integrated construction (CIC) is a new emerging research field. Its development is largely fuelled by the success of computer integrated manufacturing (CIM), a similar concept in manufacturing. CIC involves the application of computers to better manage information and knowledge in their various forms with the goal of totally integrating the managing, planning, design, construction and operation of facilities (Sanvido *et al.*, 1990). In order to utilize the computer, the users need a clear definition and architecture to organize, classify and manage the required information.

Many universities and research laboratories are actively working in the CIC area. Their work in product modelling, process modelling and developing integrated prototypes is represented in large part by the chapters in this book. It is my goal to present in this chapter a set of conceptual models and a common philosophy among the several application projects in progress. These models are presented to elicit discussion and share ideas.

17.1.1 Overview of the Penn State CIC research programme

The central goal of the CIC research programme is to define and develop better methods and computer tools to integrate the management, planning, design, construction and operation of facilities. The contribution of this research will be to define a framework for representing and integrating the key decisions and intent of the various participants in the process, and to look for totally new methods that take advantage of the computer's full capabilities. The CIC Research Laboratory houses 10 to 12 researchers with

Integrated Construction Information. Edited by Peter Brandon and Martin Betts. Published in 1995 by E & FN Spon, 2–6 Boundary Row, London SE1 8HN.
ISBN: 0 419 20370 2

architecture, architectural, civil, electrical and industrial engineering and construction backgrounds. Several faculty and industry practitioners interact with these researchers through lectures and collaboration on projects.

CIC research is divided into basic and applied research. Basic research includes process and product modelling. Much of the leading current research in product modelling was presented in Part 3 of this book. The Integrated Building Process Model – a process model of the essential activities required to manage, plan, design, construct and operate a facility – was developed from analysis of and visits to 22 successful building projects. Second, the Process Based Information Architecture – a classification of the information required to support and link the management, planning, design, construction and operation of a facility – required the collection of information from many projects over a three-year period and organization into five key categories. These models form the basis for the applied research.

Applied research develops new methods and applies several computer tools such as group technology, hypermedia, expert systems and neural networks to solve field problems. Projects completed to date include a method to select vertical and horizontal formwork systems for buildings, and a method to size hoists based on the daily labour and material requirements on a floor-by-floor basis.

Several agencies and companies sponsored the research described in this chapter. The largest sponsor is the US National Science Foundation which has funded the basic research. Specific industry applications were funded by Fujita Corporation and individual corporate sponsors, the Construction Industry Institute, the Consortium for the Advancement of Building Sciences, the Ben Franklin Partnership and the Partnership for Achieving Construction Excellence.

17.2   BASIC RESEARCH MODELS

17.2.1   Development of the process model – the IBPM

Our first step was to develop a conceptual model, or framework, of the processes and information required to provide a facility. This process model, developed from the viewpoint of a 'master builder' (Sanvido *et al.*, 1992b), identifies the essential processes required to manage, plan, design, construct and operate buildings. The Integrated Building Process Model (IBPM) (Sanvido *et al.*, 1990) identifies all major functions required to provide an operational facility to the end-user over the life of the facility. The model also identifies the information that is produced and utilized by each function. It was developed using the IDEF0 modelling methodology (Harrington, 1985) to four hierarchical levels of detail. A total of 22 building case studies were used to develop and test this model.

The IBPM in its original form had many elements and flow lines. In order to simplify the model, the elements that flow between two given subprocesses were grouped and designated as one arrow. Second, elements that were similar and generic to several subfunctions were identified. These generically similar information elements form the components of the information architecture.

17.2.2 Development of a process-based information architecture

When examining the process models, it becomes apparent that the information that is used to provide the facility can be divided into five categories (Sanvido, 1990). These are:

1. Information describing the methods used to provide the facility, or the *process elements*.

2. Information describing the physical properties of the facility, or the *product elements*.

3. Information describing the management directives to the planner, designer, constructor and operator, or the *process control elements*.

4. Information describing feedback loops from the planner, designer, constructor and operator to the manage function, or the *feedback elements*.

5. Finally, information describing the factors that impact the project, or the *constraint elements*.

These are defined in detail in Sanvido *et al.* (1992a).

This analysis of the process model defined five categories of information in the Process Based Information Architecture (PBIA). The PBIA (Sanvido, 1990) further subdivided information to three levels of detail and identified the source and destination of such packages. For example, the product information was divided into information packages such as the facility idea, planning information, design information and construction information. The planning information was divided into the program, site information, project execution plan and facility planning knowledge. Although these categories are more explicit than the parent category, they do not identify measurable pieces of information.

To further explore the content of the PBIA, the product information was divided into elements that make up a system and the systems that make up a building (Khayyal and Sanvido, 1991). This decomposition led to the development of a simple product model that divided systems into functional units. For example, the architectural system divided the building into floors, external envelope and vertical connectors. These decomposed into rooms, horizontal connectors and service spaces. Rooms are then decomposed into components. Khayyal further made a classification system to identify these elements. This was then implemented in a hyper-media environment and frames were structured to hold several types of information including function, form, cost and time data. (Evt *et al.*, 1992).

Further efforts by Sanvido and Messner (1992) decomposed the process control information into subcategories. As an example, the facility team was decomposed into subteams, e.g. management team, and further into the construction manager for the owner. This construction manager has several subordinate functions such as preconstruction, field representative or commissioning reporting back. The contract, experience and resource elements were similarly decomposed.

A third effort at decomposition is currently going on in the area of con-structability information. Prior decomposition efforts had not expressly addressed links to other data groups. This project is linking the con-structability information to the product model, the process and the process control information. It is being implemented in a multimedia envi-ronment using construction project data (Hanlon and Sanvido, 1993).

17.2.3 Process and information synthesis

'Linking levels of abstraction of a building design' (Sanvido, 1992a) achieved several important steps for our programme. First, it recognized the similarities between the relationship of the manage function to each of the plan, design, construct and operate processes. A generic relationship between the manage function and the generic 'contracted' functions was defined. The inputs, outputs, constraints and mechanisms were generi-cally defined for each of these functions.

The second important contribution of this paper was that it provided a logical tie between the processes and the various product information packages. Each product information package was loosely linked to various levels of the product model developed earlier. This was our first attempt at logically defining the product information content based on aggrega-tions from the process model.

Our modelling efforts were driven by trying to put a logical organiza-tion to all the information required to drive the process of providing a facility to an owner. In effect, we were trying to create a virtual informa-tion base similar to that of the master builder earlier this century. Sanvido *et al.* (1992b) combined the research of these three research programmes and collaboratively identified the informational, behavioural and comput-ing attributes of a virtual master builder. It is notable that several industry-wide, project-specific and technological barriers to the implementation of this concept were also identified. Conceptually, this type of integrating research may be a key step on the road towards integrated construction information.

17.3 APPLIED RESEARCH

As these models evolved, researchers tested portions of them by developing integrated solutions to their specific areas of interest. These projects try to select a key decision in, or viewpoint of, the delivery process and develop a method or tool to assist a professional in their project role. In developing the solution, we define the role or viewpoint in terms of elements in the information architecture and pose several scenarios to the model to test its robustness.

Three recent proofs of architecture projects are described. First, an information framework for facility operators (Beckett and Sanvido, 1991) defined and classified the essential information that a facility operator needs to better operate a facility. This framework (which is based on the CSI scheme) can be used to store and retrieve this essential information in multiple formats. Second, Perkinson *et al.* (1992) defined and classified the essential information that should be contained in a facility program. This framework contains the information that is the basis for a design and should be used to evaluate the design or to select alternatives. The third project defined the rules governing the selection of a project delivery decision (Vesay and Sanvido, 1992). It assists an owner in selecting the desired organizational structure and the contract strategy for a given project. It considers project requirements and risks, as well as team characteristics and uncertainties.

In addition to these procedural methods, two computer prototypes (Evt *et al.*, 1992; Kamarthi *et al.*, 1992), were developed to explore the use of hypermedia to store product information, and to compare expert systems with neural networks in selecting a formwork system, respectively. Both projects illustrated the limitations and strengths of the software. Lessons learned in both cases were instrumental in defining the scope of work and in selecting the computing tools for our first large-scale prototype described below.

Fujita Corporation of Japan and the Penn State CIC programme (Messner and Sanvido, 1992) are jointly developing a full scale computer system to assist engineers in Fujita to better select precast concrete systems for appropriate design–build projects. Fujita is a major contractor in Japan, that is, its second largest producer of precast concrete buildings. Together we are developing a system to help select two or three precast systems based on the preliminary member sizes and configurations. Manufacturing and construction plans are then developed in detail for each of these systems. For each system, the appropriate number of casting beds, manpower levels and time are presented for each phase to a decision-maker, who then selects the appropriate system for detailed design. There are three key components in the system.

The first component is StructNet (Messner *et al.*, 1993), a neural network that determines two or three most probable precast structural methods practised by Fujita for a given project. The neural network was used to overcome the need to learn the rules behind the selection process. This process was not well understood by the designers, and the language barrier made it hard to communicate. It is interesting to note that after the system was trained with the company history, the first three projects did not fit the successful profile. It was explained to us that these were done as a demonstration of company capability, and different systems would be selected given the company's experience.

The second element is the Construction and Manufacturing Planning System (CMPS), which is a concurrent planning system that plans the manufacturing and construction methods for given facilities in each of the structural systems proposed by StructNet (Messner and Sanvido, 1993a). It provides key resource requirements and cost and time data to help a planner select the best system for the site, given manufacturing space and demands on the company's plants and the required construction schedule.

The third part of this system is an object-oriented database structure, to manage the data used by these applications and the downstream computer software for detailed design and scheduling (Messner and Sanvido, 1993b). It is developed in Nexpert Object with a supercard interface.

An important part of this project was the manner in which it was conducted. Researchers spent several years working together in both Japan and the United States to understand the different construction cultures. The prototype was developed on a Macintosh platform in the United States. After testing by Japanese engineers, the software was translated into Kanji by a Japanese contract programmer under the supervision of the Fujita CIC research team. Social acceptance of the software by the large CIC research team is definitely as important as the content of the software. Fujita has wisely changed the way they do business in this area to meet the human requirements of this new system. Clearly the benefits of the new system knowledge are evident in both the technological and cultural changes to the organization.

17.4　CURRENT RESEARCH

The research philosophy in the CIC programme is to develop basic research models using a team approach. All members of the research team are currently refining the information architecture at its lower levels. The previous version of the model (Sanvido *et al.*, 1992a) has been defined to three levels of detail. Current work is extending the level of detail and identifying cross-ties between the various information categories (Sanvido, 1992b).

Each researcher also works on applying a portion of this information architecture to a critical area of interest to the industry. Currently seven projects are underway in our laboratory. Each focus on industry applications and will develop software where appropriate, to illustrate the solution in terms of the information architecture (IA).

17.4.1 A design–build project team selection system – Kevin Potter

The goal of this research is to define those qualities in a design–build team which a public sector owner can use to objectively prequalify a potential design–build project team. This system will focus on better controlling the design–build process through proper selection and evaluation of the project team. The attributes of the project team portion of the IA is being further defined and tested.

17.4.2 Contracting issues in construction – Ted Lynch

The goal of this work is to develop a logical method to identify key contracting issues so that participants can better meet the project goals and complete the project successfully. This research will define the process of selecting a project delivery method. It will provide a method for choosing a project organization, contract strategy and select particular contract forms that will properly assign duties and risks to those who are best qualified to assume them. This project defines design at the time of this decision in terms of the product model and compares that with appropriate contract models.

17.4.3 Productivity improvement information and teams – Kirby Kuntz

This research develops a set of criteria and procedures for managing site-level information that will aid site-level personnel in completing their day-to-day tasks by identifying what information they need and where it can be found. This information system should be broad enough to accept knowledge and expertise from all levels of the project and information in different formats. It should also recognize the different characteristics of the users. This project tests the lower levels of the IA.

17.4.4 Environmental efficiency management – Kathelene Bisko

The focus of this research is to define a method for environmental efficiency management for the construction industry which provides a competitive advantage, supports the future of the industry and decreases the impact on the natural environment. The research will define a method to define, promote, support and measure environmentally efficient construction practices within the industry. A life cycle view of materials which is longer than that of a facility, will identify new categories of information to be considered in the IA.

17.4.5 Strategic planning in the international construction industry – John Messner

This research develops a structured decision-making procedure to evaluate the chances of success of a company entering international construction markets. This procedure will determine whether the firm should venture internationally and identify the appropriate markets for that company. This project defines a process model that is broader than our current 'provide facility' focus. In fact, it is the next level above this model.

17.4.6 Planning and managing space for construction activities – David Riley

This project seeks to develop a space planning method to better identify and predict spatial requirements of construction activities. It utilizes past experience and expert knowledge in the space planning process. This planning experience will be used to better represent and organize planning information to support planning decisions. Finally, the new system will communicate and implement space utilization plans in the field. This system identifies the dynamic characteristics of space, a component of the IA that is typically thought of as a static model.

17.4.7 Classifying constructability information – Eric Hanlon

This project develops a generic method for classifying constructability information for the conceptual through schematic design phases of a construction project. The system should enable ease of information access and sorting for the various discipline views. This will benefit construction companies which seek to develop 'lessons learned' databases for increased competitiveness. Information is stored at various stages of the process against the product model.

17.5 TRANSFERRING RESULTS TO INDUSTRY – PACE

In our attempt to implement new ideas into the industry, we have specifically identified our partners as the construction industry, our students and the Faculty at Penn State. They are organized as the Partnership for Achieving Construction Excellence (PACE). The mission of PACE is 'to establish a working partnership between the construction industry and Penn State to achieve excellence in construction through process innovation and the development of students into leaders that shape its future'.

PACE is made up of industry leaders, students and faculty. It has several committees providing input to our education and research programme. Industry members comprising contractors, owners and designers, are innovative companies who wish to continually upgrade their processes and develop their future leaders through this partnership. PACE research activities include an annual construction roundtable where members discuss key issues facing the industry, a research programme where PACE members address problems facing the industry and an annual research seminar where members present examples of innovation.

This philosophy of developing integrated construction information in this environment has resulted in students that are well-versed in theoretical modelling and developing large-scale models in teams. In addition, these students have identified an area of critical importance to a group of industry professionals and worked with them, using the models as a basis, to develop new methods or systems to better support decision-making.

We feel that this methodology allows us to develop accurate models, while at the same time, defining small complete decision support systems. Researchers are able to contribute to a large project, yet they can fully develop their own field of expertise. Industry members are exposed to the research methods and are able to hire researchers, upon graduation, to implement these findings in their companies. This method allows all participants to gain favourably from their invested resources.

17.6   CONCLUSION

This overview of the Penn State CIC research programme has illustrated the development of a process model and its analysis to develop the resulting Information Architecture (IA). As shown earlier in the chapter, we have used the IA as the basis for all of our applied research. This gives the laboratory projects an overall goal and focus, and the success of the projects are an excellent test bed for the IA itself. Clearly we have a long way to go, but I believe that we have a basis and plan for future work.

ACKNOWLEDGEMENTS

This research programme is the result of the efforts of many researchers who have contributed to, or been a member of, the Penn State CIC Research Program. Many companies have provided access to their people and sites for data collection and feedback on our projects. The National Science Foundation in the United States provided funding under several grants to establish this programme. Members of PACE and many other agencies and companies have also funded this research. To all these people and agencies, I thank you for your continued support. Finally, I wish to thank my fellow researchers for their contributions to our programme.

REFERENCES

Beckett, J.P. and Sanvido, V.E. *INFO: An Information Framework for Facility Operators*. Paper presented at the Eighth Conference on Computing in Civil Engineering, ASCE, Dallas, TX, June 1992, pp. 57–64.

Evt, S.K., Khayyal, S.A. and Sanvido, V.E. (1992) Representing building product information using hypermedia. *Journal of Computing in Civil Engineering*, 6 (1), 3–18.

Hanlon, E.J. and Sanvido, V.E. (1993) *Classifying Constructability Information*. Paper presented at the Fifth International Conference on Computing in Civil and Building Engineering, Anaheim, June 1993.

Harrington, J. Jr. (1985) *Understanding the Manufacturing Process, Key to Successful CAD/CAM Implementation*, Marcel Dekker, Inc. New York.

Kamarthi, S., Sanvido, V.E. and Kumara, S.R.T. (1992) *A Connectionist Vertical Formwork Selection System*. Paper presented at the Eighth Conference on Computing in Civil Engineering, ASCE, Dallas, TX, June 1992, pp. 1171–8.

Khayyal, S.A. and Sanvido, V.E. (1991) *Classifying Building Product Information*. Paper presented at the 7th Conference on Computing in Civil Engineering, ASCE, Washington DC, May 6–8, 1991.

Messner, J.I. and Sanvido, V.E. (1992) A Survey of Precast Concrete Systems Used by Fujita. *Technical Report No. 27*, Computer Integrated Construction Research Program, The Pennsylvania State University, February.

Messner, J.I. and Sanvido, V.E. (1993a) Concurrent Planning for Precast Concrete Systems. Construction Congress III, ASCE, San Francisco, March 1993.

Messner, J.I. and Sanvido, V.E. (1993b) *Developing an Integrated Database Structure for the Lifecycle of Precast Concrete Projects*. Paper presented at the Fifth International Conference on Computing in Civil and Building Engineering, Anaheim, June 1993.

Messner, J.I., Sanvido, V.E. and Kumara, S.R.T. (1993) StructNet: A neural network for structural system selection. *Microcomputers in Civil Engineering*, in press.

Perkinson, G.M., Grobler F. and Sanvido, V.E. (1992) *A Facility Programming Product Model*. Paper presented at the Eighth Conference on Computing in Civil Engineering, ASCE, Dallas, TX, June 1992, pp. 41–8.

Sanvido, V.E. (1990) Towards a process based information architecture for construction. *Civil Engineering Systems*, 7 (3), 157–69.

Sanvido, V.E. (1992a) Linking levels of abstraction of a building design. *Building and Environment*, 27 (2), 195–208.

Sanvido, V.E. (1992b) *Penn State Computer Integrated Construction Research Program*. Paper presented at the CIB W78 Meeting, CIB92 World Building Congress, May 1992, Montreal, Canada.

Sanvido, V.E. and Messner, J.I. (1992) *Classifying Process Control Information*. Paper presented at the Eighth Conference on Computing in Civil Engineering, ASCE, Dallas, TX, June 1992, pp. 340–7.

Sanvido, V.E., Khayyal, S.A., Guvenis, M., *et al.* (1990) An Integrated Building Process Model. *Technical Report No. 1*, Computer Integrated Construction Research Program, The Pennsylvania State University, January.

Sanvido, V.E., Anzola, G., Bennett, S. *et al.* (1992a) Information Architecture for Buildings. *Technical Report No. 28*, Computer Integrated Construction Research Program, The Pennsylvania State University, February.

Sanvido, V.E., Fenves, S.J. and Wilson, J.L. (1992b) Aspects of a virtual master builder. *Journal of Professional Issues in Engineering Education and Practise*, 118 (3), 261–78.

Vesay, T. and Sanvido, V.E. (1992) *A Project Delivery Selection System*. CIB92 World Building Congress, May 1992, Montreal, Canada, pp. 474–5.

PART FIVE

Applications of integration

The final part of the book brings us from the need, the history and the theory to the stage of speculating about real applications. The first chapter in this section by Atkin (Chapter 18) describes, from a practical perspective, much of the difficulty faced on construction projects with information. It then describes an early attempt at addressing this problem in the form of a software product developed in Australia. It is a first step, albeit rather elementary, on the path towards integrated construction information.

This is followed by Chapter 19 from Oxman of Technion in Israel. She builds upon the technologies described in earlier sections and outlines their potential application to the subject of multiagent collaboration. She uses the ICON project, as described earlier by Cooper and Brandon (Chapter 15), to illustrate her arguments and suggests that case-based reasoning is a key application area for integrated construction information.

The third application chapter in this part of the book is from the group from VTT in Finland led by Björk. They are best known for their work in product modelling but their contribution to this volume has repeated some of the rigour for which they are known, to the application area of document management. This is an area of integrated construction information addressing more practical issues than their work normally concerns itself with.

Chapter 21 from the group in Leicester is also concerned with an applications issue. However, in this case the focus is of how to integrate construction information in the application area of environmental design. The group have used some of the product modelling technology described in Part Three in developing an integrated solution to their particular application area. This chapter illustrates well the way that the different types of contributions to be found in this book will be combined in finding practical solutions.

The final chapter in the book continues this theme, but within the application area of cost estimating. Choi and Ibbs focus on concurrent engineering technology and illustrate how this can be used in this particular application area.

As a whole, the contributions in this final part show the large range of diverse areas in which integration applications may be expected to arise. They also illustrate the variety of technologies that may be looked to as the initial integrating mechanism.

CHAPTER 18

Information management of construction projects

Brian Atkin, Department of Construction Management & Engineering, University of Reading, Reading, UK

18.1 INTRODUCTION

Information is the life-blood of a construction project. Too little and the project becomes starved; too much and it becomes swamped. Achieving a balance that satisfies the owner's objectives of time, cost and quality is crucial. Information management has to be seen as a necessary function, one that is the key to the efficient management of a project.

It could be argued that information management is at the same stage now that project management was at 20 years ago, being largely regarded as a normal part of the work of the architect or site manager. Since then, project management has emerged from the shadows and become firmly established as a discipline in its own right. Although useful similarities can be drawn between information management and project management, there are also differences. This chapter aims to examine the subject of information management by discussing some of the problems that exist on real projects and outlining solutions. It will introduce proposals for an integrated approach to the management of information on projects and provide evidence of an example, working system.

As this chapter will show, technology is not the only problem. The problems are largely in relation to people and the way they conduct themselves in managing projects. Traditional practices and customs in the construction industry have been responsible for holding back innovation in many areas, not least of all in the application of information technology. An alternative title for this chapter might have been, *there must be a better way than this*. It is hoped that it will provide a key for unlocking the potential that information technology offers and as a result might ensure the more efficient and effective management of information.

Integrated Construction Information. Edited by Peter Brandon and Martin Betts. Published in 1995 by E & FN Spon, 2–6 Boundary Row, London SE1 8HN.
ISBN: 0 419 20370 2

18.2   IMPORTANCE OF MANAGING INFORMATION

18.2.1   What do we mean by information management?

Information management is concerned with communication and covers its acquisition, generation, preparation, organization and dissemination, analysis of information and the design, implementation, evaluation and management of information resources (Hardcastle, 1982). In simple terms, information management means ensuring that information about a project is communicated to whoever needs it, whenever they need, in whatever form they need it, so that they may meet their objectives for that project. Information covers drawings, specifications, bills of quantities, schedules, programmes, financial statements and so on.

The substantial research documented in other chapters of this book, into the application of information technology shows that information management has now emerged as a serious subject, serious enough for some organizations to have invested considerable resources in developing sophisticated computer-based systems. Discovery of the need for information management has doubtless been made time and again by practitioners, yet until recently the subject has received little attention. It would be fair to say that many people fail to comprehend the size of the information problem caused by construction on a large scale and, very often, on a fast-track.

18.2.2   Larger and more complex projects

Construction projects do not have to be large to suffer from the late supply of information, but it does help! The effect of late information is usually to pave the way for a claim by the contractor for an extension to the contract period and reimbursement of costs associated with prolongation.

The need to manage information has always existed. People may not have taken it very seriously, but they have generally recognized its importance. With very large construction projects, problems are easily compounded. So, while a £1 million project might involve, at the most, 100 drawings, a major project costing £100 million might generate 10000 drawings and a difficult integration problem.

Increasing numbers of major and complex projects have helped to publicize problems of information management, yet the need to manage information exists on all projects. Information management amounts to more than handling drawing issues. The major difficulty is in getting people to recognize the scope of the problem which, on many smaller projects, may not be readily apparent. Furthermore, managing information is generally assumed to be part of the normal duties of the architect. It simply gets done, or does not, according to a number of factors.

Many types of buildings today are very different to those of yesterday, especially in terms of their engineering services and fit-out. Perhaps what is more significant is that the number of projects now in progress, as compared with, say, 10 years ago, is much greater. A consequence is that more people and organizations are having to communicate on an increasing number of fronts. Pressure from owners to complete projects quickly and competition from within the industry have forced innovative forms of procurement to appear, that is, forms of procurement that did not exist in their present state more than 10 years ago. With these forms of procurement have come new forms of contract, each with its own special way of allocating risk among the contracting parties. There may well be many individual contracts in existence: project manager to owner; designers to owner; contractor to owner; subcontractor to contractor; and so on. An inevitable consequence of these myriad contracts is that the communication chain may be broken at some point, but this may come to light only once a problem has occurred.

18.3 POTENTIAL FOR PROBLEMS

18.3.1 Conflict begins with the contract

Regular forecasting of the likely outcome of a project is essential if a manager is to attempt sensible decisions. The earlier the need is recognized, the greater will be the flexibility available to the manager to take corrective action. Unfortunately, contract forms make little, if any, mention of prevention. We are faced instead with a situation which prescribes a cure.

This cure mentality is one of the biggest problems afflicting the industry. It is responsible for people adopting the attitude that claims are inevitable and that one might as well have every avenue covered. Consider the matter of extension of time. Most contract forms require that contractors take all reasonable steps to avoid delays (even where potential delays are not of their making); but should a delay occur, the contractor must minimize its effect upon the works. A question which is likely to arise here is whether or not it is possible to prove that the contractor has taken all reasonable steps. The contractor is bound to say that this has been done.

Apart from goodwill between the parties, there needs to be some way of ensuring that the right kind of information is always provided at the right time. In a situation of confrontation, negative attitudes and the cure mentality, the contractor may well make little more than a superficial attempt to retrieve a worsening situation. There may be no incentive for the contractor to make an effort when the inconvenience or disruption purported to have been suffered might result in some further reimbursement from the owner. It might even deflect attention from the contractor's own problems, should there be any. Instead of suffering financially the contractor could benefit. Quite obviously, there must be some way of providing an incentive for the contractor and other parties to use their best endeavours and not to hide behind the contract form.

18.3.2 Typical problems

Discrepancies between tenders and final accounts have long been a source of complaint by owners and may be a manifestation of problems arising on site. In a study (BCIS, 1988), tenders and final accounts were compared across a range of building projects. Of the contracts covered in the study – data were taken from 35 contracts drawn at random from a convenience sample – 47% exceeded the tender figure. The average increase in price for these 50 contracts was, however, just 1.33%.

The study covered both cost and time overruns and it was in the latter category that the largest overruns were found: 71% of all contracts exceeded the contract period by an average of 15%. The sample size is too small to place firm reliance upon the figures produced, but the trend which the analysis indicates is probably a fair reflection of reality.

Data on time were split into delays prior to commencement on site, delays allowable within the contract and delays caused by the contractor or subcontractors. Data on cost, taken from the final accounts, were broken down to show the additions and omissions of sections covering *rise and fall*, variations and claims. A comparison of rise and fall with fixed price contracts enabled data to be analysed on a consistent basis, as increased costs were included separately. Variations were further separated into contingencies, dayworks, prime cost sums, provisional sums and measured work. Claims by the contractor, for which agreement had been reached, were analysed according to their cause. Delays were treated similarly, that is, where the contract date for completion had been postponed and reasons were identified.

Subsequent analysis revealed that 28% of all delays were due to exceptionally adverse weather, while a further 26% were caused by variations (changes and additional work) or late information/instructions. Labour shortages and strikes accounted for just 10% of delays. Materials unavailability or late delivery accounted for a further 10% of all delays recorded.

As the BCIS study has found, the greatest overrun is in time. Inferences may be drawn from this finding, but it would be unwise to place total reliance upon them without further investigation. Nevertheless, it is interesting to ponder on whether or not an increase in the price of a contract of 1.33% would be more welcome than a time overrun of 15%. These two factors may not be related and owners would doubtless prefer neither overrun to occur. On contracts where *time is money*, such as in the case of commercial developments, the cost of a delay of this order could well exceed any increase in contract price many times over.

18.4 INFORMATION IN THE INDUSTRY

18.4.1 Perceptions of information need

It is not exceptional for members of project teams to find themselves in receipt of information of little practical use, while at the same time without information that is desperately needed. Anticipating what another party might want in the way of project information is not the same as knowing what they need. Time taken to provide information of marginal value, at the expense of more valuable and much needed information, is time wasted. Admittedly, it may not always be possible to provide necessary and sufficient information that matches precisely the needs of others. Compromise may be necessary, leading to information wastage of one sort or another.

There are many potential areas of waste on a project. Non-standard specifications encourage information wastage and unnecessary cost. A standard, industry-wide specification, as is the custom in, for instance, Japan and West Germany, would mean that project documentation could be reduced to a few sheets of paper indicating the parts that apply or, conversely those that do not apply, to the project. The least unlikely culprits can also turn out to be the most guilty. For instance, the increasing use of CAD to generate more design options, which can help improve the quality of design solutions, can also create abortive work – and this is work which has to be paid for by someone. In itself, the desire to explore more design options is laudable, provided that projects are well-defined. Ill-defined design briefs, on the other hand, will cause work to be abortive. Careful control of briefs and engaging the owner in discussion during the design process will help to narrow options, enabling the design team to focus on the most workable solution sooner.

18.4.2 Volume of information

On every construction project, information must be passed among many people and organizations. There may be one owner, but there may be several designers (architect, structural engineer, environmental services engineer, fire protection engineer, cabling specialist and so on), numerous specialist and trades' contractors, as well as the main or managing contractor and quantity surveyor.

Table 18.1 was prepared using data taken from construction contracts so as to indicate the volume of information generated on projects of different sizes. Approximately 40 staff years of effort are required for every million pounds of construction cost. Over 700 people are likely to be involved in a £15 million project. The paperwork involved in administration, with respect to these people, would represent a substantial, additional burden on the project, and one that is difficult to measure.

Table 18.1 Information volume by contract value

| | Contract value (£ million) | | | | |
	350	40	25	7	5
Contracts	300	80	170	100	50
Tenders	1200	300	700	400	200
Drawings	5000	1000	30 000	4000	200
Drawings issued	400 000	100 000	150 000	48 000	30 000
Variations	10 000	1000	5000	150	300
Site instructions	30 000	3000	6000	600	1000
Rooms/areas	3000	100	11	7	50
Consultants		8	11	12	5
Cashflow/ month	£7m	£1m	£0.75m	£0.3m	£0.25m
Approvals/ week	70	26	n/a	n/a	13
Meetings/ week	20	10	30	11	6

Today's construction environment often means that additional parties such as financiers, lawyers and letting agents are employed. With the increasing number of participants, transmission of accurate information becomes a complex business. Information must be transmitted expeditiously if it is to be communicated effectively.

The figures in Table 18.1, however, represent the beginning of the information likely to be generated on a project. Each task undertaken on a project can *explode* into a requirement to generate even more information. For example, variations (or change orders) have descriptions, cost estimates, financial statements and, possibly, revised drawing issues with changed responsibilities affecting (sub)contracts. Each detail in the process depends upon another and must be communicated to many people.

18.4.3 Types of information

It is easy to talk about information in the abstract, it is another thing to define it in terms of what it is intended to satisfy. Information is not homogeneous: it is found in innumerable formats, intended to suit equally as many different purposes. A failure to understand that different information satisfies different needs, when designing a computer-based system, could easily lead to a solution that was no solution at all. If information is not the same, why treat it as though it were?

An examination of the types of information used in the industry reveals three broad types. Recognizing and accounting for these different types is of critical importance in developing integrated information systems in construction:

1. commercial;

2. technical; and

3. managerial.

Technical needs are generally determined by standards, building codes and design principles. Commercial and managerial needs are probably less well understood and, indeed, the differences between them less so. It is possible to draw distinctions between commercial, technical and managerial needs according to the criteria of accuracy, timeliness, control and auditing.

Accuracy

Commercial information, which is regulated by statute or professional accounting standards, for example, tax returns, balance sheets and payroll calculation (and which can result in penalties if the answer is incorrect) must always be accurate. Technical information also needs to maintain a high level of accuracy if it is to be useful.

Managerial information, which is used to aid decision-making, rarely has to be absolutely correct. Decisions can often be made by drawing on experience and intuition to forecast results and take preventive action.

Timeliness

Accurate commercial information inevitably requires time and checking procedures to produce the correct answer. Generally, commercial information is not time critical, but it will always follow events so as to ensure correctness. Technical information needs to be available more or less on time if it is not to lead to a delay.

As for managerial information, it is a fundamental management principle that one can control only tomorrow's events, not yesterday's or even today's. Consequently, information which is, say, 80% accurate, but on time, is more useful than information that is 100% accurate and late.

Control

Commercial information has to be accurate and therefore it is, by necessity, historical. This has a value only if it produces trends that would help in indicating future events. Implicit within most technical information are procedures which must be followed if successful results are to ensue.

Managerial information must, by necessity, forecast the future. A 100% accurate forecast is most unlikely, although a forecast will converge on actuality as time passes. Many management decisions have a long-term impact lead time and therefore varying forecast accuracies must be tolerated.

Auditing

The statutory nature of some commercial information is such that audit is a necessity. Thus, detailed records must be kept with a complete audit trail available for those with responsibility for ensuring accuracy.

There is very little need for an audit trail where technical or project-related information is concerned unless, of course, legal or contractual complications arise. Sometimes decision-making is deliberately slowed down to allow a complete audit trail to be maintained. This tends to result in missed opportunities.

These information needs can be shown relative to their accuracy, timeliness, control and auditing – see Table 18.2.

18.4.4   Relationship between information types

The relationship between the different types of information, namely, commercial, technical and managerial, and the broad objectives of time, cost, quality and (project) feasibility can also be shown diagrammatically (see Figure 18.1). The areas of overlap indicated in Figure 18.1 can be equated to the administrative aspects of information. The centre of the figure depicts the library of experience (or knowledge) which is collected on projects, for example, functional briefs, estimating price rates and lists of drawings. Information within these three sectors might include:

1. Commercial information: accounts, payroll, shareholders reporting, statutory requirements and plant accounting.

Table 18.2   Relationship between information types

Information type	Importance of various criteria			
	Accuracy	*Timeliness*	*Control*	*Auditing*
Commercial	100%	Not time critical	<100%	100%
Technical	Approaching 100%	>80%	100%	Some requirement
Managerial	80%	100%	Significant requirement	Minimal requirement

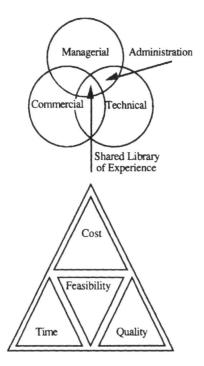

Figure 18.1 Information needs relative to time, cost, quality and feasibility

2. Technical information: engineering calculations, environmental impact analyses, functional specifications, detailed design specifications, bills of quantities and computer-drafted drawings.

3. Managerial information: time, cost, quality and feasibility aspects of the project.

The criteria used to differentiate between commercial and managerial information also suggests that administrative procedures and the use of information technology to assist in gathering and processing this information may not be compatible.

In an example provided to the author, the project team collected management information via the commercial stream so as to minimize on data collection costs. Unfortunately, the resulting management information was so late as to make it lose most of its impact upon decision-making. It had been generated as a by-product from systems designed to handle statutorily imposed controls. The accountants were found to be over-emphasizing commercial requirements and had no appreciation of the needs of the project team who were responsible for satisfying the project brief.

There are further examples of where corporate needs can overtake project needs in importance. These concern projects on which contractors and construction managers have their own accounting systems for providing cost information. These systems are generally geared to providing information which satisfies accountants, not the owner. When project managers are employed to manage a project they can very easily find themselves without the right kind of information. Consequently, they may have to resort to *seat-of-the-pants* management for which the industry is, regrettably, well known.

Some contractors have attempted to solve the problems created by their organization's preoccupation with financial accounting at the expense of project accountability. Head office accounting systems tend to treat sites as cost centres, whereas they need to be regarded as profit centres if projects are to be managed successfully. On some sites, the contract's staff have developed their own simple, but effective, reporting systems to suit the individual needs of the project, that is, the needs of the project manager and owner. This has enabled the project's time, cost and quality to be related more readily to the project team's performance.

18.5   THE PAPERLESS PROJECT: MYTH, POSSIBILITY OR REALITY?

18.5.1   More paper not less

Anyone who has thought for one moment about the increasing use of information technology will know that, if anything, it is being used to generate more paper, not less. It seems, therefore, that we might be moving further away from the paperless office and paperless project, rather than towards them.

Mistrust of computer storage media and the desire to retain earlier versions of one's work are just two examples of where the technology is helping people add to the paper mountain. People problems are not the only ones, although they do account for most of them. The ability to produce many variations on a given theme means that projects and offices are becoming swamped with paper.

A fundamental problem, viewed in information technology terms, is in providing a cost-effective means for preserving details of all previous transactions and ensuring the accuracy of the audit trail so produced. This problem area has two aspects: first there is the internal system, and second there is the external system. Audit trails are clearly useful tools to aid management within an organization, that is, the internal system. If things do go wrong, it is helpful to have some means of pinpointing past decisions. Such problems may be of little concern or consequence to external organizations or persons. However, where information has been passed externally and then acted upon in good faith and a problem has arisen, the need to have full documentation is imperative: paper satisfies this need easily and cost-effectively.

Information technology creates problems as a consequence of one of its most important features, that is, the ability to update information rapidly. Earlier versions of documents may become lost forever, with the consequence that it may not be possible to produce important, original documentation at a later date.

It would be missing the point somewhat, if we did not admit that a paperless state, both for projects and organizations, is a legitimate objective. The industry spends most of its time in communicating between various parties in the effort to design and construct buildings. Electronic, paperless communication has an obvious attraction.

18.5.2 Reducing the paper mountain

The contract forms themselves represent just a very small part of the total information base on any project. It is the storage and retrieval of the remaining information which causes the problems: correspondence, drawings, specifications, bills of quantities, tenders, architect's instructions, requests for information (and their responses), minutes of meetings, drawing transmittals, purchase orders, delivery notes and invoices, and documentation dealing with the administration of the contract (for example, progress payments, variations and claims for extensions of time).

Time and cost issues tend to require masses of information and data, underlining the need to have past records available as evidence. The storage of such information in a computer system can make for rapid search and retrieval using key words. Naturally, there is the question of time needed for entry of information in the first instance. Where a number of parties are involved, their consent to having information stored in this way is needed.

Extension of time claims are increasingly supported by great volumes of paper. The use of a computer, with project planning software, means that countless *what if?* situations can be tested by incremental changes to basic parameters. Reallocation of resources and rescheduling of activities can be greatly speeded up. Scenarios that time would not normally permit to be evaluated manually, become possible. Whereas the use of computers in this way can be a positive benefit in helping to minimize disruption and enhance decision-making capabilities, the negative side often surfaces. The computer is increasingly used to substantiate claims, which in some cases might appear all the more plausible. The ability to overwhelm people with paper is a trend that shows no sign of abating. Furthermore, the presentation of high quality, colour laser-printed barcharts, histograms, graphs and tables can add an air of respectability and credibility to a claim that might otherwise look pretty flimsy.

18.5.3 Changing attitudes

It is perhaps inevitable that the discussion in this chapter so far has focused upon the negative side of information integration. There is a positive side which would say that the use of information technology, when properly harnessed, can provide less scope for problems to arise – for instance, moving towards a situation where the project database is shared among the team. But what of the practical problems? There must be some, since it is entirely probable that the parties' agreement will be needed before documents can be entered into the system. Confidentiality becomes the next problem to solve.

It is just possible that a willingness to maintain a single source of contract documentation might instil a feeling of trust from the start and that litigation will reduce as a natural consequence. Claims will not necessarily be avoided by such developments, but might reduce significantly. The real benefits will be in bringing matters out into the open from the start of a project. This may help to avoid niggling, little problems which, although not significant in themselves, do much to destroy harmony and trust between the parties.

18.6   MANAGING OR ADMINISTERING?

18.6.1   Procedures and controls

Let us return to the distinction between management and administrative functions and the issues they raise.

First, management procedures must incorporate the controls, reports and actions necessary to ensure that time, cost and quality performance targets are met. Although it is acknowledged that some projects incorporate more detailed time controls, they still fall short of the need to cover other, important aspects of management procedures. Furthermore, some of these management procedures do not involve the contractor directly, although they do have a bearing on whether or not the building is completed on time. Thus, in the broader view of projects, we must consider management procedures in the context of the total building development from inception to occupation. By simplifying procedures and making them more explicit, we may improve communication, management reporting and decision-making. Consequently, we may expect more time to become available for prevention, even elimination, of the problems that all too often afflict projects today.

Second, administrative procedures, covering the giving of notice, approval, acceptance, valuation, certification, payment and so on, are matters of routine. They are important to a successful project outcome, but they must not preoccupy the manager to the extent that there is no time to manage the project. Passing paper around may provide evidence that people are doing something. It does not, however, indicate whether the project is meeting the performance criteria of time, cost and quality.

Part of the problem that managers face, especially when transferring to a project using a different contract form from that used on the last project, is that incorrect assumptions and inappropriate procedures might be adopted. If the manager is not careful, far too much time may be spent unravelling the administrative procedures and worrying about the contractual consequences, at the expense of managing the project. It is for this reason, and others, that management and administrative functions should be separated. Clearly, contract administration is not the same as project management. Both are aimed at different issues, they have different objectives and require, quite possibly, different types of people and computer support to make them work successfully.

Separating management and administrative functions will help to improve efficiency particularly when people transfer between projects, but these very same people still have to learn how to operate under different regimes and systems. Adequate training provides a large part of the answer.

18.6.2   Reporting responsibility

Control over management reporting must be exercised by the person accountable for the execution of the project, that is, the project manager, not someone without direct responsibility for correcting time and cost overruns. Reporting by others in a detached, *ad hoc* manner may appear, on the face of it, to make sense. Conflicts between the *controller* and *reporter* can divert attention from where it is needed most, that is, in resolving time and cost overruns. A further advantage is that the *controller* is given the chance to take corrective action before having to report the consequences, reducing the impact, perhaps, of overruns as well as *saving face*. Owners who may be concerned about the integrity of the project manager are, however, at liberty to appoint an independent project auditor.

Conflicts of interest can always surface. A time or cost overrun is inevitable when project objectives are produced by one group of people, with the final decisions on project performance being taken by another group more concerned with, say, performance than cost.

The worst example of this situation is the performance decision-maker who will be occupying part of the new facility, but who is not responsible for funding, for example, medical practitioners in hospitals, and hoteliers and restaurateurs operating on a management fee basis. The hotelier is mainly interested in maximizing profit through minimizing operating costs. Consequently, the best facilities and equipment may be wanted from the outset, with the expectation of back up by substantial spares bought out of capital budgets. This ensures that replacements from operational income are kept to a minimum. Controlling project objectives becomes crucial in such cases if capital over-expenditure is to be avoided.

18.7   INFORMATION MANAGEMENT SYSTEMS

18.7.1   Overall concept

Creating a system from scratch can be compared with selecting clothes for a child of 10 years old. The clothes must be affordable, practical and fit, not only now but when the child is 13 years old. More difficult than this would be to try to *stitch* existing systems into an integrated whole: like assembling the wardrobe for the child from items randomly selected from a group of 10-year-old children.

When designing information management systems for projects, it is important to recognize that:

1. Construction industry information covers commercial, technical and managerial needs.

2. Administrative functions are interwoven with commercial, technical and managerial needs.

3. Time, cost and quality factors are interdependent.

These might appear to point to the need for a project management system (PMS) approach but, as we shall see, this would not satisfy our objectives. Almost all PMSs fall into three categories.

1. Systems based on the critical path method (CPM) using *activity-on-the-arrow* or *activity-on-the-node* (often referred to as precedence diagramming) format with no facility to trade off time and cost, and minimal scope for handling project information generally. They are little more than tools for project planning and control.

2. Quantity surveying systems which started life handling very detailed information and which over the years have been expanded to deal with more broadly-based issues; very few are able to relate to construction activities as they are invariably tied to the elements of a building.

3. Accounting systems which are tied closely, and unashamedly, to commercial considerations, and which must embody taxation requirements.

It is difficult to say just how many so-called PMSs are in the market-place, although a conservative estimate would put it around 150. Some do little more than network analysis based on CPM, some support automatic resource levelling (among other things), and a few allow cost and time data to be handled on an activity-by-activity basis. As far as time, cost and quality factors are concerned, PMSs simply cannot cope.

Some of today's PMSs have evolved from rudimentary mainframe-based, number-crunching packages that did little more than take the effort out of scheduling. They are interactive versions of their ancestors providing the user with a rapid analysis of the project schedule in the event of change. As tools for prediction – in supporting the classic *what if?* concept – they succeed rather well. They do not, however, support much in the way of information management of projects. There are a few exceptions.

Quantity surveying systems, for obvious historical reasons, address the cost/price aspects of a project extremely well, though mostly in an accounting sense. A flaw in these systems is that they deal with just part of the problem and tend to support incredible detail on some aspects, but very little on others. For example, the proportion of cost attributable to traditional building work has decreased on certain types of building while services and fitting out have increased. By using some of the more refined versions of these systems, quantity surveyors are able to provide *more and more* detail about *less and less* of a building's content.

Many quantity surveying systems are able to address the administrative aspects of contracts, for example, variation control and cost escalation. The administration associated with issuing and tracking documents, as dictated by contractual requirements, has also found its way into some systems. Thus, quantity surveyors are being supported in their move to wrestle greater administrative control over projects by newly developed computer-based tools. The predictive element which enables management decision-making to take place is, however, generally absent. Time management is beginning to find its way into some quantity surveying systems, but they are still heavily cast in the mould of cost management.

A further issue which militates against the quantity surveyor's move into time and cost management is the scant education and training which is given over to the time aspects of construction projects. It is a fair criticism that many quantity surveyors aim for the elusive goal of accuracy regardless of the trivia that is produced in the process. For instance, 80% of the content of a bill of quantities represents 20% of the value, with measured items sometimes covering less than 50% of the total cost of the project.

Recognition of the shortcomings in PMS and quantity surveying systems has led to further research and development (Crow, 1988). A fresh approach has been adopted through a systematic analysis of the needs of project team members. The emphasis is upon information management which has, at its heart, combined techniques of time management and cost management.

A common practice when devising a computer-based solution to a problem is to solve it manually and then to convert it to run on a computer. In this case, the lack of experience of a manual system of information management has meant that little guidance was available within the industry. Predicting future problems before they have a chance to surface is not something which industry can tackle particularly well. Converting a tried and proven manual system to run on computer was, therefore, simply not possible.

Further research of the management-related needs of the project team has found that a database providing for management needs was more than adequate for supporting administration. Databases designed to serve the administrative needs of the project team cannot, however, be readily expanded to serve the broader management needs because the former are based on accuracy and auditing rather than timeliness and forecasting. It is this aspect of timeliness that represents a key difference between narrow administration/commercial systems and the broader, real needs of the project team.

18.7.2   Information management hierarchy

A partial strategy for dealing with information management involves defining levels of information which could be matched to the needs of project teams and, in particular, project managers. Thus, it is possible to consider information on a project at six levels of increasing detail.

1. *Performance indicators*: these are helpful in measuring the momentum of a project to determine whether or not an expected rate of progress is being maintained. Their use can act as a check against the contractor's own measure of performance and expectation of completion, for instance, when there seems to be too little time left in which to do a measured amount of work. Above all, performance indicators are essential if problems are to be prevented.

2. *Pareto principle*: otherwise known as the 80/20 rule, suggests that 80% of project progress is affected by 20% of its activities. Recognition of this principle enables project managers to concentrate their limited time and resources on the relatively few, vital activities that affect progress, yet retain overall control.

3. *Management exception reporting*: this helps to select those critical activities or items that management should concentrate upon. All activities or items have the potential to be selected, but only those matching defined criteria would be reported at a particular time.

4. *Action reports*: these reports are sorted according to responsibility and contain activities or items requiring attention by the person given that responsibility.

5. *Registers of information*: these generally take the form of accounting reports in the manner of detailed lists.

6. *Database or data dictionary*: this lists all items of information available to the project team, but is only of use when searched, sorted and reported. Much of these data would form the basis of an *ad hoc* enquiry by a member of the project team with access to the database.

Each of these levels represents a degree of information that can be overlaid by each of the key parameters represented in the time, cost, quality and feasibility triangle. Within each of the streams of information on time, cost, quality and feasibility are applications which concentrate on a part of the database. For instance, within the time stream there are applications such as *document control* or *production control*, and within feasibility there may be project yield and life cycle costing applications.

When designing an information management system for projects, each application should have a hierarchy of information reporting. The hierarchy helps to determine the output from each application, although it does not satisfy the complete needs of project management at any one level. This happens because managers tend to deal with a multiplicity of applications; for example, a design manager needs to know what progress has been made on document preparation and at the same time, whether there have been any changes to the brief. Thus, there has to be a way of merging the information hierarchy to suit a practical management situation.

The next challenge is to convert the information hierarchy into an information management hierarchy, with the emphasis firmly on management. This can be prepared by overlaying the information hierarchy onto a typical management hierarchy for a project. In this way it is possible to show:

1. Delegation of authority

2. Responsibility for actions

3. Membership of meetings

4. Information hierarchy

5. Information flow

6. *Who needs to know what?*

The management hierarchy can also be portrayed as a triangle, with its apex uppermost. The amount of information available is inversely proportional to the amount of responsibility at the same level in the hierarchy.

The information needed for project and corporate management controls is very simple for owners in both the private and public sectors. Experience of such enterprises reveals that the amount of information which needs to flow upwards from project/section teams to executive management level and then outwards to other divisions or corporations is minimal. However, this information is linked to:

1. Job management reports in a summary format.

2. Monthly reports analysing performance indicators and critical appraisals of actions required and taking place.

3. Commercial and accounting information.

4. Summaries of performance indicators.

These flows of information or primary communication lines can be shown schematically with the links between functional areas of administration, technical and management being linked to the executive, finance and other functional areas of the organization.

A large amount of information is needed on cost, time and quality to ensure adequate control of construction. This involves such matters as briefs, drawings, contracts and variations. It is considered that only under exceptional circumstances would there be a need for this type of information to be passed from the project team to more senior management. Moreover, such information can only be effectively utilized by people in day-to-day contact with the work involved.

General recognition of these points is evident in the expectations of senior managers who, in particular, want answers to the following questions:

1. Will the project be finished on time, within cost and to the required quality?

2. What are the critical support actions required from senior management to help the project objectives be achieved?

3. Will the function and quality of the project be as required in the agreed design brief?

18.7.3   An example information management system (IMoP)

IMoP is a modularized system developed by Crow (1988). It is built around a relational database management system and the UNIX operating system. Workstations can be located in a LAN configuration within an office or in a MAN (metropolitan area network) linking several offices. Screen forms and reports can be tailor-made to suit the needs of individual users, enabling a wider range of applications to be generated. The main modules are shown in Figure 18.2 and outlined below.

Feasibility studies

The module enables managers to build appropriate operating models of any project and obtain information on capital costs, operating costs, cash flows, financial returns and selected yield parameters, including internal rates of return.

Document control

The document control module maintains information relating to issued documents as well as an amendment register. Designers' progress is

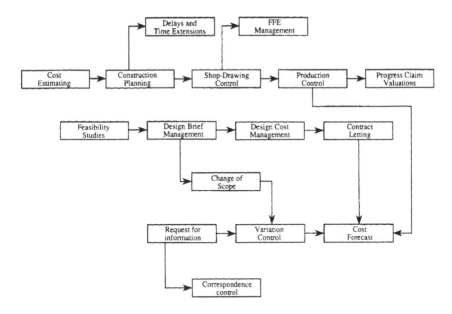

Figure 18.2 Modules of IMoP system

recorded against weekly programmes with performance tracked graphically. Transmittal slips and print orders are produced automatically with global issues by specialist, trade package or design discipline.

Variation control

Changes to a contract price require administering through an architect's instruction, variation or change order. Variations require strict management control if costs are to be kept within budget. This module tracks the approval/pricing process, where applicable, and highlights delays which can cause further costs. Contingency allowance expenditure is also monitored and reported for management control.

Production control

The production control module combines the management control techniques of programme, production, productivity and labour cost analysis to provide project managers with management exception reports on these key performance indicators. Forecasts of trade profitability assist management in preventing quality reduction, variation claims and liquidation/ bankruptcy.

Cost management

The cost management module has been designed to provide *real-time* management control of capital costs. It is based on forecasting the cost at completion, comparing it with feasibility cost limits/budgets and determining whether there are any adverse cost overrun trends requiring corrective action.

Shop drawing control

This module is used to assist multi-discipline project teams to manage the shop drawing process and ensure the project is not delayed. Exception reports highlight drawing preparation and approval delays. As drawing issues are registered, transmittal slips and print orders are produced automatically.

Design brief management

The design brief management module provides the building owner and design team with a management procedure which will enable preparation and progressive control of an initial project brief and feasibility estimates, as well as contingency budgets for owner-approved expenditure. Design variations provide a unique approach to minimizing cost overruns through unnecessary scope changes.

Delays and time extensions

During a construction project there will be delays for which time extensions may be granted. The delay and time extension module maintains a register of these delays. At any point during the project, it determines the amended contract completion date by allocating each individual delay to a particular chain of the project network. Management control graphs can then be produced to show the effect of the delays.

Tenancy management

The costs of operating a building can represent a greater investment for the owner than the capital cost of construction. Tenancy management helps owners to manage their property more efficiently and thus maximize the returns from their investments. Detailed data files enable information on tenants, leases, building maintenance agreements and servicing schedules to be retrieved for immediate scrutiny. Tenancy layouts may be displayed graphically and authority regulations checked automatically for fire safety compliance.

Construction planning

This module incorporates a bar chart approach to planning and scheduling of projects. A productivity crew/gang database can be created for a company or project to allow standard estimate rates to be easily retrieved for direct use or manipulation. This productivity database interacts with quantities taken from the building and nominated crew/gang sizes to determine activity durations. The simplified bar chart approach enables personnel to use the module without detailed knowledge of the critical path method.

Cost estimating

This module operates in a way which should be familiar to any estimator. It enables measured items with their quantities to be entered, prices of unit rates to be built up and tenders to be produced. Through the use of the library feature, a detailed database can maintain rates for different projects. Bills of quantities can be read through an optical character reader and entered into the database.

Professional practice management control

Managers frequently require information on the progressive utilization of staff on a project. This module allows the actual times worked by selected categories of staff to be accumulated by project and by staff category. Owner invoices, performance indicators of staff productivity and job cost reports are produced.

Contract analysis

The purpose of the contract analysis module is to enable the user to control the receipt of tenders, modification of tender amounts and the letting of contracts. Contract commitments are reported through the cost management module.

Project accounting

The project accounting module tracks an invoice from time of receipt to production of payment authorizations. It enables the validity of the invoice to be checked against the consultants', contractors' or suppliers' approved limit. The module produces exception reports against unauthorized invoices, thus highlighting inadequate supply of funds. Cash flows are produced automatically allowing correlation of both time and cost performance criteria.

Authority approval control

Several authorities can influence or even control a project, creating the potential for delays. The purpose of the authority approval control module is to assist the project manager/architect/contractor in managing the appropriate statutory bodies so that the approval process is not delayed.

Contractor preliminaries control

The contractor preliminaries control module allows effective control of expenditure through the progressive recording and control of actual usage of materials, equipment and labour resources. The format of reports can be tailored to suit a particular preference. Performance indicators provide forecasts of cost overruns to allow timely action.

Progress claim valuations

The progress claim valuation module provides a cost-effective method of producing accurate valuations. Detailed pricing information can be created or recalled from a library and may be stored against each bill of quantities item. Internal and external valuations can be produced, along with accurate cost/value comparisons and resource reconciliations.

FFE management

The FFE module controls the tendering, purchase, installation and ongoing inventory control of all fixtures, fittings and equipment on a project. Data can be collected automatically from CAD packages using an object-oriented methodology. Using the database, FFE items can be grouped by contract packages and suppliers selected for tendering. Purchase orders are printed automatically and reports for assisting in the installation of items by location are generated.

Instruction management control

The administration of contracts involves the issuing and *actioning* of instructions (for instance, site instructions and architect's instructions) and requests for information, some of which have monetary implications. This module issues the instructions, tracks the actioning process and, when required, creates variation records and updates cost management reports.

Information retrieval

In addition to the range of standard reports available within the system, users may also retrieve management information from the database using *ad hoc* enquiries, keyword indexing or by designing their own reports.

Other applications

The IMoP system supports links to and from other software to perform wordprocessing, mail merge, spreadsheet analysis, sketching, drafting and bills of materials extraction, contractor's accounting and professional office accounting.

18.8   CONCLUSIONS

18.8.1   Benefits of information management

The following list gives some indication of the gains that can arise from the application of information management on projects.

1. Reduced administration costs through the automation, even elimination, of drawing transmittal slips.

2. Reduced durations of meetings resulting in a saving of time and, therefore, personnel costs.

3. Decreased likelihood of building from superseded drawings.

4. Standardization of project control and reporting procedures across different projects.

5. Streamlined communication between project team members.

6. Reduced scope for further fragmentation of the project database through an ability to share information.

7. Ability to keep the owner better informed than hitherto, increasing confidence and enhancing the prospects for further work.

8. Ability to establish databases of project records for the whole organization.

9. Ability to base future management decisions on comprehensive databases, enabling strategies to be prepared and risks assessed with respect to new projects.

10. Improved quality of decisions through informed, timely briefings at all levels of management.

11. Improved documentation of progress and outstanding actions involving the owner and architect, thus fulfilling the obligation to forecast potential delays.

12. Integration of time, cost, quality and feasibility in management reporting, reflecting the interdependence of these parameters – they cannot be considered in isolation.

13. Emphasis upon exception reporting which concentrates on key performance information as opposed to reporting on *all and sundry* matters of a project.

14. Reports which can be easily tailored to suit the needs of the owner or anyone else for that matter.

15. Concentration on problem prevention through forecasting, as opposed to using historical information to cure problems – the difference between proaction and reaction.

16. Significant reduction in information system implementation time, resulting in earlier management controls from provisional project procedures, drawing checklists and user training aids.

17. Hardware used during the project can be retained for facility management, thus eliminating the need for the owner to buy a further system.

While these points represent improved efficiency and productivity, the bottom line for all parties involved in information management is increased profitability.

18.8.2   General conclusions

Successful use of IT is more evident in applications where there is a sound understanding of the procedures to be followed in practice. Often these procedures are largely sequential in nature: an accounting system is probably the best example. It is not surprising, therefore, that many computer-based systems cover much of the traditional activities that have to be performed for a construction project. The biggest problem is how these systems can be used in a dynamic way, as most of what they do is reactive. Nevertheless, controls may be built into such systems, reducing the scope for mistakes to be made. Checklists can be incorporated to prepare managers in advance of reaching a decision.

Project management systems (PMSs) have increased in popularity, but their potential falls well short of the objectives of information management. Over-concentration on theoretical time schedules, coupled with a lack of concern for communication and co-ordination issues, limits their effectiveness in practice. Most of all, such systems do not address the time/cost/quality issue. Information management requires a fresh approach; it would be unwise to contemplate converting a PMS to deal with information management. The underlying philosophies are fundamentally different and the information they are designed to handle are simply not the same.

Typical of the procedures that are involved on almost all projects are: variations, shop drawing approvals, payments, certifications (of one sort or another) and time extension claims. By involving information technology in project administration and adopting the information management concept, it is possible to release management energies for more productive work. One of the recurring problem areas is that surrounding time extension assessments. On some projects, the amount of energy that has to devoted to resolving such matters can be considerable, and it is counterproductive to some extent. Information technology, particularly the use of information management systems, can be used to make it an objective exercise, rather than an emotional activity. Anything that could get the right information to the right people at the right time must be good for the industry. Information management is concerned precisely with this issue.

ACKNOWLEDGEMENTS

The original research for this paper was supported by T. W. Crow Associates, Sydney Australia. The author is grateful for this support and for the opportunity to present the further findings here.

REFERENCES

BCIS (1988) Tender sum/final account study. *BCIS News*, (25), London, Building Cost Information Service, Royal Institution of Chartered Surveyors.

Crow, T.W. (1988) Information Management of Projects. In *Proc 9th World Congress on Project Management*, Glasgow, Internet 88, Vol. 1, pp. 309–20.

Hardcastle, C. (1982) Information management for quantity surveyors. *Quantity Surveyor*, **38**, 106.

CHAPTER 19

Multi-agent collaboration in integrated information systems

Rivka Oxman, Faculty of Architecture and Town Planning,

Technion, Israel Institute of Technology Haifa, Israel 32000

19.1 INTRODUCTION

Large projects in the construction industry characteristically involve large numbers of participants, each in a different area of specialization. Since the function of the participant is to act in a team situation, the problems associated with collaboration in the design and construction team are of great significance. In the traditional situation, collaboration is achieved through a variety of traditional media: speech, text, drawings. The achievement of systematized exchange of information and integration views of information systems in the design and construction team have been recognized long ago as significant issues in the development of information co-ordination and the development of a working model. Current related ongoing work in computer environments continues and covers new areas in information modelling such as computer-based documentation level; construction and design information modelling and computer integration of design and construction knowledge. Much of these areas of work is described in earlier chapters of this book.

Collaboration is a general term which extends existing research topics and issues devoted to the achievement of computer-based environments to support the team in collaboration. Multi-agent collaboration is complementary to current work in the area of construction information technology, which deals with the exchange of information, the construction of integrated databases, and the development of product models.

The aim of this chapter is to introduce the subject of multi-agent collaboration as a research application. The chapter presents the emerging research field of collaboration as a potential research area within CIC

Integrated Construction Information. Edited by Peter Brandon and Martin Betts. Published in 1995 by E & FN Spon, 2–6 Boundary Row, London SE1 8HN.
ISBN: 0 419 20370 2

which complements current ongoing work, and suggests a preliminary agenda for future research and development in the field. It provides a general introduction to the research area, describes subareas, discusses relevant research issues and proposes certain specific research topics as being particularly relevant to CIC.

19.2   BACKGROUND: THE SHARING OF DATA AND INFORMATION IN CIC

Sharing of information among the various participants in design and construction is based upon mutual understanding. The sharing of information by electronic means requires the development of formats, standards and models for the description of the various representations that agents employ (speech, text, drawings, etc.) and the capturing of the domain logic underlying interagent processes of collaboration.

19.2.1   Low-level approaches: exchange formats

Current approaches at the low end are based on the development of standardized electronic formats for data and information interchange. For example, the transferring of data using DXF format among application software, such as CAD packages, is considered to be an example of information transfer at a fairly low semantic level. Low-level approaches differ from high-level approaches in the semantic level in which they require standardization.

19.2.2   High-level approaches: conceptual models

Exchange and sharing of information at a higher semantic level requires a higher level model. This approach differs from the low-level approach, since it does not deal with the format of the document which describes the building, but provides a conceptual model and structures the information about the building to which these documents refer. Such a model depends on our ability to define, capture, integrate and model information in the various disciplines in a specific domain. A conceptual model provides for modelling of the semantic information by specifying the categories of information used in a specific domain. Conceptual models can be built through languages which employ certain modelling mechanisms. Current work in the field deals with process, information and product models as described previously. A *generic information reference model* for AEC (IRMA) has been proposed in a recent workshop organized by the Finnish research institute, VTT.

19.2.3   Product data models – entities and relations

A product data model is a conceptual description of a product which is capable of structuring all the information necessary for design, manufacturing and use. Current research work in ISO has concentrated on product modelling. The CAD/CIM research community has been developing an international standardized modelling language, for the Exchange of Product Data (STEP) (ISO, 1988). Work on data modelling in design and construction such as RATAS (Björk, 1992) is related to work on product models. These proposals interpret product models as the modelling of physical building components. Currently, various building product models in the domain of construction and building are being developed employing the EXPRESS-G language, which is used in the STEP product modelling standardization effort. Examples are RATAS, COMBINE DIM and GSD. A comprehensive comparative analysis of the various models is provided by Björk (1992). The building product data model which is currently developed at VTT, which is based on earlier RATAS prototypes, is particularly advanced and provides for the structuring of information about the topological relationships between building components and the spaces they bound.

19.2.4   Integrated information models – entities, activities and processes

Recently, there has been an attempt to develop a generic model which will satisfy the different information needs of the various actors by capturing and extracting high level abstractions, in terms of both components and processes in the different domains. The ICON project (Aouad *et al.*, 1992) is an integrated information model in which all the participants can share and access information according to their contextual interpretations. This is described in Chapter 15 by Cooper and Brandon. Object-Oriented Case tools provide various useful mechanisms for the representation of entities and processes to be employed.

The ICON (Intelligent Integration of Information for Construction) project has among its objectives, to demonstrate an intelligent information system for multiagent integrated activities such as architectural design, engineering design and construction management, and which support multi-agent communication during all phases of the project life cycle. One of the key issues in this project is how to capture and model the semantics of the diverse views, or perspectives, of the different participants. When communication takes place among multiple users with different perspectives, certain typical problems of communication, co-ordination and understanding occur. These are outlined in the following section.

19.3   FUTURE RESEARCH ISSUES IN INTEGRATED INFORMATION
SYSTEMS: MODELLING COLLABORATION IN CIC

19.3.1   Information systems and collaboration

When participants communicate with one another and exchange data or
knowledge, problems of disagreement and conflicts may arise. When co-
operating in order to perform collaborative tasks participants are required
to negotiate and resolve conflicts by employing their knowledge and
experience.

Current CAD is limited in supporting collaborative tasks, since it generally provides only for the sharing of data. CAD lacks an intermediary environment for the support of collaboration through the communication of
ideas, the negotiation of solutions and the resolution of conflicts. Generic
conceptual information models described above, capture the information
needed by various agents. Such generic models provide for a shared underlying representation of design and construction generic elements, with relations and processes upon which all agents agree. Such a representation
models the internal processes in order to ensure consistency. However,
while there is agreement about the model, it does not assure agreement
about its information and semantic contents, and its use in the processes of
design and construction. When data is distributed among a number of different intelligent agents through electronic communication, these agents
must co-operate with one other in order to perform a collaborative task.
Collaboration models must be developed in order to support computer-based
collaboration. While most AI in design research has considered design so far
as an activity carried out by an individual designer, research in other fields
of AI are currently addressing models of collaborative problem-solving for
multi-agent activities. These fields in AI have the potential for contribution
in the field of design and construction collaboration.

19.3.2   The AI perspective

Recently, the subject of multi-agent collaborative work has gained attention from various subfields in AI. *Collaboration* is a general term which covers various research topics and issues in AI devoted to the achievement of
computer-based environments to support the team in collaboration. Current
research focuses on the development of comprehensive models in AI for
collaboration, co-operation, negotiation and conflict resolution in various
domains. In the following sections, the relevant fields and recent developments in AI in collaboration are reviewed.

Distributed artificial intelligence (DAI)

Distributed artificial intelligence (DAI) is a field which deals with the design
of automated agents which can interact effectively in order to

co-operate in collaborative problem-solving. Blackboard systems (Engelmore and Morgan, 1988) have been the most commonly accepted paradigm for the development of different control architectures in such systems.

Co-operation in DAI is the design of automated agents which can interact effectively in order to co-operate in collaborative problem-solving. Current research focuses on the development of comprehensive models for collaboration in various domains. Negotiation is recognized as an important means by which interagent co-operation is achieved. This is currently a research area in distributed artificial intelligence (DAI) which is concerned with distributed problem-solving and task allocation by providing negotiation protocols among autonomous agents. That is, DAI is concerned with the design of agents which are able to communicate in such a way as to enhance the possibility of reaching mutually beneficial agreements concerning problems among the agents (Rosenschein, 1992).

A central aspect of co-operative problem-solving by groups is the avoidance, identification and resolution of conflicts among the various participants. In order to achieve implemented conflict management systems for human and computational agents, theoretical research and empirical studies are of current interest among the AI community. This is currently another significant area in distributed artificial intelligence (DAI), which begins to outline a single general theory of conflict management that works across multiple domains.

In order to apply this theory to practical applications in the fields of design and construction the difference between the generic and the domain-specific aspects of conflict management need to be explored.

Computer-supported collaborative work (CSCW)

Computer-supported collaborative work (CSCW) deals with the theory and application of how people work together and how the computer may enhance group behaviour. Research areas in this field deal with the modelling of group meetings and agents' interaction, such as communication (Pankoke-Babatz, 1989; Smith *et al.*, 1991)and capturing group decision rationale (Maclean *et al.*, 1991). In order to enhance such teamwork, different techniques for long distance communication, both textual and graphical, are employed in networking for the development of CSCW applications systems.

Case-based reasoning (CBR) in collaboration

Case-based reasoning is a computational paradigm which employs past experience of similar problems in current problem-solving (Riesbeck and Schank, 1989). Based upon past experience of similar problems, Case-based reasoning (CBR) presents an approach for the resolution of conflicts which may emerge among the various agents. It has already been applied in this way in various domains such as in the field of law (Sycara, 1989). In the CBR approach, cases may provide a way to solve conflicts without the necessity of deriving a new solution. In order to advance this significant field, there is a need for theoretical and applicative research in the area of multipurpose and multiagent case-based systems for collaboration.

Design rationale (DR)

Work in design rationale and design rationale capture (Lee, 1990) has potential relevance to case-based collaboration, and to CIC in general. Among other things, design rationale capture attempts to formalize and represent the deliberations underlying a design process and to employ the captured rationale in order to support better design and construction. A current direction proposes that a model of design rationale may support subgoaling as a result of decision-making and the propagation of the effects of decision revision through a distributed dependency network.

19.3.3 Conclusions

Distributed artificial intelligence (DAI), computer-supported collaborative work (CSCW) and case-based reasoning (CBR) are currently active research fields in the AI community. CSCW has the potential to contribute to research in multi-agent case-based systems by providing communication techniques. Case-based collaboration is proposed as a new research field which emphasizes research in computer-supported human collaboration through the use of previous experience (cases). Design rationale also provides a potential contribution to the advancement of this research area by providing techniques for design case rationale capture as a basis for collaboration.

19.4 COLLABORATION ISSUES IN ICON

19.4.1 Information modelling issues in ICON

The objective of this section is to identify specific research issues in collaboration for CIC. The relevance of these issues to the field of CIC can be demonstrated in the framework of the ICON research programme. In the beginning of this section, issues and approaches to information-sharing in ICON will be presented. In the rest of this section, future directions for extending these approaches to information-sharing in order to deal with collaboration will be identified, proposed and discussed.

The ICON project is currently concerned with the creation of a generic information model for the construction industry which will enable the multiviews of different agents in various domains to be accommodated within a single information model. The ICON project investigates the way in which information is passed and understood by the different agents. ICON is currently dealing with three characteristic domains in CIC, the professions of architecture, quantity surveying and construction management. In order to provide for multi-agent communication, semantics should be transferable across disciplines. This raises the following problems with which ICON research is dealing.

The first significant problem, is the problem of *diverse perspectives*. An aspect of communication among diverse perspectives is that of diverse semantics. Different agents in different domains may relate to the same object, semantics which differ from one agent to another. That is, while employing the same objects, they do not necessarily share the same interpretations of these objects. For example, the object 'wall' for the architect means a 'spatial partition' and is a part of a higher level system which consists of a set of corridors, hallways, rooms, etc. For the structural engineer, the same wall is a 'structural load-bearing component' which is a part of a structural system consisting of a set of structural elements such as columns and beams. For the quantity surveyor the object 'wall' means a 'costed element'.

Another related problem is that of multiple *domain vocabularies*. Agents employ different accepted conceptual vocabularies which are conventionalized in their specific domain. In order to understand one another, domain agents usually share certain categories at higher levels of abstraction. Therefore, in order to achieve a shared vocabulary and shared semantics for multidomain interaction in conditions of diversity of representations, shared levels of abstraction are required. The problem in modelling is how to capture and provide mappings and consistencies among multiple domain vocabularies.

An additional problem in multi-agent systems is that of *diverse information requirements*. The kind of information each requires in dealing with the same object may vary from one domain to another.

19.4.2   Approaches to information modelling in ICON

Presently, issues of information-sharing have been addressed in ICON. Information in ICON is represented at the knowledge level, which will allow for the interpretation of different contextual domains. ICON has utilized a concept termed 'perspectives' to solve these problems. Domain perspectives are concerned with the modelling of information as viewed from a top–down hierarchical description of processes capturing the vocabulary when dealing with design objects. Shared information structures accommodate the perspectives of different users at different hierarchies and different abstractions. In order to represent requirements of different agents, modelling of the data provides the basis for establishing the semantics of the shared data. The method that has been applied in

ICON is to construct a number of separate information models each describing a different perspective on the information. ICON is using object-oriented modelling methods and techniques which are supported by the CASE tool approach.

The theoretical approaches to information-sharing in ICON may be summarized as follows:

1. *Domain model*: provides for an information model which satisfies the needs of one agent in a particular domain. Captures the concept vocabulary of that domain.

2. *Multi-domain generic model*: integrates diverse domain specific models in a single generic information model which can be accessed by all agents.

3. *Integrated model*: integrates the different domain models by modelling shared conceptual categories in different domain vocabularies.

4. *Multi-agent semantics in integrated models*: models the overlapping of concepts which occurs between the various domains and which provide for the passing of information from one domain model to another.

These modelling approaches within the ICON project already provide for the sharing of information. Beyond this, in collaboration, we are interested in the design of multi-agent systems which are able to communicate and interact with shared information in order to clarify rationale, negotiate and resolve conflicts.

19.4.3   Collaboration research issues in ICON

In order to apply current work on collaboration in AI to practical applications in the fields of design and construction, the relations between the generic and the specific aspects of the domain should be explored. In collaboration, we are interested in the design of multiagent systems which enhance communication and interaction with shared information in order to clarify rationale, negotiate and resolve conflicts. These issues are elaborated below.

Rationale

Information models such as ICON contain the various description views of an artefact. In collaboration, we are interested in going beyond the semantics of these different perspectives in order to achieve an understanding of the underlying rationale. The use of design rationale is very appealing in multiagent collaboration. It is a promising direction because it may be used as a basis for the integration of multi-agent perspectives in design and construction. In applying design rationale to collaboration, explicitly represented rationale can serve as a basis for communication and agreements among the different agents. It may also provide a method for the resolution of unresolved issues which might occur during design and construction collaboration.

In recent work we have investigated how rationale can be communicated between multiple agents. We have proposed a *representational vocabulary* which appears powerful enough to capture design rationale (Oxman, 1993).

Protocols for negotiation

This approach suggests collaboration through negotiation protocols among the agents. That is, protocols provide detailed consideration in order to enhance the possibility of reaching mutually beneficial agreements concerning trade-offs among the agents. While classical research in DAI is mainly concerned with the view of shared common goals, research in CIC and design should emphasize the role of agents endowed with their own individual goals, knowledge and experience particularly when pursuing competing goals.

Conflict resolution

The identification and resolution of conflicts among various agents is a main issue. CBR presents an interesting approach. In such an approach the early identification and warning of potential conflicts through accessing a case library is possible. Cases can provide a successful previous solution in which trade-offs are part of the case itself. Currently, research work is being carried out in the design domain (Oxman *et al.*, 1993a,b) in which the conditions of multi-agent conflicts, typical problems requiring negotiation and trade-offs are defined. A case schema for the representation of information on multi-agent conflicts and their associated methods of conflict resolution is currently being investigated. In such an approach, a case library may be indexed and relevant information retrieved relative to a general schema of conflicts between the diverse perspectives within particular collaboration tasks.

19.5   CONCLUSIONS

This chapter has attempted to outline the new research field of collaboration, and to suggest its potential relevance to CIC. Current work in AI in the areas of collaboration have been reviewed, and significant general contributions of this work has been identified. It has been proposed that this area is complementary to current international research efforts in the integrated modelling of construction information, since it advances this work by identifying computational contributions to collaboration support between multi-agents. Since problems of understanding, communication and conflict resolution within the environment of CIC are particularly significant, the information exchange emphasis should be expanded to accommodate approaches to computer-supported collaboration in CIC. The chapter has sketched certain of the problems, issues and potential research directions in order to suggest the content and potential of this area.

REFERENCES

Aouad, G., Brandon, P., Brown, F. *et al.* (1992) *ICON Half Yearly Internal Report*, Salford University, Salford, UK.

Björk, B.-C. (1992) A conceptual model of spaces, space boundaries and enclosing structures. *Automation in Construction*, **1**(3), 193–214.

Engelmore, R. and Morgan, T. (1988) *Blackboard Systems*, Addison-Wesley, Reading, MA.

ISO (1988) Industrial automation systems – exchange of product model data – representation and format description.

Lee, J. (1990) *SIBYL: A Tool for Managing Group Decision Rationale*, in Proceedings of the Conference on Computer Supported Cooperative Work (CSCW-90), Los Angeles.

Maclean, A., Bellotti V., Young, R. and Moran, T. (1991) Reaching through analogy: A design rationale perspective on roles of analogy, in *Proceedings of the Conference on Computer Supported Cooperative Work*, Robertson, S.P., Olson, G.M. and Olson, J.S. (eds), ACM, New York.

Oxman, R.E. (1993) Precedents in design: a computational model for the organization of precedent knowledge. *Design Studies*, in press.

Oxman, R.E., Aouad, G., Brandon, P. *et al.* (1993a). The Semantic of Collaboration: Capturing Semantics among Agents in Design. *Working Paper*, Faculty of Architecture and Town Planning, Technion, Israel.

Oxman, R.E., Aouad, G., Brandon, P. *et al.* (1993b). Multi-agent Model: Modeling Entities and Processes of a Phase-Agents-Activity in Design. *Working Paper*, Faculty of Architecture and Town Planning, Technion, Israel.

Pankoke-Babatz, U. (1989) *Computer Based Group Communication: The AMIGO Activity Model*, Ellis Horwood, Chichester.

Riesbeck, C.K. and Schank, R.C. (1989) *Inside Case-based Reasoning*, Lawrence Erlbaum Associates, Hillsdale, New Jersey.

Rosenschein, J. (1991) *A Theoretic Approach to Distributed Artificial Intelligence*. Paper presented at the *8th Israeli Symposium on Artificial Intelligence and Computer Vision*, Tel-Aviv, Israel.

Smith, H., Hennesy, P. and Lunt, G. (1991) An object-oriented framework for modelling organization communication, in *Studies in Computer Supported Cooperative Work: Theory, Practice and Design*, J. Bowers and S. Benford (eds), Elsevier Science, Amsterdam.

Sycara, K.P. (1989) *Argumentation: Planning Other Agents' Plans*, in *Proceedings of IJCAII-89*, Michigan.

CHAPTER 20

Integrated document management

*Bo-Christer Björk, Department of Building Economics and
Organisation, Royal Institute of Technology, Stockholm, Sweden
Pekka Huovila and Sven Hult, Technical Research Centre of
Finland, Espoo, Finland*

20.1 INTRODUCTION

Several authors have recently discussed the characteristics of a future construction process where information technology is used extensively both in numerous data processing tasks and for data transfer (Howard *et al.*, 1989; Dupagne, 1991; Sanvido, 1992). The term which has been used for such a state is computer integrated construction (CIC) which is analogous to the computer integrated manufacturing (CIM) concept used in other branches of industry. Some authors place more emphasis on the data exchange aspects of CIC, others focus on developing the overall process, including both construction activities and information processing activities.

One of the main purposes of the use of computing in construction, in addition to automating tedious information processing work, is to facilitate the use and reuse of information, thus integrating the information and knowledge available in different parts of the overall process. Using computing is only one possible way of achieving these goals as Part Two of this book, on historical classification work, showed. Another way would be to gather whole design teams co-operating on a project in the same physical location, thus facilitating rapid information access. Vertical and horizontal integration, where firms try to incorporate as large a part of the design and construction process as possible inside single organizations, obviously represent a further way of achieving integration goals.

Data transfer in digital form between the different applications and the different actors which participate in the construction process is the key element in CIC. Many research projects are currently trying to define methods and standards which would facilitate and automate such trans-

Integrated Construction Information. Edited by Peter Brandon and Martin Betts. Published in 1995 by E & FN Spon, 2–6 Boundary Row, London SE1 8HN.
ISBN: 0 419 20370 2

fer. Organizing data transfer on the physical level is a trivial problem compared with the problems of getting different applications to understand the semantics of each other's data structures.

20.2 TWO TYPOLOGIES OF DATA TRANSFER

Physically, data transfer can take place in many ways. Today, predominant modes are the mailing of copies of written documents and drawings, telephone conversations and face-to-face communication. The use of electronic media such as the telefax or E-mail is rapidly increasing. CAD data is often exchanged by the handing out of diskette copies of drawing files. Almost all of these activities depend on either the producer of the information actively sending it out after it has been produced, or on the producer interpreting a request for information and manually going through the steps of retrieving and sending the information (which in itself may be as a file in a computer).

In a more developed CIC process, the most interesting mode of data transfer will be in the form of well-defined information requests, issued by human operators or computer applications, which are answered automatically without the need for human intervention. Today, this type of interchange is rare and mostly limited to the management of CAD databases with shared access to the data via computer networks. Another mode of data transfer is where the sending of the information is highly automated, as in sending E-mail messages to sets of predefined recipients using single commands. Table 20.1 tries to summarize the categories discussed above. In today's practice the emphasis is on manual data transfer. As networks and methods for data transfer develop, the emphasis will more and more shift towards the automated handling of both categories of automated transfer.

A second typology deals with the information content of the transferred data, viewed from the receiving end. Looking at things from the viewpoint of the application receiving information, there are two fundamental levels on which data interchange can take place. The more advanced is the information level. This means that the application receiving data can directly interpret the information for further processing,

Table 20.1 A typology of data transfer based on the degree of automation

	Manually controlled	*Automated*
Producer-generated	i.e. Mail, telefax, mailed diskettes, E-mail	E-mail
Request-driven	i.e. Telephone, mail telefax, mailed diskettes, E-mail	On-line access to distributed databases

using its knowledge of the semantics of the data itself or possibly the conceptual model of some data transfer standard into which the data has been translated prior to the transfer. An example would be an EDI message, where the receiving application can decode the message concerning an order for a particular component. Another example would be data in object form from a product model, where the receiving application would fetch information about the exact geometry and location of a component in the building.

A less ambitious data transfer level, which in research seems to have received much less attention than the information level, is the document level. On this level the party requesting information is able to locate in digital form a document which contains the relevant information and retrieve it, without knowing the internal data structure of the information. The receiving system may, at the most, be able to move the data around as a digital file or display it as a raster graphics image. Human interpretation is needed to understand the document, but no human intervention is needed to transfer it. Digital data exchange on the document level is closer to enhancing today's practice using the possibilities of modern workstations and networking, and would therefore be relatively easy to develop. It is at a different point on the map of types of integrated construction information, put forward earlier by Betts, Fischer and Koskela, than much of what has been discussed earlier.

Another dimension of data exchange which we can discuss at this point is the homogeneity of the information in the applications or databases on the two sides of the exchange. Data exchange between, for instance, word processing systems or between draughting systems concern homogeneous information, at least from the software-type viewpoint. Another criteria for determining the homogeneity of the information is the scope. Geometrical design information is fairly uniform across design disciplines and differs significantly from, for instance, tables of cost data or contractual information in text form.

In the more complicated types of data exchange, which should be characteristic of fully developed CIC, quite a lot of exchange between hetero-

Table 20.2 A typology of data transfer based on the nature of the exchanged information

	Between homogeneous applications	*Between heterogeneous applications*
Information level integration	EDI, CAD data	Product models, full scale CIC
Document level integration	—	Integrated document management

geneous applications would be needed. An example where both the software types used and the information types differ would be the cost calculating or project planning applications of a construction firm utilizing information from the CAD database of the designers. If we cross-tabulate these two classifications we get Table 20.2.

Three of the four types of integration which result from this cross-classification are discussed one at a time. The category document level integration between homogeneous applications seems somewhat artificial, and it is difficult to find practical examples of it.

20.3 INFORMATION LEVEL INTEGRATION OF HOMOGENEOUS INFORMATION

In *information level integration of homogeneous information* the system requesting the information knows the detailed data structure of the accessed file or database and utilizes this information both in making the query and in the postprocessing of the information. The information processed on both sides of the transfer is fairly homogeneous in scope and usually the same type of software or even exactly the same software are used.

Standards such as EDI standards for a number of different types of commercial messages deal with this type of integration. In Finland such EDI messages have been developed for purchase orders, delivery notes, etc., of construction materials. Other examples of integration at this level concern the exchange of written documents using electronic mail. Often neutral formats such as plain unformated ASCII text or quite sophisticated formats for formated text such as SGML are used. Bilateral conversion software between different types of market-leading word processors are becoming standard features in most systems.

The exchange of CAD data, possibly enhanced by techniques such as layering and reference files, is another good example of this type of integration. In the case of different CAD systems, standards such as IGES or DXF can be used for the transfer.

20.4 INFORMATION LEVEL INTEGRATION OF HETEROGENEOUS DATA

Information level integration of heterogeneous data is the most difficult type of integration, since it implies that the application issuing a query or receiving information must understand the semantics and syntax of quite dissimilar categories of information treated during a construction project. Product models implement this type of approach for the limited scope of data describing the building. Thus the systems originating the data are rather homogeneous (CAD systems), whereas the applications needing the exchanged information will be of a larger variety (in addition to CAD systems, a large variety of calculation software, etc.).

In product data models an object-oriented approach for structuring design data is used. One of the basic tools used in product models for structuring information about a product is an abstraction hierarchy, starting with an object describing the product in its entirety and going down through several layers of decomposition to single components. This is made possible by the facility of objects to contain information about their relationships to other objects. Techniques such as inheritance of data structures are also used extensively.

The main advantage of this way of structuring information is that it offers good possibilities for all kinds of different applications to extract exactly the information needed using database queries. Automated access to design data from calculation applications is a typical example.

More detailed discussions of the techniques of building product models are beyond the scope of this chapter. For overall presentations see (Gielingh, 1989; Björk and Penttilä, 1989).

An interesting aspect of the current work on product models is the use of formalized tools. In particular, the tools used in the international product model standardization effort under ISO (the STEP standard) are quickly becoming *de facto* standards for presenting product model definitions.

A number of proposals for extending the use of conceptual modelling techniques used in product modelling, to other categories of information issued and utilized during a construction project, already exist (Luiten and Tolman, 1991; Froese and Paulson, 1992). An earlier paper by the author of this chapter has proposed a possible generic framework model usable for supporting this type of integration (Björk, 1992a). In this framework model a number of generic information categories which form the backbone of existing traditional construction classification systems, have been formalized using conceptual modelling techniques. In addition other categories have been added. Within single generic categories traditional classifications can easily be modelled using subtyping. What is even more important is that such conceptual models also incorporate the relationships between information from the different categories, which traditional methods have great difficulty coping with. Figure 20.1 shows an EXPRESS-G diagram of the main information categories in the model.

The advantage offered by extending formal conceptual modelling to categories beyond product definition data, is that it offers the possibility for linking information in widely heterogeneous applications. The definitions of such models and the development of prototypes is, however, only just emerging as a new research topic, and a long time will pass before practical applications can be expected. It goes without saying that standardization will be extremely difficult to achieve in this area. The need for work in this area has, however, recently been recognized by the ISO

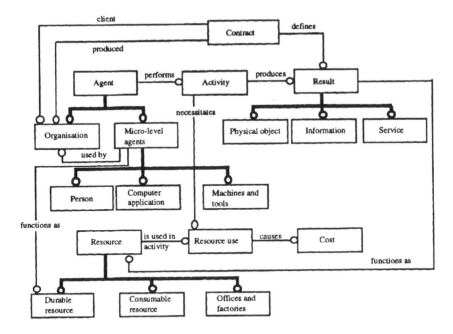

Figure 20.1   A proposed generic model of construction project information

working group dealing with the further development of traditional building classification systems (Karlsson and Allott, 1991).

20.5   DOCUMENT LEVEL INTEGRATION OF HETEROGENEOUS DATA

Document level integration of heterogeneous data has recently been studied in a project at VTT in Finland, by developing a demonstration system for integrated document management in construction projects (Björk *et al.*, 1993). The purpose of the system is to offer integrated access to all the documents (regardless of the scope or the type of originating software) that have been issued during a construction project. The granularity of this system is the individual document which, using the object-oriented approach, is the fundamental object class of the system. For each document object, the system contains enough information to locate it, based on the specifications in a query, and to transfer it in digital form to the recipient. The internal format of the document depends on the program which has issued it and is, to the document management system, just a string of illegible characters. It is not a concern of the document management system to be able to decode the document; it will only contain enough reference information to make this possible if suitable conversion programs are at hand at the receiving end.

The demonstration system was defined using formalized tools familiar from product model development. A process model of the activities of the document management system was defined using the SADT activity modelling technique (Marca and McGowan, 1987).

This activity model provided a system specification for the programming of the generic tasks of the document management system. In parallel with the activity modelling, a conceptual schema of the information that is needed to describe each document was developed. This conceptual model forms the basis of the design of the database containing information about the documents.

Development of the above ICPDM system was based partly on the analysis of documents from an empirical test case: a two-storey office and exhibition building for car sales. The main criteria for the choice was to find a completed construction project with all documents available (not only drawings) and where as large a share of these as possible had been produced using computers. The chosen documentation consisted of more than 800 documents covering a variety of drawings, different types of lists, schedules, calculations, calls for tenders, contracts, invoices, etc. The documents have been produced by the architect, the different design subdisciplines, the client, the main and subcontractors as well as public authorities.

20.6   FUNCTIONALITY OF THE DEMONSTRATION ICPDM SYSTEM

The demonstration ICPDM program was developed for personal computers (PCs) running Windows version 3. This configuration was selected because it is the *de facto* standard in the construction business today. A PC is flexible enough to be easily transferred from one site to another.

In the ICPDM system the document data itself and its management data are separated from each other. The management data will be sent to every party actively, that is, the creator of a document sends messages to other parties, but the data must be retrieved by the other parties from the file servers of the data transfer network. The ICPDM consists of local management programs running on each party's workstation and a message handling system provided by a telecommunication company. The file transfer can be arranged separately using alternative means of communication.

The user interface is defined in conformance with the standard Windows user interface. An example menu is shown in Figure 20.2. The management program performs the following tasks:

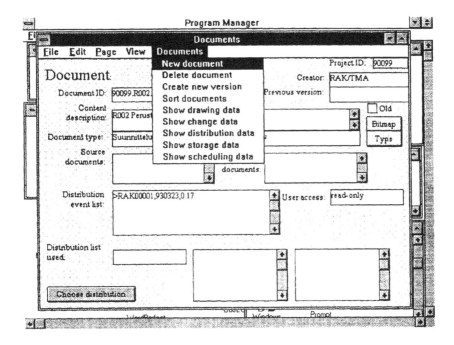

Figure 20.2 An example of the user interface of the demonstration system

1. Creating, editing and deleting document management data

2. Sending, receiving and handling document data management messages

3. Making reports from the document management database

4. Acting as a platform for accessing other programs used to edit the actual contents of documents.

The last function is quite useful from the user's viewpoint. Ideally they would like to be able to open the program needed to edit the document just by double-clicking the icon of that document. Features of this type are now becoming available through the object link and embed techniques used in the Windows environment.

The activity model was developed using the SADT technique and consisted of five main tasks (activities). Each of these main activities was zoomed using separate SADT diagrams which form a hierarchical structure. An example diagram is shown in Figure 20.3.

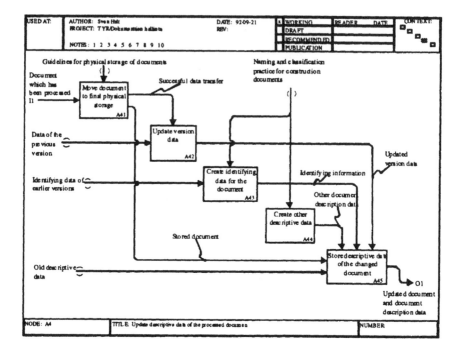

Figure 20.3   IDEFO diagram AO

20.7   THE CONCEPTUAL SCHEMA OF THE ICPDM SYSTEM

The conceptual model, which forms the kernel of the ICPDM system, was developed using the EXPRESS language (CEN, 1991). EXPRESS is an object-oriented data definition language which has become more or less a *de facto* standard in product model work. The ICPDM schema consists of a number of basic object classes or entities.

1.   Document object

2.   Project participant object

3.   Document distribution objects

The document is the main object of the ICPDM conceptual model (Figure 20.4). A document can be independent or it can belong to a construction project. It can consist of design data, project management data or contractual data. If a document is revised, the new version will be a totally new document. The new document will have a relation to the older version (the relation type is 'previous_version').

   A document has an identifier. If a document is a new version of a previously stored document, it has a textual description of changes made after

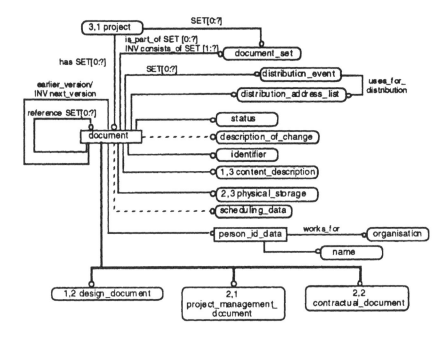

Figure 20.4 The document object and its attributes

the previous version. A document has a content description that can be textual or/and a scaled-down bitmap picture. This description can be used for quickly browsing through documents without actually accessing them in full.

The document has information concerning the way it is stored. If the document is stored electronically, the object description includes the file name, the directory path, definition of the program and the file format used when the document was created. If the document exists only on paper (for instance, in a drawing archive), the description of its storage can be inserted into the management base.

The document's scheduling data consists of planned and occurred completion dates, document's state data (a document can have, for example, the following states: proposition, draft, accepted for construction, as-built, etc.).

A document has a type. The type defines what kind of document it is. The three main types are:

1. Design document

2. Project management document

3. Contractual document.

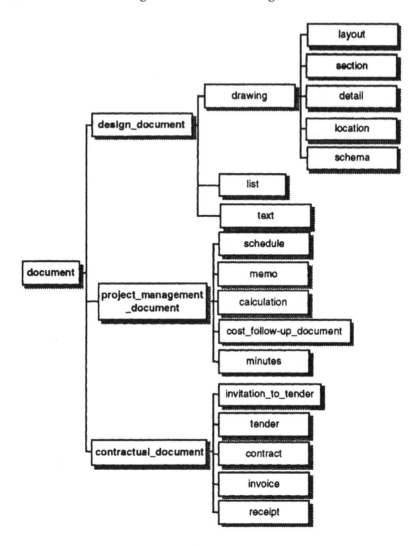

Figure 20.5 Document-type tree

The type attribute is a hierarchical list of types, as shown in Figure 20.5. The first level defines the main type and the lower levels define the document more specifically.

Design documents have certain additional management information lacking from the other document types. They always have design discipline information that defines the scope of documents. The design discipline can be, for example, 'Architect' or 'HVAC Designer'. This attribute allows queries using the design discipline as a search key. One important subclass of design documents is a drawing. It has additional attributes

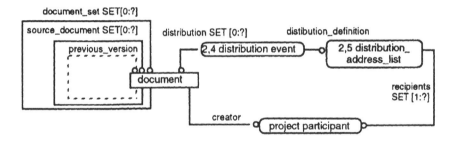

Figure 20.6   A document's relations to other objects

concerning the scale of the drawing, and the artefact or subject described by the drawing. The user of the system can define which floors, sections or building parts are defined or shown in the drawing.

A document can have relations to other documents (Figure 20.6). There are three types of relations.

1. A document_set relation defines a part-of relation between an object describing a set of documents and its part documents. A good example of a document set is provided by all the documents needed to define a contract's specifications, or all the documents needed to apply for a building permit. A document set is, in many ways, analogous to the folder concept in user interfaces. A very important difference is, however, that the same document may belong to several document sets at the same time.

2. A source_document relation defines which documents were used as sources when a new document was created. With this relation one can find all documents that need some sort of inspection if the source document is changed.

3. A previous_version relation shows which document was the previous version of a document.

A document has a creator. The creator is defined by a relation to the object 'project participant'. A project participant can be an organization or a person working in an organization. When document management data is sent to another party, the system creates a distribution event entity and a relation 'distribution' between it and the corresponding document. A distribution event is an object that defines one specific distribution event in which a document is sent from its creator to another party. A distribution event has an id, time of the event and recipient data as its attributes. The EXPRESS code for this is shown as:

ENTITY Distribution_event;

event_id : STRING;

distribution_time : Date;

recipient : Project_participant;

distribution_criteria : Distribution_address_list;

INVERSE

distributed_document : document FOR distribution;

UNIQUE

event_id;

END ENTITY;

Recipient data is actually a relation to a 'Distribution address list' object. A distribution address list has an identifier, and items of a list. Each item has a relation to a project participant object and user access rights data. User access right defines how the recipient can access the actual data of the document. There are three different values: *read_and_write* gives the recipient the right to make a new version of the document, *read_only* gives the recipient rights to browse the document's contents, and *exists* gives the recipient only the information that a document exists. In the last case the recipient has no right to browse the document. This kind of user access data can be used, for example, if two parties want to inform a third party that they have made a contract but do not want to distribute the details of the contract.

20.8 CONCLUSIONS

Research studying integration methodology and techniques is a high priority area in many of the leading research units dealing with the use of IT in construction. Among the subjects of research are product models, knowledge-based construction management systems, concurrent engineering, etc. Looking at this research from the viewpoint of industry, there seems to be too big a gap between the medium term needs of industry and the subject area of much of the research. Standardized product model descriptions of buildings may prove extremely useful in the long run, say within 10–15 years. Enterprises are today mostly involved in design and construction projects where many of the participants produce drawings using CAD systems, while others still draw manually. Some participants may occasionally have E-mail, but the most common denominator is that almost everybody has a telefax. In such a hybrid environment less sophisticated methods for managing information are needed, methods which could lead to improvements in management procedures and productivity within 2–3 years.

During the duration of the project described above, the researchers have received encouraging signals that developments along these lines are starting. In Finland a large engineering consultancy (FINNMAP Oy) has recently developed a full-size prototype of a document management system. The system is especially meant for concurrent design in a draughting system environment. In Sweden, a similar system has been developed and tested in a real project. The system was centred around a server workstation which contained copies of all document files. From this server all files could be printed out and mailed based on E-mail requests. Developments due to the impetus of the US Army CALS project and the commercial development of groupware software are also starting. An obvious strategy for building full-sized prototypes would be to utilize some commercial software and tailor it to the needs of the construction industry rather than start from scratch.

The research described in this chapter draws on earlier work at VTT concerning overall strategies for CIC and product modelling. The author believes that it points out a new and slightly differing approach to integration of the total information produced during a construction project. It is particularly important within the context of this book as an example of an application of integrated construction information.

ACKNOWLEDGEMENTS

The research described in the paper is based on ideas developed by the author during earlier projects focusing on product modelling and on defining an overall framework for construction computing. An earlier version of the first part of this paper was presented in a workshop at VTT in 1992 (Björk, 1992b). Pekka Huovila managed the ICPDM project. Sven Hult co-ordinated the definition of the activity and conceptual models, and was responsible for the software implementation of the ICPDM. The research was funded by the Technology Development Centre of Finland.

APPENDIX A: EXPRESS SCHEMA

```
SCHEMA ICPDM-model;
ENTITY Document;
SUPERTYPE OF (Design_document);
     identifier : STRING;
     description_of_change : OPTIONAL STRING;
     content_description : Content_data;
     physical_storage : Storage_data;
     document_scheduling : Scheduling_data;
     document_type : Type_data;
     creator: Project_participant;
     distribution : SET [0:?] OF Distribution_event;
     previous_version : OPTIONAL Document;
     source_document : OPTIONAL SET [0:?] OF Document;
     document_set: OPTIONAL SET [0:?] OF Document;
     UNIQUE
     identifier;
END ENTITY;
ENTITY Design_document;
SUBTYPE OF (Document);
SUPERTYPE OF (Drawing);
     design_discipline : STRING;
END ENTITY;
ENTITY Drawing;
SUBTYPE OF (Design_document);
     scale : INTEGER;
     sections : OPTIONAL SET [0:?] OF STRING;
     floors : OPTIONAL SET [0:?] OF STRING;
     building_parts : OPTIONAL SET [0:?] OF STRING;
END ENTITY;
ENTITY Content_data;
     textual_description : STRING;
     picture_of_document : OPTIONAL BITMAP;
END ENTITY;
ENTITY Storage_data;
storage_type : SELECT (Paper_document, File);
location : STRING;
END ENTITY;
ENTITY Scheduling_data;
planned_start : OPTIONAL DATE;
planned_end : DATE;
actual_start : DATE;
actual_end : DATE;
completion_level : REAL;
```

```
completion_state : SELECT ('draft', 'proposition', 'inspected', 'ready');
END ENTITY;
ENTITY File;
    file_name : STRING;
    program : STRING;
    file_format : STRING;
    backup_file : SET [0:1] OF File;
END ENTITY;
ENTITY Paper_document;
END ENTITY;
ENTITY Distribution_event;
    event_id : STRING;
    distribution_time : Date;
    recipient : Project_participant;
    distribution_criteria : Distribution_address_list;
INVERSE
    distributed_document : document FOR distribution;
UNIQUE
    event_id;
END ENTITY;
ENTITY Distribution_address_list;
    list_id : STRING;
    distribution_list_item : SET [1:?] OF List_item_data;
INVERSE
    user_event : SET [0:?] OF Distribution_event FOR
    distribution_criteria;
UNIQUE
    list_id;
END ENTITY
ENTITY List_item_data;
    user_access_right : STRING;
    recipient : Project_participant;
END ENTITY;
ENTITY Project_participant;
    participant_id : STRING;
    function_in_the_project : STRING;
    address: STRING;
    works_for : Project_participant;
INVERSE
    address_data : SET [0:?] OF List_item_data FOR recipient;
    document_created_by : SET [0:?] OF Document FOR creator;
UNIQUE
    participant_id;
END ENTITY;
END_SCHEMA.
```

REFERENCES

Björk, B.-C. (1992a) A unified approach for modelling construction information. *Building and Environment*, **27**(2), 173–94.

Björk, B.-C. (1992b) *Information Versus Document Level Data Transfer in Computer Integrated Construction*. Preproceedings of the international workshop on *Frameworks for Computer Integrated Construction*, VTT, Espoo, Finland, 5–9 October 1992.

Björk, B.-C. and Penttilä, H. (1989) A scenario for the development and implementation of a building product model standard. *Advances in Engineering Software*, **11**(4), 176–87.

Björk, B.-C., Huovila, P. and Hult, S. (1993) *Integrated Construction Project Document Management (ICPDM)*. Paper presented at EUROPIA '93 Conference, Delft, The Netherlands, 20–23 June, 1993.

CEN (1991) Express Language Reference Manual. *CEN/CLC/AMT/WG STEP N48*, Association francaise de normalisation, Paris, 160 pp.

Dupagne, A. (1991) Computer Integrated Building. *Strategic Final Report*, Esprit II, Exploratory Action No. 5604, CE Commission D.G. XIII, December.

Froese, T. and Paulson, B. Jr. (1991) An Object-Oriented Approach for Integrated Project Management Software. *Working Paper Nr. 11*. CIFE, Stanford University, USA, 15 pp.

Gielingh, W.F. (1989) *Computer Integrated Construction, a Major STEP Forward*. ARECDAO89 Conference Proceedings, ITEC, Barcelona, pp. 29–48.

Howard, H.C., Levitt, B.C., Paulson, B.C. *et al.* (1989) *Computer integration: reducing fragmentation in the AEC industry. Journal of Computing in Civil Engineering*, **3**(1), 18–32.

Karlsson, H. and Allott, T. (1991) Classification of Information in the Construction Process. *ISO TC59 SC13*, Discussion paper, 25 pp.

Luiten, B. and Tolman, F.P. (1991) *A Conceptual Modelling Approach to the Development of Integrated Building Project Models and Systems*. 4th International Conference on Computing in Civil and Building Engineering, Tokyo, Japan, p. 7.

Marca, D.A. and McGowan, C.L. (1987) *SADT–Structured Analysis and Design Technique*, McGraw-Hill, New York.

Sanvido, V.E. (1992) Linking levels of abstraction of a building design. *Building and Environment*, **27**(2), 195–208.

CHAPTER 21

Environmental design of buildings

Kevin J. Lomas, Stewart Thornton and Rowan Shulver, De Montfort University, The Gateway, Leicester, LE1 9BH UK

21.1 INTRODUCTION

Prompted initially by the oil crisis of 1973 and, more recently, by concerns about the well-being of the global environment and the quality of the environment within buildings, there is considerable world-wide interest in predicting the environmental performance of buildings. This interest has coincided with the rapid evolution of computers, from the mid-1970s mainframe giants to, in the mid-1990s, extremely powerful workstations and personal computers. Building environmental performance (BEP) methods capitalized on this technology beginning first with thermal analysis, for which computer power was necessary, and later lighting design and airflow analysis for which computers provided rapid and accurate solutions to problems which previously could only be resolved by experimentation.

The most sophisticated BEP tools try to mimic the thermophysical and luminous processes which occur in buildings by modelling the underlying physics from first principles. These sophisticated programs are termed simulations and are the main focus of this chapter about one application of integrated construction information.

Simulations for thermal analysis (the predictions of temperatures and energy use), airflow analysis and lighting design have developed opportunistically with developers being responsible for the design of data storage, data input and data display functions of the software as well as the physical modelling. There has been little, if any, attempt to incorporate intelligence to resolve 'what if' questions.

It is now widely recognized that more efficient, productive and cheaper design advice could be provided if the functions of data input, data storage and data interpretation are separated out from the physical model-

Integrated Construction Information. Edited by Peter Brandon and Martin Betts. Published in 1995 by E & FN Spon, 2–6 Boundary Row, London SE1 8HN.
ISBN: 0 419 20370 2

ling. These three functions could then become generic to a wide range of application areas, design, environmental analysis, cost estimation, etc., thus presenting users with a familiar interface, relieving developers of the task of creating them and enabling data to be stored in such a way that other members of a design team can make use of it directly. Thus integration covers both the integration of interfaces and data storage with specific modelling programs as well as integration of different professionals working on a particular building (architects, environmental designers, quantity surveyors, etc.). The key to such integration is the use of common data structures to describe the building – a product model.

This chapter maps out the problems to be overcome in order to achieve integrated systems, and reviews existing work in the area of building environmental design with a view to identifying the most favourable features of this research. It gives a more detailed description of one project being undertaken by the ECADAP Group at De Montfort University to develop a low cost Integrated Building Environmental Performance System (IBEPS).

21.2 BUILDING ENVIRONMENTAL PERFORMANCE SIMULATION

21.2.1 Detailed thermal simulation programs

Computers have been used for building energy analysis since the late 1960s; however, the programs focused on sizing heating, ventilating and cooling plant using simple, steady-state assumptions for the building envelope losses and gains. They were largely computerized versions of calculations which hitherto had been undertaken by hand and very little attention was paid to the rigour, usability, or validity of the programs.

With the oil embargo of 1973 came a raised consciousness of energy issues and a serious interest in reducing the use of non-renewable resources. By 1976–77 more attention was being devoted to passive solar and innovative design strategies, both in the United States and the United Kingdom. The attempt to accurately evaluate such designs fostered an entirely new generation of building energy analysis programs called variously building envelope, or thermal loads, programs. The most sophisticated of these programs are dynamic simulation programs (DSPs) which are capable of predicting the hour-by-hour variations of internal temperatures, heat fluxes and energy usage in response to occupancy patterns, plant schedules and weather conditions for real zoned buildings. They worked by trying to solve, at each time step, the physical equations governing heat transfer via transient conduction, convection, infiltration, ventilation and long and short wave radiation, etc. Notable public domain examples which emerged in the late 1970s and early 1980s, and which are still in widespread use, were, in the USA, BLAST (Control Data

Corporation, 1980), DOE (Lawrence Berkeley Laboratory, 1980) and SERIRES, formerly SUNCAT (Palmiter and Wheeling, 1983), which had a private domain twin (SUNCODE) and used a finite difference approach. In the UK, ESP (Clarke, 1982) was developed and a little later HTB2 (Alexander and Lewis, 1984) emerged. Tas, a response factor program, also emerged in the UK (Jones Cassidy Mellor Ltd, 1982) as a commercial venture.

Most of these packages consist of a number of FORTRAN programs which operate as a closed system. They were originally developed for mainframe computers which accept input via comma or space delimited ASCII files. In particular, the geometrical building description required by these programs took this form. Many packages can access data libraries (of building material properties, for example) but their structure is also unique. Similarly, the output interface is specific to each package, as are the internal data transfer formats which are used. Developments in desktop computing power have enabled these programs to run on powerful desktop workstations and PCs. This has enabled more complex problems to be analysed and users have demanded more sophisticated interfaces. Specifically, some systems are claimed to be able to import data from other CAD packages, e.g. ESP+ (ASL, 1992), and some DSPs are part of larger systems which contain a central building description file capable of serving more than one analysis program, e.g. BEANS (Oasys/Arup, 1991).

21.2.2 Airflow prediction

Physical modelling has been the traditional approach to assessing the airflow in and around buildings for critical conditions. However, the cost and timescale of such evaluations means that it is often not possible and computers offer one possible solution.

Computer-based predictions of fluid flow were spurred on by the requirements of 'safety-critical' industries (in particular, the nuclear and aerospace industries) to evaluate performance when no other (physical) modelling method was possible. Thus computational fluid dynamics (CFD) codes evolved in the late 1960s for use on mainframe computers and attempts to apply them to buildings were made in the early 1970s (Nielsen, 1974).

CFD programs work by dividing spaces into a number of discrete 2- or, more usually now, 3-dimensional volumes using a mesh. A well-designed mesh is crucial to obtaining accurate results. Where the flow is turbulent, as it often is in buildings, additional equations defining the turbulence model must also be solved. Further equations are introduced for turbulent flows when temperature predictions are needed and/or if variables representing the concentration of containments are involved.

The partial differential equations are highly non-linear and can only be solved by iteration and, since thousands of such equations must be solved for a reasonably detailed description of the velocity and temperature field in a space, substantial computing power is needed. The extensive output describing the values of key parameters across space and time requires the use of good graphics-based post-processors for its interpretation.

Initially, these considerable computer demands constrained the use of CFD to simple problems on mainframe computers (e.g. Reinartz and Renz, 1984; Alamdari *et al.*, 1984). However, the emergence of powerful UNIX-based desktop workstations and, very recently, powerful PCs offered the potential for the wider uptake of the technique. Programs which had originally been developed for other applications areas such as Phoenics (Spalding, 1981) migrated to these platforms for use in buildings (e.g. Holmes *et al.*, 1990) and numerous other CFD packages began to emerge, e.g. (Flow3D, 1992; Fluent, 1992) and the buildings-oriented package FloVent (Flomerics, 1989). However, the need for close interaction between a very skilled user and the program is a barrier to the use of CFD by all but the most experienced operators. The resolution of this issue is a specialist activity for CFD experts.

21.2.3   Lighting design

Daylight levels in simple spaces can be estimated from simple equations, charts or tables, such as the BRE daylight factor protractors and nomograms (CIBSE, 1987); the exact times when sunlight will penetrate into a space and the surfaces upon which it will fall can also be assessed manually. Isolated PC programs have also been produced to address these simple lighting issues, the DAYLIGHT program (Birch and Frame, 1991) and Superlite (Lawrence Berkeley Laboratory, 1985) are widely used programs.

True lighting simulation has, however, only emerged recently in the form of workstation packages using either a radiosity or a ray-tracing approach. In this field, the RADIANCE system is unique since it appears to be the only physically valid ray-tracing program available world-wide (Ward, 1992). RADIANCE is capable of accurately predicting the daylight factors, surface illuminances and surface luminances within complex spaces containing any type of surface finish, furnishings, glazing, etc. It can predict the lighting levels produced by daylight, sunlight, artificial light or any combination of these. RADIANCE also accounts rigorously for specular reflection and the reflection and retransmissions of light from one space to another. This is most important in the design of complex daylit spaces such as atria, and the spaces surrounding them. The most impressive feature of RADIANCE is that it combines these capabilities to produce photo-realistic images which convey accurately the appearance of the illuminated scene. Because of their numerical accuracy RADIANCE images are much more convincing than those produced by CAD rendering packages, e.g. 3D STUDIO (AUTODESK, 1992) or by radiosity-based methods.

RADIANCE has been developed since the advent of the desktop work-station. It is written in C, runs efficiently in the UNIX environment and utilizes the powerful graphical capabilities of workstations. It has recently been ported to run on the PC.

21.2.4   Comparison of BEP Tools

It is evident from the foregoing that the analysis engines of the three types of BEP tools are highly specialized, complex and unique to their application domain. However, it is also apparent that in all three areas there are features which offer hope for the development of a low cost integrated system.

1. All three operate on both workstations and PC platforms.

2. In all three areas, the data input interface can be separated from the analysis engine and geometrical data could be supplied, in principle, by a CAD system. This would reduce both the time taken to import the data and the likelihood of making mistakes.

3. Much of the remaining data are related to the properties (thermal, lighting) of the materials which bound the space, thus extensive use can be made of databases.

4. This building-specific data could be stored for use in other application domains in the form of a product model.

   There are, however, a number of challenges to such integration.

1. In all three areas there are elements of applications specific data (e.g. convergence criteria or meshing requirements, CFD). This implies that, to a certain degree, the input interface must be able to accept applications-specific data.

2. The results output and analysis interfaces are highly domain-specific and may thus be difficult to standardize.

3. The level of modelling detail demanded by the tools is dramatically different (e.g. in visualization a very detailed description, in thermal analysis and CFD much sparser representation).

4. Very little effort has been directed towards the encapsulation of knowledge in such systems so that they can resolve 'what if' questions.

The approaches which can be taken to integrate these sophisticated tools by capitalizing on their beneficial features and overcoming their problems are explained further in the next section.

21.3. A STRATEGY FOR BEP INTEGRATION

21.3.2 Overview

As we can see in the other chapters and sections of this book, considerable research effort is being focused on the problem of integration in general, and the integration of BEP tools is but one aspect (Hertkorn, 1993). The 3 staff year effort at De Montfort University (DMU) can only address a small part of the problem but, by building on the efforts of others, it will complement work being undertaken elsewhere.

The following are key features of the DMU project.

1. A low cost solution which can operate on both PC and UNIX platforms.

2. The use of AutoCAD (AUTODESK, 1993) to supply geometrical information.

3. A core product data model represented in an object-oriented database management system. Extensions will be possible without impacting on the current users of the product model. These extensions would include the addition of new data and the provision of existing data in the format required by new evaluation programs.

4. Transfer of data from the CAD system to the product model and vice versa.

5. Retention of the native BEP tool interfaces.

6. The ability to support multiple users, which implies a mechanism for ensuring the security of data within the product model and change control.

7. The provision of a facility for data exchange using the STEP standard.

ESP will be the target for a prototype system because it requires a full geometrical description of the building being modelled and it needs a detailed description of the thermophysical properties. At all stages however, the needs of RADIANCE and CFD programs will be considered to facilitate further development of the system.

21.3.2 The CAD System

AutoCAD has been chosen as the CAD system to integrate in the prototype because of:

1. Its market-place dominance

2. Its *de facto* standard for data exchange DXF

3. Its availability on a wide range of platforms (DOS, Windows, UNIX)

4. The range of third party software available for it

5. The powerful customization facilities provided.

Customization is possible using AutoLISP and C when direct communication with the host operating system is needed. The ability to store extra data on graphical entities – extended entity data – is provided (from Release 11). Object identity can be retained from one editing session to the next (from Release 10) and a GUI interface is provided (from Release 12). Entities can also be separated by being drawn on a different layer and level.

These features enable semantics to be added to the graphical entities (lines, polygons, etc.), which give physical meaning (as walls, doors, etc.). For example, polygons are being used to represent wall surfaces with the distance between them defining the wall thickness. A combination of polygons which forms a closed mesh defines a room. By assigning the entity handle of one of these objects to the extended entity data of another, a referential link can be established between the objects.

21.3.3   The product data model

The building data model is the key to the system and will be the main focus of attention. It will not be possible to produce a single generic conceptual data model that will allow for all the future data and their interrelationships. Consequently, the ability of the administrator to handle the complexity of changing requirements is vital. This data independence can be provided by using a database management system (DBMS) rather than a file system to store the data and a key choice in choosing a DBMS is the data model that the system incorporates. The relational model (Codd, 1970) is the basis for many commercial DBMS. However, current relational database management systems (RDBMS) create difficulties when used to support building design applications.

1. Normalization tends to cause design objects to be stored across multiple tables which leads to slow retrieval/update times.

2. The simplicity of the modelling constructs makes common design relationships difficult to represent, e.g. inheritance, aggregation.

3. The data types provided are too simple to handle common design situations, e.g. graphical data, ordered sequences.

4. The language SQL (Date, 1989) is not computationally complete so that it must be embedded in a host language.

The object approach implemented in object-oriented database management systems (ODBMS) offers a potential solution to these problems (Bertino and Martino, 1991); it allows

1. the encapsulation of methods with the data;

2. the sharing of commonality via inheritance;

3. close integration of an object-oriented programming language;

4. simplified representation and manipulation of persistent design objects;

5. provision of version control facilities; and

6. provision of long duration transactions.

These are necessary features of an integrated BEP system.

Encapsulation and inheritance are important in providing a technique for handling the complexity of the product model. Commercial ODBMSs are now available and some RDBMS vendors are extending their products with selected object model features. The benefits of the RDBMS are well established, although it is not yet clear whether ODBMS products will become established as mainstream databases or their functionality will become merged with extended RDBMS products. At present, the issues associated with conforming to the emerging STEP standard (Clark, 1990) have not been addressed explicitly, partly because the standard is not fully developed and partly because the purveyors of some ODBMSs are considering these problems.

A prototype is being developed as a set of C++ classes. The classification of the classes, the relationships between these classes and the functionality within them can then be tested within a shorter time frame. After this has been achieved, an implementation of these C++ classes will be mapped to an ODBMS. This approach is made possible due to the prominence C++ has within the ODBMS world, being the supported host language for most products.

21.4 OBJECT MODELLING

The Object Modelling Technique (OMT) (Rumbaugh *et al.*, 1991) has been chosen for developing the conceptual data model. Other projects in this area have chosen to use modelling techniques such as NIAM (Björk, 1992), however OMT was chosen because it supports the concepts of inheritance, association and aggregation that are central to the conceptualization of object-oriented systems.

The proposed object model (Figure 21.1) has been strongly influenced by that proposed by Björk (1992), which in turn derived from a review of models produced by several other research groups. A space/room is the basic enclosing volume used in the model. It is bounded by several physical space boundaries (PSBs) which are planes lying on the face of the enclosing structure. The structure may contain a number of different static

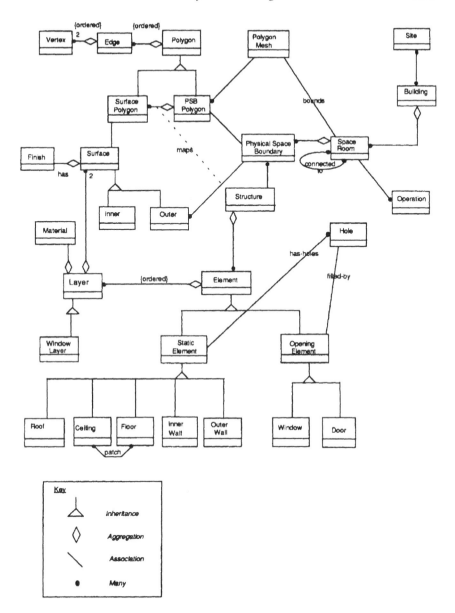

Figure 21.1　Proposed object model

elements (such as walls) or opening elements (such as windows and doors). These elements are composed of layers of material (brick, glass, air, etc.). The layers in the element will be ordered in a different direction depending on which PSB is current (i.e. the side from which the PSB is being viewed). The surfaces have applied finishes (paint, paper, etc.), which differ from the natural properties of the materials. The finishes of the outer surface layers of an element are associated with a PSB. At present space/room has an explicit association with itself to show which rooms are adjacent to each other. As an alternative, the information already included in the model (via the inner walls which will bound two space/rooms) can be used.

One noticeable difference between the model and that proposed by Björk is the use of delegation rather than inheritance in the key relationship between Structure and Element (called 'enclosing structure' and 'enclosing entity' by Björk). Delegation is a much safer way of sharing behaviour between classes that have noticeable differences. By including Element as an attribute within structure, delegation has been used in this model. Only meaningful operations are delegated, so there is no danger of inheriting meaningless operations by accident. By defining Structure as a subclass of Element (i.e. inheritance) there is a danger of inheriting methods not needed in the subclass because Structure is quite a different semantic object from Element.

In thermal models, the space/rooms may be aggregated to form a space assembly so that very large buildings can be simulated more efficiently. The complete building will thus consist of an aggregation of space/rooms or space assemblies or both. The Site class enables the meteorological and other site features to be defined, and the Operation class enables occupancy characteristics and system schedules, etc., to be described.

The data fed into thermal simulation programs will contain many space/rooms or space assemblies defined by a few polygon meshes. In contrast, a lighting simulation program such as RADIANCE will define one space/room with many polygons, which forms a complex mesh, and other shapes such as cones, spheres and cylinders may also be needed. Thus the complexity, or richness, of the model will increase as extra application programs are fed from it. Similarly, the model will be extended to enable the full complexity of a particular building to be described as it evolves during the design/construction/refurbishment/demolition process.

21.5 CONCLUSIONS

1. The evolution and current status of the most sophisticated building environmental performance (BEP) analysis tools for thermal design, airflow prediction and lighting design/visualization has been outlined.

2. These advanced tools are closed systems; however, they each run on a range of platforms and require a common core of geometrical building attribute data. This could usefully be supplied from a building product model. There are, however, many domain-specific demands which cannot easily be addressed via such a data model.

3. An outline product model, tested in one project aimed at integrating advanced BEP tools, has been described.

4. There are many other projects, throughout the world, which are grappling with the issue of integration in general and BEP tool integration in particular. By pooling knowledge and sharing effort, data models with the widest possible application will emerge.

REFERENCES

Alamdari, F., Hammond, G.P. and Melo, C. (1984) Appropriate Calculation Methods for Convective Heat Transfer from Building Surfaces. Proceedings of the 1st National Heat Transfer Conference, Leeds. *IChemE Symposium Series no. 86*, **2**, 1201–11.

Alexander, D.K. and Lewis, D.T. (1984) HTB2 – A Model for the Thermal Environment of Buildings in Operation. *Technical Reference Manual*, Rev. 0.0 (draft), 32 pp.

ASL (1992) *The ESP+ Training Manual*. Version 1, Release 1. Abacus Simulations Limited, Glasgow.

AUTODESK (1992) *3D STUDIO Release 2, Reference Manual*. Autodesk Inc.

AUTODESK (1993) *AutoCAD Reference Manual* Release 12. Autodesk, Inc.

Bertino, E. and Martino, L. (1991) Object-oriented database management systems: concepts and issues. *Computer*, 33–47.

Birch, S. and Frame, I. (1991) *Computer Generated Analysis of the Daylight Performance of Internal Architectural Spaces*. Proceedings of Building Environmental Performance '91, BEPAC, Canterbury, UK, pp. 89–102.

Björk, B.-C. (1992) *The Topology of Spaces, Space Boundaries and Enclosing Structures in Building Product Data Models*. CIB W78 Workshop on Computer-Integrated Construction.

CIBSE (1987) *Window Design. Applications Manual* Chartered Institution of Building Services Engineers, London, pp. 66.

Clark, S.N. (1990) The NIST Working Form for STEP. *NISTIR 4351*, NIST, Gaithersburg, MD 20899.

Clarke, J.A. (1982) *ESP Manual*. Architecture and Building Aids Computer Unit, University of Strathclyde, Glasgow.

Codd, E.F. (1970) A relational model of data for large shared data banks. *Comm ACM*, **13**, 6.

Control Data Corporation (1980) *Cyber BLAST 3.0 User Information Manual*, Control Data Corporation, Alexandria, VA, USA.

Date, C.J. (1989) *Guide to the SQL Standard*, 2nd edn. Addison Wesley.

Flomerics (1989) *FloVent User Manual*. Flomerics Limited, Kingston-upon-Thames, Surrey.

Flow3D (1992) *User Manual*. CFDS, AEA Industrial Technology, Oxfordshire, UK.

Fluent (1992) Fluent Europe Limited, 146 West Street, Sheffield.

Hertkorn, C. (1993) *Report on Projects Dealing with Building Design Support Environments*. IEA Annex 21 Subtask D, Institut für Industrielle Bauproduction, Universitat Karlsruhe.

Holmes, M.J., Lam, J.K.-W., Ruddick, K.G. and Whittle, G.E. (1990) *Computation of Conduction, Convection and Radiation in the Perimeter Zone of an Office Space*. Proceedings of International and Conference ROOMVENT 90, Oslo, Norway.

Jones Cassidy Mellor Ltd (1982) *Tas Handbook*, Cranfield Institute of Technology, Cranfield, UK.

Lawrence Berkeley Laboratory (1980) *DOE-2 Reference Manual*, version 2.1. LBL, Energy and Environment Division, Berkeley, CA, USA.

Lawrence Berkeley Laboratory (1985) *Superlite 1.0 Manual*. Windows and Daylighting Group, LBL, Berkeley, CA, USA.

Nielsen, P.V. (1974) Flow in Air Conditioned Rooms. PhD thesis, Technical University of Denmark, Copenhagen.

Oasys/Arup (1991) *BEANS User Manual*, Oasys, 13 Fitzroy Street, London.

Palmiter, L. and Wheeling, T. (1983) *Solar Energy Research Institute Residential Energy Simulator* version 1.0, Solar Energy Research Institute, Golden, CO, USA, 365 pp.

Reinartz, A. and Renz, U. (1984) Calculation of the temperature and flow field in a room ventilated by a radial air distributor. *International Journal of Refrigeration*, 7, 308–12.

Rumbaugh, J., Blaha, M., Premerlani, W. *et al.* (1991) *Object-oriented Modelling and Design*. Prentice Hall.

Spalding, D.B. (1981) A general purpose computer program for multi-dimensional one and two phase flow. *Mathematics and Computers in Simulation XXIII*, pp. 267–76. London.

Ward, G.J. (1992) *The RADIANCE Lighting Simulation System*, Lawrence Berkeley Laboratory, Berkeley, CA, USA, 16 pp.

CHAPTER 22

Project historical information system

K. C. Choi, Bechtel Corporation, San Francisco, CA, USA and
C. W. Ibbs, University of California at Berkeley, CA, USA

22.1 INTRODUCTION

The importance of early project planning is widely recognized for engineering construction projects. (CII 1986; Barrie and Paulson, 1984). Company historical data is a major source of information for developing project scope and cost estimates during conceptual planning when only limited amounts of project-specific data are available. Despite the recognition of its importance, however, the compilation, storage, retrieval and use of project historical data in many engineering companies is still not effective. Furthermore, the current historical data cannot support integrated conceptual planning and design systems. Work by Serpell-Bley (1990) concluded that the improvement of capturing and using historical data is one of the most critical improvement areas in conceptual planning currently.

This chapter identifies the current problems and needs associated with managing project historical cost data and develops, from a user perspective, a complete set of functional specifications for a new kind of historical cost information system for a process plant project. Our long-range goal is to develop integrated concurrent engineering systems that can be used to develop accurate scope and cost baselines from an historical database.

22.2 METHODOLOGY

First, we analysed five existing project historical reports in a turnkey contracting company to identify their contents and format. In addition, we conducted site interviews with 10 users of historical data in the same company to identify the improvement needs. The interview participants were from Marketing and Business Development, Project Controls,

Integrated Construction Information. Edited by Peter Brandon and Martin Betts. Published in 1995 by E & FN Spon, 2–6 Boundary Row, London SE1 8HN.
ISBN: 0 419 20370 2

Engineering, Procurement, Construction and Start-up. We asked the following questions:

1. What are the problems and needs with current historical data?

2. Describe the information flows in your function whether they flow on paper or electronically.

3. What are the data items that you need to capture for the following projects?

4. What are your views of historical data?

5. What are the reports (data queries) that you need from that view?

6. What are the levels of detail for the items in each view?

7. What would you like to see in a new historical information base that is designed to support concurrent engineering? (Users were introduced to concurrent engineering concepts during an interview.)

Once the existing ways of storing historical data and problems with them were identified by Question 1, the input from interview participants to Questions 2–6 were analysed to identify the requirements for supporting concurrent engineering. The requirement analyses include high level data flow diagrams, data matrix and entity relationship diagrams. These tools were then used along with responses to Question 7 to develop the functional specifications presented in the last section of this chapter.

22.3 OVERVIEW OF CURRENT HISTORICAL REPORTS AND DATA

Typically, corporate historical data are gathered from various project historical reports prepared by the project participants at the end of a project. The compilation of various data generated on a large project into such an historical report is commonly a major task consuming several thousand labour hours, and tens of thousands of hours for mega-projects. Multiplied by the number of projects completed each year, historical report preparation cost can be a major expenditure item. As more historical reports become available, more hours are expended to extract data from them, and to make them useful and available to various project planners including engineers, construction planners and cost estimators. Clearly, a disorganized information system is a major source of overhead expense and provides a major improvement opportunity by applying appropriate information technologies. Project historical reports typically include the following major data components:

1. General project data (description, location, scope, photos, maps, etc.)

2. Management data (execution plan, schedule, organization charts, etc.)

3. Key design document (plot plans, general arrangements, piping and instrument diagrams (P&IDs), etc.)

4. Material vendor and delivery data

5. Construction progress and production data

6. Subcontractor data

7. Cost data

8. Labour hour and productivity data (home, office and field)

9. Technical and cost study data (trade-off studies).

Though used for other purposes as well, historical project reports and corporate databases are primarily used for estimating project scope, cost and other resource requirements. The estimating department extracts various data from project historical reports to maintain company estimating databases. They also develop and maintain various cost and scope (quantity) models based on project historical data. Regression is the most common parametric estimating technique used for such models. The definitions of cost items have to be consistent in all historical reports. Otherwise, it will be difficult to combine data from various projects.

To develop meaningful regression coefficients, the historical data have to receive a rigorous statistical treatment, called data normalization. For example, material and equipment costs have to be normalized for time and place of purchase, currency, transportation, scale of economy by purchase quantity, delivery schedule, etc., before they can be added to a cost-capacity regression model. Likewise, labour productivity has to be normalized for various considerations including crew mix and size, weather, work volume, location, work schedule, overtime hours, physical and logical (schedule) congestion. Other qualitative information is also recorded, such as the overall project and materials management effectiveness that impacts labour productivity. The data normalization process can be quite complex.

22.4   PROBLEMS WITH CURRENT HISTORICAL DATA

The responses from the interview participants indicated problems in the following areas:

22.4.1   Data storage, access and distribution

1. All project historical reports are on paper, which is difficult to use and easy to lose or damage.

2. Most data structures were determined based on what was typically available and possible to capture manually at the end of project, rather

than what is needed at the planning phase of subsequent projects; e.g. actual cost data by the engineering system is not captured due to the difficulty in manual cost allocation.

3. People who collect the data may not necessarily understand user needs. For example, cost data are collected by cost engineers. They may not fully understand the needs (data views, details, types, etc.) of design engineers and cost estimators during the conceptual planning phase.

4. Most historical data capture only 'snapshots' at the end of project; e.g. the last monthly cost report and schedule. People who have knowledge about the project execution history are likely to have left for another project by this time and their input is not captured.

5. Qualitative data are captured separately from quantitative data. Understanding the project context of a specific set of data (e.g. why the labour productivity differs from other projects) is not practically possible for people who are not familiar with the reference project.

6. The raw data in an individual project historical database or reports are not connected to the normalized data in a company historical database. The detailed information in the project historical reports is not accessible from the company historical database.

7. Searching for relevant historical reports and data is time-consuming.

8. Project historical reports are difficult to use by users who are not familiar with the overall organization of reports.

9. The corporate historical database is not tied to the integrated project planning systems. Ideally, users should be able to access information without being required to know cumbersome access details. An engineer who wants to know the cost history of a heat exchanger on previous projects should be able to query that information by navigating through user-friendly commands.

10. Lessons learned are not organized well and buried in the details. It is difficult to compile and disseminate them to other projects.

22.4.2   Information extraction processes

1. Much useful data is lost during the extraction and summarization processes. The level of detail in the historical reports is inadequate to support robust cost and scope models required for conceptual planning.

2. Converting this useful information into electronic form is time-consuming and error-prone.

3. Since people always shuffle within and between companies, it is difficult to reach the historical report authors who understand the hidden meaning of project historical data. The historical data should provide rich data context so that it can be used with minimum or no consultation of authors.

22.4.3 Data views (model)

1. When companies started collecting historical cost data several decades ago, they did not need a data model to capture data manually in the paper-based reports. Therefore, the amount of data and the ways in which the data are represented and organized are severely limited. Usually, only summary level data at the end of the project is captured using a rigid data structure, such as a company standard cost code of accounts. To support concurrent planning, data should be captured with a robust data model that can support multiple views and abstraction levels.

2. The relationship between data entities, which is an important knowledge element for data to become information, is not captured in both project and company historical data. This is attributable again to the lack of a rich information model. This limitation is one of the main motivations of this research and discussed in detail later.

22.4.4 Data quality

1. Since projects are not identical in type, size, location, technology, time of installation, etc., historical data have to be normalized extensively to make them useful. This data normalization is manually performed, which is time-consuming and error-prone. Updates are not timely.

2. Even the limited amount of qualitative data captured in the historical reports is not consolidated into a company historical database. Most historical data contexts and semantics are lost during data summarization and normalization. It is like compressing three-dimensional space into a two-dimensional plane. We refer to this as 'flattening the historical data dimensions'.

3. Data semantics (entities, attributes and relationship) change over time. The data structures used to store the data semantics in a company database are presently not easily extendible.

22.4.5 Data maintenance

1. Project data have a limited useful life. For example, when new technology (material, process, control systems, etc.) is introduced, the

old technology cost data should be eliminated from the database. This is difficult though because the data details are lost during the manual data normalization processes. Thus, cost and scope models are maintained based on the mixture of both good and bad data.

2. Companies usually maintain a few 'historical data experts'. They analyse and publish a closely circulated historical information bulletin. Those historical data experts maintain the knowledge themselves, and that knowledge is vulnerable to the loss when they leave the company.

3. Historical data are considered company confidential and not shared between companies. This places an expensive burden on individual companies since each company has to maintain its own historical information.

22.5 FUTURE HISTORICAL INFORMATION SYSTEM TO SUPPORT CONCURRENT ENGINEERING

The concurrent engineering principle advocates that the cost impact of early design decisions on downstream project phases should be evaluated *simultaneously* with the design process (Creese and Moore, 1990). The information systems should provide the capability to perform such concurrent evaluation, compressing sequential steps in the traditional approach. This process is labelled as concurrent project planning and is depicted in Figure 22.1.

The central idea in concurrent engineering is design-to-value. The value is determined by the facility owner (i.e. customer of engineering or turnkey

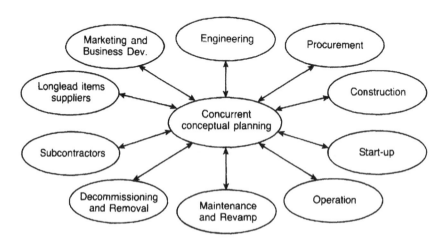

Figure 22.1 Concurrent conceptual planning

contractors) in terms of cost, schedule, quality and safety. Since delivery date is such a critical constraint in the industrial plant projects, the response time to analyse these value parameters for a design alternative must be kept short. Simultaneous evaluation of design and its value implication will greatly enhance planning capability and the chance for overall project success. The conceptual planning information system should support such needs for rapid design optimization. To assist in the team work approach, each team member involved in concurrent conceptual planning needs rapid access to historical information on the similar projects.

During concurrent project planning, designers, planners and cost estimators depend on their personal knowledge, reference materials and historical data to make design decisions and interpret their cost and schedule implications on downstream project phases. The data model has to support different data views to be useful for concurrent planning. For example, the cost and quantity data are typically collected by the company standard commodity code of accounts. However, construction planners or engineers may want to view project cost and scope data in different ways from commodity code of accounts such as engineering systems and construction work packages.

As identified in the previous sections, company project historical information plays a key role in conceptual planning. Figure 22.2 shows the use of an historical information base to support concurrent project planning throughout the project. The information captured during design, procure-

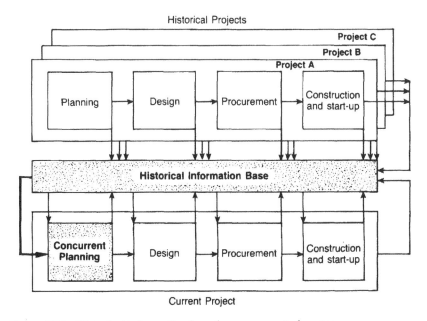

Figure 22.2 Historical information base for concurrent planning

ment and construction phases is used for concurrent planning of the next project. Conversely, information captured during the concurrent planning phase can be used in downstream phases of the same or other projects.

22.6   ANALYSIS OF INFORMATION VIEWS FOR CONCURRENT ENGINEERING

3D CAD-based detail design and its database structure are centred on the bottom–up composition of design objects. In this bottom–up concept, design progresses by reverse hierarchical composition of physical components. Design components are typically represented at the lowest level of detail and stored in the database with their attributes (e.g. piping components, structural components, etc.). Building a 3D model of a large process plant requires the use of several hundred thousand such components.

This bottom–up composition using a large component library is not appropriate for conceptual design that progresses in top–down decomposition of functional systems. The information abstraction during conceptual design is dynamic because the design continues to develop from an abstract (summary) level to a concrete (detail) level. This top–down design should be able to support various types of information views (e.g. equipment modules, process systems, areas, work packages, activities, etc.) and data abstraction needs, i.e. generalization and aggregation.

An information model should be able to represent and relate data in the dynamically evolving and changing project environment. It should provide a framework for integrating various forms of data, procedures and knowledge. Eventually these design objects have to be represented graphically in a CAD environment, and the data model has to be able to support this requirement effectively. Moreover, this model has to support knowledge representation needs to incorporate expert system concepts that are becoming increasingly useful in engineering–construction.

Current planning and design systems and supporting databases do not provide that flexibility. Typically cost is captured in a cost database by commodity cost account or construction work packages. Design information is captured separately in drawings and CAD files. Work execution information is captured by organizational structures, work breakdown structures, construction and subcontract work packages, and CPM schedule activities. This results in fragmented information systems with inconsistent data representation. If the information model is designed to represent all such views and the information can be captured to support that model, planning can be improved significantly.

One desirable feature to support concurrent planning is the ability to abstract project planning information in many different planning views (i.e. scope definition, cost estimates, work plans, etc.) while maintaining consistent relationships among them. That is, conceptual planning information should provide consistent and integrated data abstraction for scope definition, estimating, work execution planning and subsequent project control functions. Information systems integrated using a uniform data representation can provide such powerful data abstraction capabilities.

To analyse the information to be captured in the historical information base that supports concurrent conceptual planning, the data flow in a typical process plant project was modelled at the highest level based on the input from the interview participants; see Figure 22.3

The information contents in each data flow were then itemized and categorized into Table 22.1 from a systems engineering perspective, i.e. input, process, output. Project and project environment categories have been added to identify all necessary project attributes that are not covered in other categories, but necessary for data access and use. Within each category, the data types are created from the viewpoint of a 'history writer', who may specify what, who, when, where, how and why. Table 22.1 represents the ultimate scope of the historical information base.

22.7 ENGINEERING VIEW

Conceptual engineering requires building and analysing various functional models of a constructed facility. Such engineering models include mathematical, statistical and process simulation models. The functional models and the supporting data views for conceptual engineering (design) are primarily oriented by engineering systems. Design development typically evolves top–down along the various system and component assembly hierarchies. Many engineering disciplines (e.g. mechanical, structural, electrical, piping, etc.) interact with each other to develop design into more detail by exchanging various information from various engineering system views.

Although the volume of data to be integrated is much larger in the detail design phase, the data model for integrating conceptual planning functions is conceptually more complex. The design of a system or a component affects other systems or components successively. The depth and magnitude of this successive influence is much larger during the conceptual planning than during detail design. It is one reason why improved concurrent planning capability during the early project phase is critical for the project success. To analyse the relationships accurately between design data entities by systems, the information flow during the conceptual planning phase is modelled in a simplified information flow diagram; see Figure 22.4. This figure is a subset of Figure 22.3, and shows how the various engineering and design functions are affected by their upstream functions. This engineering data can be grouped into two broad categories: process systems and building systems. The information details are discussed next.

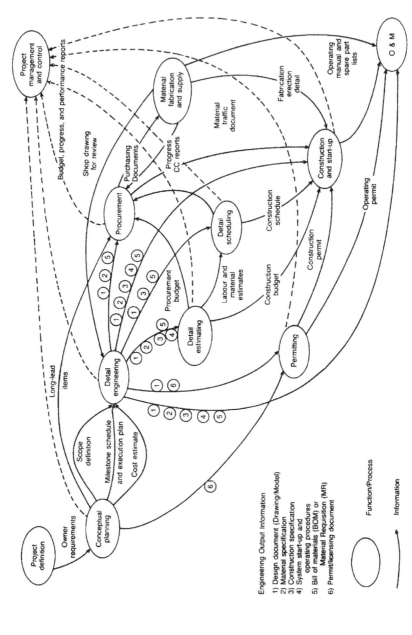

Engineering Output Information

1) Design document (Drawing/Model)
2) Material specification
3) Construction specification
4) System start-up and
   operating procedures
5) Bill of materials (BOM) or
   Material Requisition (MR)
6) Permit/licensing document

◯ Function/Process

───→ Information

Figure 22.3 Information flow diagram on EPC project

Table 22.1 Project historical information

	Project data	Project environment data	Input data	Process data	Output data
What	Project number Project name Project type Project description Scope of services Project bid/competition	Physical site conditions Weather data Site access transport Local economy Social environment Labour availability Currency	Process technology Resources – Home Office labour – Field labour – Material/equipment – Cash (cost) – Budget vs. actual	Eng. work pkg/activities Procur. WP/ activities Const. WP/ activities Start-up WP/ activities	Design Document (as-builts, calc shts. submittals, tests Bid Tabs/POs Installed quantities Test reports and spare part lists
Who	Owner Competitors	Regulatory agencies Local government Local suppliers Labour unions Permit agencies	Engineer/consultants Supplies/fabricators Contractor Subcontractors Labour Unions	Organization Charts – E, P, C – Start-up – Project Mgmt/ Controls – Subcontractors	Operator
When	Project milestone dates	Project milestone dates	Man loading chart Staffing Plan Installation progress Equipment chart Cost/commitment chart	Schedules Progress summaries Plan vs. actual	Commercial Oper. Date Maintenance schedule
Where	Project location	Area map	Resource supplier locations	Prefabrication sites	Project information Archive locations
How	Financing arrangements Contract type		RFP/bidding document Logistics/traffic Collective bargaining	Project execution E, P, C methods* Project Mgmt/Controls	O&M procedures
Why	Feasibility studies Project economics	Site selection criteria	Quality Cost Schedule Safety Resource availability	Quality Cost Schedule Safety Resource availability	Engineering calculations Cost studiies Alternatve selections Design considerations

*The E, P, C method examples include 3D modelling, Just-in-time delivery and modular construction.

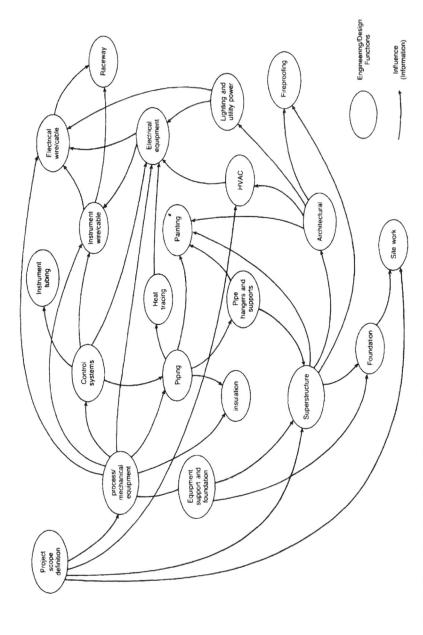

Figure 22.4 Information flow diagram for engineering process

22.7.1 Process systems

Process systems usually consist of process equipment, interconnecting piping and control system components that change the characteristics of fluid flowing through them. System design at this stage considers the chemical and physical characteristics of fluids such as state (gas vs. liquid), molecular structure, temperature and pressure. Such systems in a process plant include hydrocracking, reforming, steam generation system, etc. A major grass root process or power plant can have over 100 such systems. These process systems require auxiliary items such as power supply, heat tracing and structural support.

The engineering work for process systems typically starts from various process definitions, usually represented in PFDs (process flow diagrams) or preliminary piping and instrument diagrams (P&IDs). These schematic diagrams identify process equipment, major piping and control system requirements. P&IDs are the most important documents produced in the early design phase. They define all major components and their logical configuration by process system. Detailed design of process systems commences after P&IDs become available. The design of equipment, piping, instrument, electrical and some structural components progresses in a logically sequenced way as depicted in Figure 22.4. Based on the information flow diagram in Figure 22.4 a simplified E–R data model for process systems was developed, and is shown in Figure 22.5 (Barker, 1989).

22.7.2 Building systems

Building systems support or enclose process systems. They also support and protect the operating and maintenance personnel, e.g. offices, control rooms and other personnel support facilities. Figure 22.6 shows the simplified E–R diagram for building systems. The data relationships in building systems are simpler than those of process systems. Once the historical data are collected and stored to support both process and building system views, conceptual engineering can be done analysing the potential impact on costs and quantities *concurrently* from the engineering systems perspective.

Though the design of building systems should be directly influenced by the physical characteristics of process systems (e.g. volume, weight, etc.), fast-tracking precludes the detailed consideration of individual process system components during building systems design. For example, foundation design should be based on the weight of the superstructure and process system components it supports. However, the foundation is usually placed long before detailed design of process components is completed. Thus, this foundation design is based on the approximate weight of the superstructure and process systems, and is independent of detailed design of process systems.

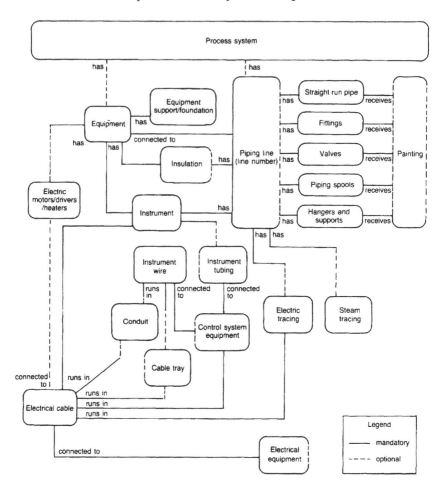

Figure 22.5 Simplified E–R diagram of process systems

When process and building systems definition is almost completed, detailed design proceeds by engineering discipline. The overall dimensions and general equipment arrangement are the design basis of superstructure. Individual pipe and equipment supports, on the other hand, are directly related to process systems and considered as a part of process systems. The amount of information interaction is significantly reduced at this stage. The design progresses mostly to fine-tune the discipline-specific items while avoiding physical interferences with other commodities. The impact of design change in one component is not as significant at this stage since most major system design decisions have been made.

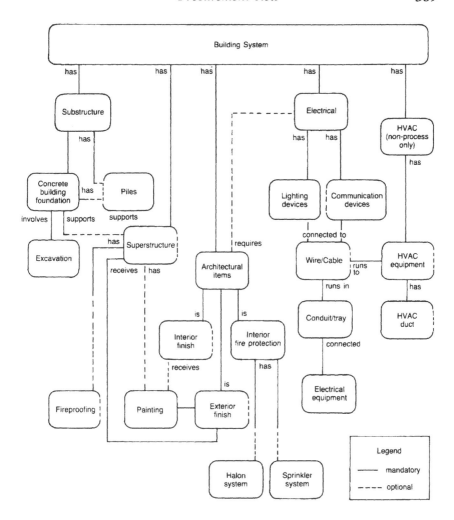

Figure 22.6 Simplified E–R diagram of building systems

22.8 PROCUREMENT VIEW

Material procurement needs to be planned by commodity type and area to support start-up and construction schedule. The cost should be traceable by purchase order as well as by item since major vendors usually have more than one purchase order with one company. In that case, procurement personnel need to view cost data by supplier to examine their delivery performance.

Besides material cost data, other procurement-related cost data (e.g. transportation, custom, tax) need to be captured in the historical

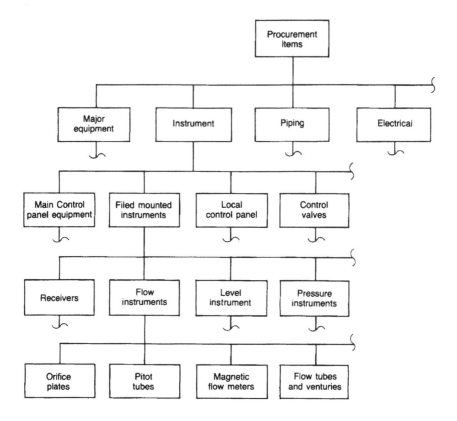

Figure 22.7   Procurement historical data view

information base. Another important data item in the historical information base is procurement lead-time information. The commodity view to support procurement planning can be modelled by a hierarchical tree structure as shown in Figure 22.7.

22.9   CONSTRUCTION VIEW

Construction views the field labour and construction support data primarily by construction work packages. The construction work packages are influenced by location, labour craft involved, commodity type, schedule logic and organizational breakdown structure identifying management responsibility. Construction work packages can be modelled as another hierarchical tree structure. Historical data items unique to construction have to be captured in this model also. They include temporary facility, indirect labour services and construction equipment not associable to

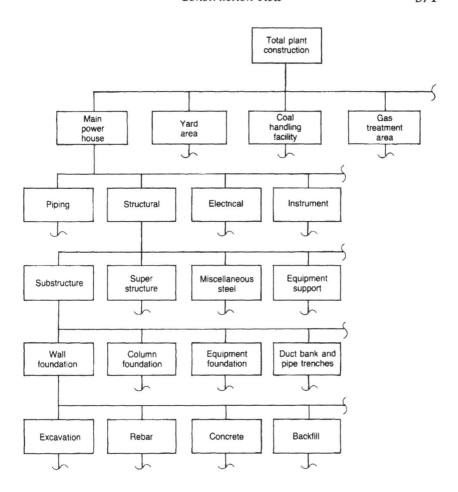

Figure 22.8 Construction historical data view

individual work packages such as cranes. Figure 22.8 shows an example of a construction work package hierarchy.

The items in the construction and procurement views have many-to-many relationships between the two views. For example, the concrete work in Figure 22.8 can appear in many foundation work packages as well as other work packages, e.g. superstructures in other areas or buildings. As far as material procurement is concerned, however, concrete may be viewed as one data item if it is purchased from one vendor and the material characteristics are the same, e.g. the same 28-day strength, slump and gravel size. On the other hand, many procurement work packages can be involved in one construction work package. For example, the local mounted instrument work in a main powerhouse can be one construction

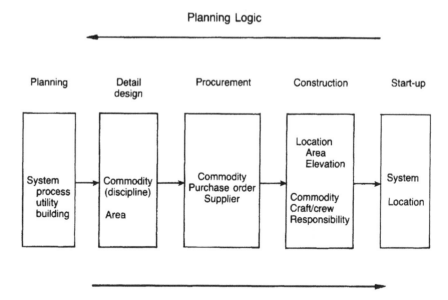

Figure 22.9   Information views by project phases

work package if done by one crew. However, the materials can be procured by many procurement work packages issued to many different instrument suppliers.

22.10   START-UP VIEW

As construction reaches about the 70% completion point, the focus of field erection work shifts from construction work packages to start-up systems. The historical data view by start-up systems is similar to the one by engineering systems presented earlier. In addition to those data captured to support the engineering view, historical data items unique to the start-up view need to be captured. They include the costs of operating, manual preparation, operator training, residue disposal and the testing equipment and services required for start-up services. Spare parts costs also need to be captured by start-up systems.

22.11 ISSUES IN CAPTURING HISTORICAL DATA TO SUPPORT MULTIPLE VIEWS

Figure 22.9 summarizes the user views by project phase. When the historical data are captured to support these views at any level of detail, users can concurrently perform cost optimization by viewing historical cost data from their viewpoints. The trend base quantity and cost details can be established at the same level as the detail estimates so that change controls can be performed more effectively based on the consistent levels of detail between the baseline and subsequent changes.

One important issue is the practicality of capturing the data to support these many views at a fairly low level of detail. The information base should be able to handle any difficulty related to data capture by creating data details automatically. Data capture issues are discussed in the following sections of this chapter.

22.11.1 Budget quantity and cost

Capturing budget quantity data by four different views is becoming easier because material quantity data can be extracted from CAD files by system, area and commodity views. Budget costs from the estimate can be easily allocated to support those views by using the quantity breakdown by different views.

22.11.2 Actual quantity and cost

Capturing actual material costs by commodity code of accounts can be done easily since material costs can be allocated by different quantity views available from material management systems. What remains as a challenge in gathering historical data is capturing actual construction and start-up data by work packages and systems. Traditionally, field costs have been captured primarily by commodity code of accounts. Though critical for planning future projects, construction management teams are usually less interested in gathering data by process systems since it is not their immediate concern.

Sometimes gathering actual construction cost data by systems is not possible if many different systems are included in one construction work package. For example, a pipe rack that carries many different pipes for different process systems may be installed by one crew under one construction work package. The field project controls system is unlikely to track the installation hours by a different pipe on the same pipe rack just because they are different systems. One solution to this problem is to include an automatic cost allocation and aggregation mechanism in the historical information base as procedural knowledge. Using quantity breakdown by components, the construction costs can be allocated to components first, then aggregated by systems. Considering the accuracy range of actual data collected and the required accuracy level required for conceptual planning, the potential error from this allocation and aggregation should be tolerable.

22.12   FUNCTIONAL SPECIFICATION

A set of functional specifications of a historical cost information base have been developed based on the analysis in previous sections. Accordingly, we recommend system designers consider the following:

1. The system should provide the capability to access historical project data for estimating project scope (quantities), costs and other resource requirements. The historical data should include:
    (a) Plant Scope

    > (i)   Output capacity by plant and system
    > (ii) Equipment list and bulk quantities

    (b) Home office costs and resources

    > (i)   Labour by departments/disciplines
    > (ii) Other resources by home office cost code of accounts

    (c) Field costs and other resources

    > (i)   Equipment and bulk materials
    > (ii)   Manual labour hours and costs
    > (iii) Construction equipment and facilities
    > (iv)   Subcontract hours and costs
    > (v)   Other indirect construction services costs (e.g. mobilization, general support, etc.)
    > (vi)   Contractor overhead and profit
    > (vii) Spare parts

    (d) Other historical data in Table 22.1 that are relevant to the prototype system scope.

2. The system should provide the ability to organize the historical data by systems. In addition, construction and start-up costs and resource data captured by construction work packages should be automatically allocated and aggregated by engineering systems. Specifically, the historical database should provide information for the following user views:

    (a) Process and building systems (systems engineering work packages)
    (b) Commodity code of accounts
    (c) Engineering discipline and tasks
    (d) Engineering area/building
    (e) Procurement work packages (material requisitions)
    (f) Purchase order/Subcontracts
    (g) Material supplier/Subcontractors

    (h) Construction work packages
    (i) Start-up systems (work packages)

3. Historical data should be automatically generated from the project databases as a by-product. They include design, cost, schedule and progress data.

4. The system should provide a summary of historical project information to define plant attributes with only general plant characteristics.

5. The system should provide a rapid access to the database in the following ways:

    (a) Scope of services (e.g. engineering only, turnkey, construction only, etc.)
    (b) Contract type
    (c) Size of project (e.g. total cost, labour hours, key quantities, duration, etc.)
    (d) Type and size of plant (capacity)
    (e) Customer
    (f) Location
    (g) Process type and technology
    (h) Milestone schedules (e.g. start and finish dates)
    (i) Procurement and contracting strategies (e.g. Just-In-Time)
    (j) Construction strategy and methods (e.g. modular construction, execution details)

6. The system should provide the capability to automatically extract standard corporate estimating databases. This includes:

    (a) Cost factors and parametric estimating model coefficients
    (b) Cost indexes
    (c) Productivity factors and wage rates
    (d) Crew and craft mix
    (e) Major equipment pricing curves
    (f) Material pricing basis and unit rates
    (g) Construction equipment rental rates
    (h) Procurement source and delivery data
    (i) Codes and regulations
    (j) Environmental constraints
    (k) Market trend analysis
    (l) Indirect costs, overhead and profit data

7. The allocation of actual costs and labour hours to support various views should be done automatically in the database.

8. The data should be normalized by the appropriate adjustment factors including:

    (a) Escalation
    (b) Location

(c) Economy of scale
(d) Productivity factor
(e) Schedule
(f) Management effectiveness
(g) Materials availability

9. The system should link 'integrally' with other databases. They include:

   (a) Expert knowledge base for standard systems design
   (b) CAD
   (c) CAE (e.g. process simulation programs)
   (d) Estimating systems
   (e) Database for scheduling systems
   (f) Materials database

10. The system should support on-line knowledge bases for corporate-wide standard systems design by type of system function and environment.

11. The system should provide the capability to support exception reporting to compare progress data on a current project against the historical progress on similar past projects.

12. The system should provide the following database management system functionality:

    (a) The information base (i.e. data entities and attributes) has to be both expandable and extensible to support future needs.
    (b) The information base should be flexible to support additional user views easily. In other words, the database has to support dynamic schema (view) evolution after the database is implemented.
    (c) The information base should support data abstraction at multiple levels effectively.
    (d) The data model should be captured without losing semantics, i.e. relationships between data entities.
    (e) The generic behaviour of data (i.e. data allocation and normalization) should be encapsulated with the data so that users need not deal with them in application programs.

22.13   SUMMARY

Project historical information is a critical resource for defining, pricing, bidding, planning and controlling subsequent projects. The ability to capture and share this critical corporate resource is becoming more important as we enter the information era. As the AEC industry automates more work processes and generates more electronic data, it is timely to investigate what and how to capture and retrieve from the historical information system.

The new historical information system should support paperless, integrated, concurrent engineering information systems. Users in various project phases should be able to retrieve historical information at different levels of detail and from different views, e.g. by process systems, by commodity account, by construction work packages, etc. This final chapter of the book presents the results of a detailed analysis for developing user requirements of such an information system. Other published work by Choi (1992) describes our design and implementation of a historical information base, which meets these requirements using state-of-the-art database technology.

ACKNOWLEDGEMENTS

The authors acknowledge the support of many managers of Bechtel Corporation who participated in the user interviews for this research, including Keith Burrowes and Maurice Heley.

REFERENCES

Barker, R. (1990) *CASE*METHOD-Entity Relationship Modelling*, ORACLE Corporation, Addison-Wesley.

Barrie, D. and Paulson, B. (1984) *Professional Construction Management*, McGraw-Hill.

Choi, K.C. (1992) An Object-oriented Historical Information Base to Support Concurrent Engineering. PhD thesis, Civil Engineering Department, University of California, Berkeley.

CII (1986) *Constructability, A Primer*, CII Publication 3-1, July.

Creese, R.C. and Moore, L.T. (1990) Cost modeling for concurrent engineering. *Cost Engineering*, **32**(6), June.

Serpell-Bley, A.F. (1990) Improved Conceptual Estimating Performance Using a Knowledge-Based Approach. PhD thesis, Civil Engineering Department, University of Texas, Austin.

Index